SCIENCE, EVOLUTION AND CREATIONISM

ORIGINS AND EVOLUTION OF PLANTS ON THE EARTH AND THE DESCENDANTS OF ANITA

SCIENCE, EVOLUTION AND CREATIONISM

Additional books and e-books in this series can be found on Nova's website under the Series tab.

SCIENCE, EVOLUTION AND CREATIONISM

ORIGINS AND EVOLUTION OF PLANTS ON THE EARTH AND THE DESCENDANTS OF ANITA

SUBIR RANJAN KUNDU

AND

PRITHA BISWAS

nova science publishers

New York

Library of Congress Cataloging-in-Publication Data

ISBN: 978-1-53614-065-1

Published by Nova Science Publishers, Inc. † New York

To my parents Sova Kundu and Subodh Ranjan Kundu. The incessant inspiration, love and empathy of them during writing of this book helped me a lot in every walk of life to fnally finish the book.

CONTENTS

ABSTRACT

Evolutionary biology may still be a complicated arena for many but in Subir Ranjan Kundu's "Origin and Evolution of Plants on Earth and the descendants of ANITA," concepts are presented in a simplified manner, with a focus on Angiosperms. The book walks its readers through a series of events that caused Evolution in the different spheres of life on Earth over the past 4 billion years. The theory explores the transformation of the pre-existing "Dark Planet" to the "Blue Planet" focusing on areas such as spatiotemporal changes, aqua life, and organic and inorganic evolution. The mystery of evolution has impacted everything from the shape of a flower to sliding continents, the writer substantiates his hypotheses with relevant examples. This non-fictional piece of work can potentially change the commonly held notions of life and evolution, leaving its readers curious to delve deeper.

Keywords: origin, plants, evolution, angiosperms, ANITA

ACKNOWLEDGMENTS

I express my deep sense of gratitude and indebtedness to Peter K. Endress, Emeritus Professor, Institute of Systematic and Evolutionary Botany, University of Zurich, Switzerland; Shinobu Akiyama, Professor, National Museum of Nature and Science, Department of Botany, Tsukuba, Japan; Giovanni Cristofolini, Professor, Department of Biological, Geological and Environmental Sciences, University of Bologna, Bologna, Italy; Paramjit Singh, Director, Botanical Survey of India, Ministry of Environment and Forests, Government of India, Howrah, India; Lokos Laszlo, Emiretus Professor, Hungarian Natural History Museum, Botany Dept., Budapest, Hungary; Rothin K. Chakraverty, Former Additional Director, Botanical Survey of India, Ministry of Environment and Forests, Govt. of India, Kolkata, India; P. K. Sarkar, Ex.Deputy Director, A. J. C. Bose Indian Botanic Garden, India; Sean Thomas, Professor Faculty of Forestry, University of Toronto, Toronto, Canada; Todd Boland, Research Horticulturist, Memorial University of New Foundland, Canada; Zoltan Barina, Chief Curator, Hungarian Natural History Museum, Botany Dept., Budapest, Hungary; Mahbuba Sultana, Senior Scientific Officer, Bangladesh National Herbarium, Dhaka, Bangladesh and Sudipa Das, Principal Investigator, SERB Young Scientist, Assam University, Assam, India for their suggestions and guidance in this present writing.

I am especially pleased to Ashwini Srivastava, Former Scientist "SG," Birbal Sahni Institute ofPaleobotany, India; Sujana K. Arjunan, Scientist "SD," Central Botanical Laboratory, Botanical Survey of India, India; Dr. Himadri S. Debnath, Principal Scientist, Japan International Cooperation Agency; Ger Van Den Beuken, Dutch Alpine Garden Society, Holland; Asma Khatun, Former Chief Scientific Officer, Bangladesh Jute Research Institute, Bangladesh; Nate Yuen, Conservation Chair, Sierra Club of Hawaii, United States; Andrey Dedov, Barnaul, Russia; Preeti V. Phate, Asst. Professor, Botany Dept., Mumbai University; Prasad M. Govindan, Fellow, International Association of Plant Taxonomy, CSIR-National Botanical Research Institute, India; Adrian Young, Waterperry Gardens, Blackpool, United Kingdom; Alexander Sartori, Restoration Ecologist, Viva

Papa Tree Initiative of Congo, Germany; Reshmi Chatterjee, Ex-Research Fellow, Birbal Sahni Institute of Paleobotany, India, who helped me with the different part of the manuscript required reviewing consultations of literature, facilitating the studies in Herbarium, library consultation and discussion etc.

I am also thankful to Ms. Susan Zadek, Librarian of the Albert Campbell Library, Toronto Public Library Network, Toronto, Canada and Mr. Uday R. Biswas, Assistant Library Information officer, Central Library, Central National Herbarium, Botanical Survey of India, India for extending Library facilities and providing necessary help in locating literature required for this project to complete on Ms. Sraboni Modak, Alumna, Department. of South and South-East Asian Studies, University of Calcutta, Kolkata, for her graphical presentation and Ms. Esha Mitra, Department of English Literature, Presidency University, Kolkata for paying especial attention in the final phase of drafting and editing the manuscript.

I am expressing my gratitude to thank all authors, editors, publishers of books and journals, colleagues for sharing the printed and digital materials, providing valuable suggestions and tendering necessary clarifications during the last few years, selflessly helped me out to complete this task.

Chapter 1

Origin of Life on the Primitive Earth

Formation of the "Blue Planet"

As we try to envisage our future as the dominant species on Earth, a grey picture comes to mind: an expansive, desolate graveyard with a little cenotaph weighs down on our souls. Nevertheless, a stoical quest for our home gives us an optimistic pleasure as the mystical picture of the "blue planet" appears in our minds. Naturally, its origin is a soul-searching enigma of putting together all scientific innovations related to the origin of the primitive Earth; combining the fields of cosmogony, cosmology, astronomy, and geoscience.

On the basis of the collective observations of scientists from these inter-disciplinary fields, the proponent of the "Big Bang Theory," Astrophysicist Georges Lemaitre (1931), claimed that the primitive Earth was formed along with the other planets in our solar system, approximately 10-20 b.y.a. (billion years ago).

The "Big Bang Theory" deals with the formation of the "universe" or "cosmos," of 100,000 million planets or galaxies (a planetary organization comprised of numerous planets, satellites, stars, asteroids, and gaseous clouds of stellar dust bounded to each other by gravitational pull). On the scale of stellar distance, the universe has occupied almost a span of 10 billion light years of spatiotemporal area in space.

To have an idea of the stellar distance we need to look at a twinkling star in the night sky, emitting rays of light at an unbelievable rate of 1,86,000 miles/second. This light crosses billions of light years in distance for millions of years before it reaches the Earth. 1 light year is approximately 6 trillion miles; so, looking at the twinkling starlight may be considered akin to travelling in time to witness a historical event, traversing through spatiotemporal galactic pathways.

Scientist Georges Lemaitre (1931) postulated that the formation of the "Universe" occurred as a result of an explosion of energized, high density, minute particles, and that the planet Earth belongs to the "Milky Way" galaxy.

In his "Nebular Hypothesis," Kant (1755) claims that the formation of the universe took place around 4.6 b.y.a., which is less than the estimation made by Lemaitre. In "Nebular Hypothesis" Kant (1755) suggests that the universe evolved from a huge gaseous cloud in space known as the "Solar Nebula."

According to scientist Kant (1755), "Solar Nebula" took the shape of a rotating disc under the influence of gravitational pull in space. He further argues that the center of the "Nebula" contained an extremely hot gaseous mass known as the primitive "Sun" (Cassen, 2001), around which the solid masses (that we today call "planets") began rotating, encircling the Sun in "orbits" as they are known today.

Though there may be controversy regarding the formation of the Universe it has been discerned unequivocally by cosmologists, astronomists, and geoscientists, on the basis of their collective observations, that the Sun forms the center of the planetary organization known as the "Solar System" (Cameron, 1978). This began with the conception of the Sun as a huge ball of heterogeneous mixture of gas, or as a dust cloud. It was then understood that the Sun was made up of Hydrogen and Helium gases, and some heavy elements that suspended on the outer side of the core of the earliest solar system, the "Proto-star System."

Based on the analysis of the available data on the Meteorite Age, the formation of the "Proto-star System" was initiated 4.5-4.6 b.y.a. The "Proto-star System" was considerably colder than the present-day solar system (Cameron, 1978). The gradual accumulation of heat in the gas dust-cloud sphere contracted in volume and formed the centrifugal gaseous core (the H-He gas mixture). This was followed by the formation of rings on its mantle with denser gases and solid particles.

The collusion of heated solid and molten particles formed the earliest planets, called "planetesimals," which remained suspended in the gaseous matrix. The gradual accumulation of larger volumes of "planetesimals" due to the continuous collision of planetary solid particles lead to the formation of gradually larger sized planets (Asphaug et al., 2006); centripetally, in the planetary orbits from inside out.

The astrophysicists and geoscientists opine that a combination of different factors such as electrical shock waves, solar explosions, nuclear radioactivity and collision force among planetesimals, provided the heat-induced energy which raised the temperature beyond 1000 degree Celsius, melting the solid particles and forming larger meteoroids. Among all the different sources of heat energy produced during the formation of planets in the solar system, nuclear fusion was the key source which led to the emission of elements such as Uranium (^{235}U), Potassium (^{40}K) and Aluminum (^{26}Al).

The geological analysis of meteorites (fragments of meteoroids) and remnants of asteroids from space, which landed on the Earth's surface led to this conclusion. From the different gradients of concentration of hydrogen and helium gaseous clouds around the planetary orbits of the Sun, scientists deduced that the planets which were found in close proximity to the Sun formed the inner group of planets (e.g., Earth, Mars, and Venus) and the rest formed the outer group of planets (e.g., Jupiter, Neptune, Saturn, and Uranus), whose planetary orbits are relatively far from the Sun.

It is easy to comprehend that the gravitational force of larger planets on outer orbits is greater than that of the smaller planets on inner orbits. The planets are endowed with solid objects such as mineral-rich asteroids, comets, frozen carbon dioxide ice (also known as dirty snowballs), dispersed in the gaseous clouds of intra-planetary areas.

The slow but gradual contraction of the gaseous ball (earliest Sun) increased the internal temperature of the core of the gaseous sphere to over 1 million degrees Celsius inducing nuclear fusion and producing a "solar wind" (Stern and Peredo, 2001) which created a centripetal force driving the solid particles away. As the inner group of planets is directly exposed to the solar winds, the hydrogen and helium gases in the surrounding gaseous clouds are thinned out by the excessive force of the solar wind. This is lower than the retaining capacity of gravitational force. The bigger planets are comparatively less exposed to solar-wind forces and have a greater gravitational pull; hence the planets in the outer orbits are found to be enriched with hydrogen and helium gasses.

Therefore, it may be concluded that the outer surface of the earliest Sun was much cooler than the present-day Sun and as a result, energy emissions (light and heat) from the earliest Sun were also lesser. The geochemical analysis of different meteorites on different locations across the Earth (which were formed due to the collision of planetesimals and planets and landed on Earth due to its gravitational pull) revealed that they were primarily carbonaceous chondrite type rocks (water containing, carbon-enriched rocks). Furthermore, the existence of iron-magnesium silicates and iron-nickel silicates evidences the heat-induced melting and fusion of solid metallic dust particles in the gaseous cloud.

The strategic position of the Earth on the planetary orbits of the Solar System allows there to be ample space between the Earth and the Sun to sustain the vortex of solar-winds. The extra barriers of comets act as a preventive safeguard against the direct exposure of solar heat that induces desiccation of the Earth's surface. Hence a slow but gradual process of melting of ice on Earth (in the proto-planet condition) created a huge volume of water. The location of the Earth with respect to the Sun provided the optimal surface temperature to maintain this liquid state since its prehistoric formation. It is quite interesting to visualize the evolutionary process of life on Earth in the backdrop of the timescale proposed by cosmologists that the formation of the Earth and the entire planetary system around the Sun happened around 4.6 b.y.a. (Cameron, 1962; Ezer and Cameron, 1965).

According to the "Proto-star systemHypothesis", cosmologists explain the formation of our "blue planet" and its evolutionary foot prints from its earliest form, the "proto-earth." This evolved around 4.5 million years ago as a part of the "Proto-star system" (Hayashi, 1966). The process of formation, evolution, and metamorphosis of the "blue planet" is said to have transpired over the course of 1 billion years (Hadean Eon). The initial phase of fifty million years had the "proto-earth" go through the phase of "melt-down," resulting in the formation of the "core" of the Earth with heavier metallic elements such as iron-nickel and a crust of outer layer of lighter metallic elements; Iron-Magnesium silicates formed the mantle, also known as "proto-continents," along with the early ocean, "Magma Ocean," formed as a result of weathering. These were continuously impacted by incoming bombardments of meteorites and asteroids on the surface of the proto-continents (Ernst et al., 2016).

The geological analysis of the samples of rocks from "Halemaumau" crater of the "Kilauea lki" volcano in Hawaii further substantiates this "Melt-Down Hypothesis". The nature of the igneous rock at surface level has been identified as "basalt" and "pyroxenites" were identified at deeper levels.

These presumably occurred around 3.8 billion years ago. The astrophysicists and cosmologists discuss a collision of a planet or asteroid (also known as planet-x), similar to the size of Mars, with proto-earth, as a result of which a part of planet-x was further fragmented and meltdown, adding-up to the volume and mass of the Earth. A large chunk made up the moon, rotating around the proto-earth. The collision also shifted the Earth's axis of rotation from 30 degrees to 23 degrees (Laskar et al., 1993; Lissauer et al., 2012).

The distance between the early moon and the proto-earth according to scientists, was one-tenth of the present distance, and the meltdown and gradual increase of the volume of the early moon also happened during this phase. In due course of time, the moon slowly moved outward due to rising tidal force counteracting the gravitational force of the Earth (Laskar et al., 1993, Lissauer et al., 2012). The sample analysis of lunar rocks further ascertains that the core of lunar rock is enriched with iron, whereas the mantle is made of feldspar enriched rocks.

Analysis of the history of high paced rotation of the proto-earth on the proposed 30-degree rotational axis suggests that the proto-earth experienced 6 hours of daylight and there was a lesser scope of the Earth's surface being exposed to the Sun for a prolonged period of time. The rotational pace has since slowed down due to the shifting of the axis of rotation to 23 degrees. The Earth now experiences 12 hours of daylight (approximately); resulting in higher chances of prolonged exposure of the Earth's surface to sunlight (Laskar et al., 1993; Lissaeur et al., 2012).

Scientists also suggest that the cooling phase of the proto-earth was initiated after the completion of the melt-down phase, which lasted for 50 million years of the first Hadean Eon (1 billion years), when the metamorphosis of the early Earth occurred and greater exposure to sunlight was possible (Brown and Jenvey, 2012). The extent of the rate of bombardment of the asteroids and comets on the newly built surface of the Earth and moon was higher then, than it is today.

While direct evidence of sample-based analysis of meteorites on the Earth's surface is not available, lunar chronology has revealed a heavy bombardment around 3.9 b.y.a, on the Earth's surface.

In the course of metamorphosis from "proto-earth" to "blue-planet," if the first shift of unique form of changes on the Earth is the long time exposure of the Earth's surface to sunlight due to a rise in daylight hours; then the second unique form of change is the changing of the proto-earth from a solid, rocky, hot planet to a larger, voluminous, water-containing, cooler, "blue planet" (Valley et al., 2002).

According to cosmologists, around 150 to 200 million years after the formation of the proto-earth, the cooling phase of the metamorphosis of this dry, solid, dark, rocky, primitive Earth to the "blue Planet" took place. This was the result of the cumulative effect of bombardments, melting and addition of fragments of meteoroids, asteroids, and comets, that contained frozen water, which was transferred to the surface of the Earth (Brown and Jenvey, 2012). Geoscientists have considered that factors such as the initiation of the hydrological cycle (evaporation, rainfall, and etcetera) and the maintenance of temperature, ranging from 0 degrees to 100 degrees Celsius of the primitive Earth in the "cooling phase," helped the melting of the solid ice to water and also helped retain the water in the liquid state in the primitive "Magma Ocean." The hydrolysis of water produced H^+ (Hydrogen ion) and OH^- (hydroxyl ion) but it was found that water is a stable compound and hydrolysis of water is not a usual phenomenon (Halliday, 2004; Lunine, 2006). The existence of Zircon rock, formed between 4.4 b.y.a. to 4.0 b.y.a., and the formation of sedimentary rocks in Greenland (3.9 b.y.a. to 3.8 b.y.a.), also substantiate the above transition.

According to the cosmologists, this transition process happened between 4.6 b.y.a. to 3.7 b.y.a., which covers the "First Hadean Eon," and was subsequently followed by "Proterozoic Eon" (2.5 b.y.a. to 1.9 b.y.a.) on the geological timescale.

The crust of igneous rock enriched with iron, magnesium, calcium and little silica content (basalt rock), supposedly formed the raft of the ocean, popularly known as "simatic subcrust." Geologists believe that "proto-crust" of the primitive Earth was modified to form the simatic subcrust over a due course of time, which exists in the lithosphere till date. Simultaneously the sedimentary rocks enriched with silica, aluminum, and sodium minerals of the first Hadean Eon metamorphosed to form the "sialic crust" (of granite rock) forming the "nuclei" on top of the simatic crustal mass and forming an outer boundary around it (Halliday, 2004). These further ascertain that the plate tectonics in the early stages of the proto-earth formation during the cooling phase, lead to this action.

The analysis of the sample, collected from sedimentary rocks in Western Australia has been found full of zircon crystals but carbon-dating indicates that the age of formation ranges from 3.3–3.7 b.y.a. (Brown and Jenvey, 2012). This creates confusion as to whether zircon is part of the igneous rock. The presence of zircon crystals in sedimentary rocks further justifies this possibility of its origin. The formation of the gemstone zircon ($Zr\ SiO_4$) has been conceivably created from granite rocks, a type of sialic igneous rock, due to weathering processes that subsequently lead to its incorporation into sedimentary rocks (Saxena, 1966). This was a slow process.

Later, carbon-dating of different samples of zircon from different locations showed that the age of zircon ranges from 4.4–2.0 b.y.a. The analysis of oxygen isotope, known as O^{18}/O^{16} ratio of any sample of rock, was performed to determine the potential water level in it. A level of more than 5 indicates the presence of water in the liquid form in the atmosphere. A Higher value of O^{18}/O^{16} indicates an overwhelming presence of O^{18} (Peck et al., 2001), which is a heavier form of oxygen than O^{16}.

In the meltdown phase, the ice crystals of rock started changing their physical state from solid to liquid, but the existence of oxygen and water vapor was less due to the high temperatures of the proto-earth due to the absence of the protective blanket of the ozone layer. At the end of the cooling phase the ratio of O^{18}/O^{16} exceeded the level 5 cut-off mark, which indicated a higher volume of the heavier O^{18} than O^{16}; presence of enough moisture was evident (Scott and Krot, 2001). The presence of oxygen in the atmosphere and its gradual cooling lead to the condensation of water vapor into its liquid form in the "Magma Ocean" of the proto-earth.

The 7.4 value of O^{18}/O^{16} ratio of zircon (e.g., Archaean Greenstone of the rocks of Pilbara, Australia) in the silica enriched sialic layer of igneous rocks (like basalt and granite) and in meta-sedimentary rocks (like metamorphosed basalts, Archean granites, Gneiss) prove the transition of the proto-earth from the melt-down phase to the cooling phase (Peck et al., 2001). This supposedly took place around 100 million years ago, determined by analyzing the geological age of zircon sample on Earth. According to the geoscientists, this sialic metamorphosed igneous rock formed the nucleus of the earlier continents of the proto-earth and the phase of metamorphosis from igneous rock to sedimentary rock went through a series of actions such as bombardment, melting, weathering, and sedimentation. zircon can be considered as the "geological footprint" of this transitional phase. Most likely, in the process of sedimentation, the water current of the primitive ocean played a key role (Mojzsis et al., 2001). The condensation process was designated as a "rapid cooling action" as it took only 100 million years compared to the actual age of zircon, formed around 4.4 bya.

The nuclei of the early continents of the proto-earth formed around the core of basalts were interspersed by granites. Those nuclei remained buried under heaps of outer layers of metamorphic rocks (granites, gneiss etc.) and formed a number of huge "continental mass," also known as "continental islands," the tips of which were far above the primitive ocean.

These "continental islands" existed in the first Hadean eon. The tidal force and water current of the primitive ocean lead to the weathering of metamorphosed rocks followed by sedimentation and hardening to form the mantle of the early Earth. Under the influence of rotating convection currents of the hot proto-earth, the mantle layer ruptured into a few "early continental blocks" (also known as "proto-continents") floating above the ocean (due to low density) and moving laterally (Petit, 2010). These movements are popularly known as "plate tectonics" and it's assumed that the maiden plate tectonics first occurred in the Hadean eon.

According to the culminating formation of multiple hypotheses of scientists, the formation and distribution of proto-continents took place in the first Hadean Eon, while the sea-floor expanded in the Archean Era. The combined effect of convection currents and the sea-floor spreading propelled the lateral movements of the blocks of proto-continents, which continued until they collided into the super-continent known as "Ronidia." Ronidia has since been fragmented, forming the new continents (Stanley, 1999). The geologists suggest that the merging, collision, and fragmentation process happened repeatedly until the present shape of continental masses on the Earth was obtained.

In the meltdown phase, volatile gases (e.g., helium and hydrogen) of the proto-earth's atmosphere were thinned out by the combined actions of solar wind and heat formed from molten rocks (identical to lava of volcanoes). This resulted in a hydrogen and helium gas deficient atmosphere in the cooling phase.

During the formation of the primitive sea, which happened in the cooling phase, the smokes from volcanic eruptions led to the accumulation of non-volatile gases: carbon dioxide, nitrogen, ammonia, methane, sulfur and water vapor, in the Earth's atmosphere (Sleep and Zahnle, 2001; Tobie et al., 2006). While the zircon footprint analysis ascertained the presence of oxygen 4.4 billion years ago, it also suggests that oxygen was present in a nascent condition (dominance of O^{16}) in the melt-down and cooling phases of the early Earth. This could be interpreted as a "reduced environment," more precisely, an atmosphere where the free form of O_2 was deficient. Hence, naturally, the chances of evolution and existence of life were lesser. This condition persisted for the initial 2-2.3 billion years of the 4.4 billion years of the "oxygen transition regime" (a gradual shift from O^{16} to O^{18}).

Most of the oxygen, coming out of volcanoes, combined with metallic iron deposited along-with silicates in the metasedimentary rocks. The latter 2-2.5 billion years of the 4.4 billion years was the O^{18} dominant regime, or the secondary phase of cooling[1]. The existence of the primitive ocean ensured the huge volume of water in the atmosphere, and the abundance of non-volatile gas like nitrogen, ammonia, methane, carbon dioxide and the presence of free oxygen in the atmosphere in this period formed a suitable environment for the evolution of life and has been recognized as the "oxidized environment."

The existence of free oxygen in the atmosphere is the synergistic result of the evaporation of water from primitive oceans and the flow of lava through hydrothermal vents present on the ocean floor, which helped release oxygen along with the other gases from the ocean water (Stein and Stein, 1992). The basic requirements to support life were now present: water, CO_2, O_2, N_2, SO_2 (in the cooling phase of the Earth).

The palaeoclimatic conditions of low temperature, longer daytime, presence of non-volatile gases in atmosphere, lightning, volcanic action, etc., along with the metamorphism of rocks and the huge volume of water in the ocean, provided the conditions necessary for physicochemical reactions to take place on the primitive Earth utilising the free atomic ingredients to form primitive molecules and simple inorganic compounds. This was followed by the formation of simple organic compounds and subsequently by the formation

[1] Which indicates that "rapid cooling" was not an instantaneous event and took more than a 100 million years, as predicted earlier.

of complex organic compounds, which ultimately led to the maiden appearance of the earliest form of life also known as "Biopoesis" (Voet and Voet, 2004; Scharf et al., 2015). There can be no conjecture about the incidences in the "Biopoesis" as it is an integral part of the evolution of the universe. It is palpable that the appearance of living organisms on Earth in the melt-down phase was most unlikely given that the free oxygen, water, and low temperatures were not enough for aerobic, unicellular, prokaryotes (which are considered to be the maiden form of life). Even if we consider that the anaerobic thermophilic microbes were the earliest living beings in the melt-down phase on the primitive Earth, the direct exposure of these microbes to deleterious cosmic rays would induce higher mortality. With the help of age-dating, it was determined that the primitive form of life on Earth occured between 3.85-3.55 b.y.a. in the form of simple, structured, unicellular bacteria. Alhough the palaeoclimatic data is in favor of the formation of life around 4.55-4 b.y.a., foot-prints of the evolution of life have not been traced back to rocks beyond that time. Few hypothetical models have been proposed by different scientists from time to time to track the footprints of "Biopoesis" (Scharf et al., 2015; Voet and Voet, 2004), some of these are as follows:

1) The theory of special creation: According to the spiritual beliefs of different religions, some unknown, sacred, supernatural power created the living beings on Earth. For instance, according to Hindu mythology, the God "Bramha" is the creator of the universe and all its living organisms. Furthermore, according to 'Genesis' in the Holy Bible, Christians believe that "God" created this Universe and its different living organisms within the period of six days. However, no conclusive scientific explanations can be found in favor of these theories.
2) Theory of 'spontaneous generation' ("Abiogenesis Theory" or "Theory of autogenesis"): According to this theory, proposed and supported by the philosophers Aristotle, Anaximander, Plato and Plutarch, Thales and Xenophanes, living organisms evolved spontaneously from dead organic matter. The philosophers Anaximander and Empedocles stated that primitive living organisms (e.g., Toads, Mice, Snakes, and Crocodiles etc.) evolved as a result of exposure to sun rays on the hot muddy soil of river Nile in Egypt. In the early days of 16th Century this theory received appreciation in China and Egypt. According to the philosopher Van Helmon, the mixture of wheat and human sweat could also give rise to living organisms such as mice (Oparin and Morgulis, 1938). This hypothesis has been discarded due to a lack of supporting evidence. Since the end of the 16th century till the middle of 18th century, eminent scientists like Louis Pasteur, Lazzaro Spallanzani and Francisco Redi proved that the evolution of living organisms is not possible from dead organic matter, rather, new living organisms have always been evolving from pre-existing living organisms (Omne Vivum Ex-

vivo). This theory is known as the "Theory of Biogenesis" (Harvey, 1651; Huxley,1870).

In this context, an experiment conducted by Spallanzani (1765) should be mentioned. A certain type of nutrient medium was boiled in eight glass jars in aseptic conditions. Four jars were left open for a certain period of time; and the rest were in an air-tight condition. After a period of time, microbial growth was noticed in the nutrient medium in the four open jars but there was no trace of microbial appearance in the nutrient medium in the closed jars. From these results, Spallanzani inferred that microbial population need air as well as contact with nutrient media for growth. No appearance of microbial population in the closed jars proved that the theory of spontaneous generation or "Abiogenesis" is invalid. On the other hand, the theory of biogenesis supports the view that life originated from a "pre-existing life form," but a lack of evidence meant that this theory was also rejected.

3) Theory of catastrophism: According to the renowned paleontologist Georges Cuvier, there was a mass extinction of all living organisms on the primitive Earth due to a sudden catastrophic accident, which led to the end of an era. The scientist Orbigney, a proponent of the "Theory Cataclysms," supported this view (King, 1877). The beginning of a new era was marked by the appearance of a range of diversified living organisms. This theory can be thought of as a modified form of the "Theory of Special Creation" and hence not enough scientific justification of this hypothesis exists.
4) Theory of Eternity of life or Steady State Theory: This theory was proposed by Preyer (1880), who proposed that diverse forms of living organisms existed forever. These living forms could change their forms and shapes in time. The analysis of different pieces of evidence that has proved the formation of the Earth happened at a certain point of time; so, an eternity of life is not an acceptable proposition as there is no acceptable evidence that life existed in a period when universe or planetary formations did not even exist.
5) Extraterrestrial Origin: this can be studied under the following categories-
   I. Cosmozoic Theory: According to the Cosmozoic theory of Richter (1865), pollen or seed-borne "Protoplasm" came from a distant location in the cosmos or universe, known as "Cosmozoa," and reached the surface of the primitive Earth to initiate life. Another supporter of the "Cosmozoic theory," Helmholtz (1884), suggested that the meteorites carried the "protoplasmic part" to the Earth from another part of the universe.
   II. Theory of Panspermia: In this theory, scientist Arrhenius (1908) proposed that the primitive molecular entity, known as "Panspermia," was carried by a comet from a distant planet to settle on the primitive Earth to form a living organism.

In support of the extra-terrestrial theory of life, scientists cite the existence of amino acids on Murchison meteorites, collected from Australia (Engel et al., 1990). In 2011, scientist Hoover reported the existence of bacteria in these meteorites found in Alaska. Except for Mars, there was no sign of the existence of water on any other planet, so the notion of the existence of biological reproductive units like spores and seeds in distant planets was not supported later.

PHYSIOCHEMICAL EVOLUTION – THE PENULTIMATE PHASE OF ORGANIC EVOLUTION

1. The theory of Physiochemical Evolution or the "Primary Abiogenesis Theory," in terms of the origin of life on Earth, is popularly known as the "Oparin-Haldane Hypothesis" and is the most feasible explanation (Gordon-Smith, 2012). This theory was proposed by the scientists I. Oparin (1924) and J.B.S. Haldane (1929), later proven experimentally by Miller, and substantiated by Urey and Fox (Miller, 1953). In their mile-stone literature "the Origin of Life," scientists Oparin and Morgulis (1938) postulated that "abiogenesis first, but biogenesis ever since." This further clarified that the inorganic compounds of non-living entities on the primitive Earth went through gradual chemical changes to form organic compounds or living forms. This is also recognized as "Physiochemical Evolution" which was followed by the continuous appearance of new organisms from existing living organisms (known as "Organic Evolution"). It has been accepted by the scientist Haldane with a few modifications. The "Theory of Physicochemical Evolution" has further been classified into two phases: Chemogeny and Biogeny.

i. Chemogeny: Chemogeny is classified under the following headings-

a) Origin of the primitive Earth and the primitive atmosphere: It has been stated that the origin of the Earth was in a huge volume of a gaseous cloud, known as "Nebula," as a result of condensation around 400-500 million years ago (Cassen, 2001). Due to extremely hot conditions, with temperatures as high as 5000^0 Celsius, inorganic components existed on Earth in atomic forms during the melt-down phase. The onset of condensation in the early cooling phase of the Earth segregated the free atoms according to their atomic masses. The heavier metallic atoms of Iron, Nickel, Copper etc. formed the core of the planet Earth, and the outermost crust-mantle was formed by the lighter atoms of carbon, oxygen, nitrogen, and hydrogen, with a lower atomic mass. The lighter metallic and non-metallic atoms with intermediate atomic masses like Sodium, Magnesium, Potassium, Sulfur, Silicon, Chlorine etc., formed the first layer of the crust or inner mantle (Ganguly and Bose, 1995). Eventually, the outer mantle formed sedimentary rocks and soil and the outer layer of the crust formed the palaeoclimatic atmosphere.

b) Origin of early molecules and formation of simple organic compounds: In the second half of the cooling phase, the free atoms of the outermost crust like hydrogen, carbon, oxygen, and nitrogen were modified (as a result of intra-atomic reactions) to form molecules and inter-atomic reactions yielded carbon dioxide, water, acetylene, ammonia, methane, and ethylene-etc

The palaeoclimatic analysis of the atmosphere of the primitive earth revealed that the free form of hydrogen existed in the primary stage of cooling and gradual decrease in temperatures from 5000^0 Celsius to 100^0 Celsius triggered a pivotal change on Earth; the physicochemical bonding of free hydrogen with free oxygen to form water vapor, which created an oxygen deficient environment, due to the exhaustion of free oxygen.

Further cooling of the Earth's surface enhanced the rate of condensation of water vapor to form a cloud and hot rain filled up the primitive oceans and aquifers with hot water, which was referred to as the "Warm Little Pond" by Charles Darwin. Owing to the liquid state and stable form of water, it was found to be a suitable medium for biochemical reactions. As a good solvent, water contains a number of electrolytes or polar electrolytes disolved in it. Moreover, heterogeneous mixtures of molecules created from the condensation of atmospheric gases were present in the hot water of the primitive ocean. Volcanic eruptions and weathered meta-sedimentary rocks induced the formation of simple inorganic compounds such as:

$H + H = H_2$ (Hydrogen molecule),
$N + N = N_2$ (Nitrogen molecule),
$C + H = CH$ (transitional form of hydrocarbon),
$CH + CH = C_2H_2$ (Acetylene),
$CH_2 + CH_2 = C_2H_4$ (Ethylene),
$CH_2 + CH_2 = CH_4$ (Methane) + C (free carbon).

c) Origin of simple organic compounds: The gradual cooling of the Earth's surface from 100^0 Celsius to 60^0 to 70^0 Celsius, led to the second pivotal change of in the palaeoclimate. The reciprocal reactions of inorganic molecules and compounds (e.g., hydrogen, nitrogen, water, carbon dioxide, methane etc.), formed earlier, reacted to form simple forms of organic compounds, like simple carbohydrates (e.g., ribose, glucose, galactose), various fatty acids, different type of amino acids, and two types of nitrogen bases (purines and pyrimidines).

Scientist Haldane has designated this aqueous admixture of simple organic compounds which filled-up the primitive ocean as the "Prebiotic soup" or "Hot Dilute Soup." Water has played a key role in the hydrolysis of a number of complex chemical compounds (Shapiro, 1987). The catalysis of those physicochemical reactions was supposedly driven by the oxygen deficient environment of the Earth, high temperatures, the penultimate level of radioactivity, a high voltage of lightning sparks of gaseous clouds, cosmic and ultraviolet rays, and etc. Some of the end-products or simple organic compounds (also known as monomers) are as follows:

$CH_4 + CO_2 + H_2O$ = simple carbohydrate + fatty acid
$CH_4 + NH_3 + CO_2 + N_2 + H_2$ = purine + pyrimidine
$CH_4 + NH_3 + CO_2 + H_2O$ = amino acid

In support of the Oparin-Haldane Hypothesis, scientist Miller (1953) cited the results of the great experiment, popularly known as the "Swan Neck Flask" experiment, where they used different inorganic molecules and compounds like methane, carbon dioxide, ammonia, hydrogen gas, and water vapor in a glass container called the "Swan-necked flask." They then applied an electric spark of 75 kiloVolts through it and heated it up to 800^0 Celsius for 18 days (as a simulation of the palaeoclimatic environment of the primitive Earth) and observed the liquid (soup) suspended in the container after condensation. The scientists' trio then examined the condensed form of the liquid with a chromatographic analysis and noticed the presence of 15 amino acids, a ribose sugar, a nitrogen base (e.g., adenine), and a number of organic acids (e.g., Lactic acid, Acetic acid etc.). This validates that organic compounds evolved from inorganic compounds on the primitive Earth.

Furthermore, in support of the Oparin-Haldane Hypothesis, the findings of the scientist Melvin Calvin (Garrison et al., 1951) can be cited. He used a mixture of carbon dioxide and hydrogen gas in the laboratory to produce Succinic acid and Formic acid which are simple organic acids.

d) Origin of complex organic compounds: The solar heat induced vaporization of the "pre-biotic soup" increased the density of the aqueous solution. Consequently, the density of simple organic compounds also increased gradually. Solar radiation and high voltage electric sparks from lightning accelerated the special kind of chemical reactions such as the polymerization of simple organic compounds or monomers to form complex organic compounds like disaccharides, polysaccharides, fat, nucleotides, polypeptides etc. It must be noted that enzymes did not take part in these chemical reactions. Some of the end-products were as follows:

(Mono-saccharide + Mono-saccharide) …n times = Di-saccharide. Polysaccharide

Fatty acid + ……………………………………………… Glycerol = Fat

Amino acid + Amino acid ……..n times = Protein (Polypeptide)

Nitrogen base + Carbohydrate + Phosphate = Nucleotide

Nucleotide + Nucleotide ……….n times = Nucleic acid

Adenine (Purine base) + Phosphate + Phosphate + Phosphate = ATP (Adenosine Tri-Phosphate)

In support of the second part of Oparin-Haldane Hypothesis, scientist Sydney Fox (Fox and Foster, 1957) performed another simulation experiment with a mixture of 18 amino acids. He heated them upto 200^0 Celsius for a few hours. At the end of the experiment, he got a protein (polypeptide), which he called a "Proteinoid." Furthermore, he synthesized nucleotides, porphyrin, and ATP in the laboratory as well. The "Proteinoid" also came in contact with water to form a colloidal (i.e., viscous solution), and a uniform and stable form of molecules covered with two membranes called a "Microsphere" (Hayakawa et al., 1967).

Due to insolubility of protein and nucleic acid in water, water is the ideal medium for sustaining protoplasm and hence the ideal medium for the origin of life.

In support of this part of the hypothesis, the work of scientist K. Bahadur (1954; 1965) should be considered. He exposed a mixture of ammonia, ferrous chloride and formaldehyde to a high intensity of light to induce a photochemical reaction and yield a mixture of different amino acids known as "Jeewanu"; likewise scientist Carl Sagan (Ponnasperuma et al., 1963) used the mixture of purine and pyrimidine to synthesize nucleic acids, NAD, and ATP in the laboratory.

ii. Biogeny: Biogeny can be explained under the following headings-

a) Formation of Coacervates: The complex form of organic compounds (like proteins and carbohydrates) from Chemogeny settled down in the clefts of rocks in the sea-bed as well as the narrow spaces between boulders in coastal regions. The intra-atomic attraction between these complex organic compounds lead to the formation of colloids, suspended at the bottom of water-bodies to form unique form of molecules known as "Coacervates," in varied sizes (Priftes and Tirrell, 2012).

   Having a larger mass and being transitional in nature, colloidal suspensions or coacervates were designated as "Prebiotic forms." They have been considered as the precursors to life. Interesting features of vegetative reproduction of prokaryotes, like "binary fission" and "bud formation" (Schmidt and Turgeon, 2011), have been noticed in coacervates.

   Contextually, the experimental observations of the scientists Oparin and Morgulis (1938) need to be cited her. They successfully produced coacervate molecules in laboratory conditions by mixing the opposite charges (+, -) of colloids: 'gum arabica' and gelatin.

b) Formation of Eobionts: Geoscientists say that one or more coacervate molecules, encapsulated within protein and fatty acid (Cephalin, Lecithin) membranes, absorbed protein and nucleic acid (RNA) on the primitive Earth to yield anaerobic heterotrophic organisms with a metabolism.

It was noticed that lipids, in the presence of water, formed a bilayer membrane which has been considered as the precursor to the cell membranes of living organisms. The protein content of the membrane of "protobionts"[2] have a functional resemblance to bacterial chromatophores involved in the absorption of light energy.

The combined action of electrical gradients and enzymatic action of RNA (Ribo Nucleic Acid) induced the various chemical reactions inside the "protobiont" (Chen and Walde, 2010) resulting in the formation of different organic compounds with replicable proteins. The modified "protobionts" as well as replicable organic compounds have been found to possess the characteristic features of living organisms, such as growth and cell division, and some resemblance to virus particles, and are referred to as "Eobionts'. The "Eobionts" of the primitive ocean were considered to be the first living entity or the penultimate formation of the prokaryotic cell, called "Protocell" (Chen and Walde, 2010), which most likely appeared 3.7 b.y.a. (in the Pre-Cambrian Period) in the primitive ocean.

The "RNA world Hypothesis" (Joyce, 1989) asserts that as a reproductive and hereditary component, and functionally as an enzymatic material, RNA was the primitive form of nucleic acid found in living beings of the primitive Earth. This is designated as the "RNA world" or "RNA first," as it could retrieve and functionally execute genetical information as well as synthesize proteins.

The role of RNA or ribozyme (also known as a molecular fossil) as the key ingredient of replication of the virus and its presence in prokaryotic living forms, supports the "RNA World Hypothesis." The critics of this hypothesis also admitted that due to its poor catalytic power and weak stability it could not support the secondary phase of evolution of living organisms. This has been supported by the DNA (Deoxyribo Nucleic Acid), which formed as a result of the replication of single strand RNA followed by a sugar substitution.

According to the renowned biochemist who created a protocell in his laboratory, J. Szostak (2012), DNA was created by the replication of RNA and there is supposed to be an intermediate form where the combinations of the complementary nucleotides of the RNA strand template and deoxyribose sugars exist together. However, it was noticed that the chimeric formation of DNA and RNA in the laboratory were found to be quite unstable to sustain together. This indirectly challenged the validity of the "RNA World Hypothesis". This proved that RNA would not have had the stability necessary to form DNA in the due

[2] Protobiont: An aggregate of membrane bound organic molecules, acting as precursor of prokaryotic cell.

course of evolution. An alternative theory; Co-evolution of DNA and RNA molecules at the same time, was thus proposed.

Bernhardt (2012), a strong critic of the "RNA first Hypothesis", proposed the "Protein first Hypothesis" that suggests that protein preceded RNA, and alternatively, that RNA and protein co-evolved to initiate the journey of life. This hypothesis wasn't adequately substantiated.

c) Origin of the first cell: In the cooling phase, the temperature of the primitive ocean gradually decreased to 500 Celsius. This transformed the nucleoprotein (of the free gene), modified "protobionts" or "protocells" (Szostak, 2012) to form the "Primitive Cell" containing cytoplasm and naked DNA. These cells were aapable of synthesizing protein, producing energy, and utilizing it in a productive manner. Evolutionary scientists also suggest that the early form of cells were anaerobic and prokaryotic, and were classified as heterotrophs. These cells absorbed organic nutrients from the surrounding waters of the primitive ocean. Contextually, a 3 billion years old epigraph of 'Eobiont' from the African continent "*Eobacterium isolatum*," should be mentioned here.

iii. Cognogeny: This is an extension of the "Oparin-Haldane Hypothesis," which narrates the characteristics and evolution of the primitive call as under the following headings:

   a) Origin of Chemosynthesis: The analysis of the nutritional uptake of primitive prokaryotic cellular organisms helped classify them as "Chemoheterotrophs" due to their pattern of assimilation of nutrients from organic compounds in their aquatic ambiance. The gradual decrease in the temperature of the paleoclimatic atmosphere lowered the temperature of the ocean water which in turn affected the production of organic compounds required by the simple form of prokaryotes. The continuous absorption of organic compounds by the prokaryotic heterotrophs from the ocean water reduced the pool of organic nutrients available in the ocean water. This dire need induced the evolution of the "Chemo-heterotrophic organisms" from "Chemo-autotrophic organisms." This kind of a paradigm shift in the absorption of nutrients, readily available from the organic compounds in the surrounding habitat, to achieve the power of the primary stage of self-reliance by developing a physiological mechanism, is known as "Chemosynthesis." The efficacy of this mechanism is related to the high temperatures on Earth. The existence of simple prokaryotes (without a true nucleus) like Iron bacteria and Sulfur bacteria, further prove this phase of evolution.

Paleoecologic evidence made it conceivable that the ancient Earth or proto-earth collided with meteorites causing a solar wind which created a huge quantum of heat energy resulting in extreme hypoxia, dry and hot conditions for millions of years on Earth. In those extreme conditions of the paleoclimatic atmosphere, the successor of "coacervates," a group of living organisms known as "hyperthermophiles" or "extremophiles" evolved. They could survive under conditions of extreme atmospheric pressure, exposure to radiation, in a microenvironment with a high pH, and hypoxia, which was acidophilic, alkaline, and had high salinity, of hydrothermal vents (the submerged crater of volcanoes that molten lava or gaseous compounds admixed in the ocean water) under the sea floor. In such extreme climatic conditions, the extremophiles comprising of different chemo-autotrophic, prokaryotic archaebacteria, e.g., *Pyrolobus fumaria,* and some of the chemo-autotrophic, eukaryotic insects (grylloblattids), worms (e.g., Pompeii worm), and crustaceans (Antarctic krill) etc.), evolved, diversified, and played an important role in the formation of the ancient biosphere, markedly different from the present one (Goswami and Das, 2016). These unique groups of primitive living entities on Earth could survive and reproduce in the saturated brine of the benthic level of the ocean, -12 degrees to +100 degrees Celsius, could sustain up to 1400atm of hydrostatic pressure, and were capable of synthesizing energy from inorganic minerals like iron, sulfur, manganese and etc. These extremophiles have been found to carry nucleic acids, produce proteins, enzymes (extremozymes), and can sustain in extreme weather conditions. These traits were passed on to the next generation in the course of evolution. Extremophiles provide an idea of primordial life on Earth and on other planets in the solar systems.

b) Origin of Photosynthesis: In the next phase of evolution, the temperature reduced further and the availability of free oxygen gas induced the prokaryotic simple form living microorganisms to go through structural modifications and acquire a complex form. This structural (morphological and anatomical) variation lead to the upgradation of the physiological activities of these organisms. Eventually, some of the "Chemo-autotrophs" (non-transducing prokaryotes) of the primitive ocean absorbed magnesium porphyrin from the surrounding water to form the photoreceptor pigment known as "bacteriochlorophyll," the precursor of chlorophyll of the eukaryotic transducer. This initiated the mechanism of "photosynthesis" among the transducing prokaryotic cells.

Around 22 amino acids were traced from the 3000 million years old rock crevices (Schopf, 2001). The prokaryotic, single-celled, photosynthesizing micro-organisms, Cyanobacteria, also known as "Blue-Green algae", evolved on the primitive Earth and are still existing. The earliest microfossil of cyanobacteria, *Archaespheroides barbertonensis,* evolved around 3.3–3.5 b.y.a. 2.9 billion years old fossils of cyanobacteria from India have been kept in the archives of the Birbal Sahni Institute of Palaeobotany (Schopf, 2012).

It was noted that millions of bacterial populations combined together to form bead-like structures called stromatolites- a complex structural formation of microbial fossils

(Nutman et al., 2016). They are found among the layers of calcium carbonate and anatomically comprise of three layers; the outer layer of stromatolites is made up of cyanobacterial mats and the innermost layers are made up of chemo-organotrophic bacteria, interspersed with anoxygenic photosynthesizing bacteria.

The epigraphic existence of stromatolites was further affirmed in the hot springs and hot hydrothermal vents of the ocean floor, which are 3 to 3.5 billion years old. As water is one of the most important ingredients of photosynthesis, it has played a key role in the origin and evolution of life. The primitive form of the rare stromatolite colonies was discovered by paleontologists from two locations (Hamelin pool and Lake Clifton) in Australia (Nutman et al., 2016).

c. Origin of the transducing of the Eukaryotic cells: A combination of different factors in the paleoclimatic atmosphere, like the elevated levels of free oxygen, the gradual decline of the Earth's surface temperature, the occurrence of thunderstorms, and the electrical sparks of lightning clouds in and around the atmosphere of the Earth, set up a fundamental change in the Earth's atmosphere. The dissociation and re-association of 0_2 molecules (as a result of physicochemical reactions) lead to the formation of the O_3 molecule, also known as Ozone. The accumulation of O_3 gaseous molecules in the Earth's atmosphere helped form a safety blanket, known as the "Ozonosphere," which prevented the exposure of the Earth's surface to the incoming Ultra Violet rays of the sun.

The Ozonosphere stopped the incidence of high solar radiation from reaching the Earth's surface and also helped reflect the heat energy back into space. This helped maintain an isothermal condition around the Earth.

Epigraphic analysis ascertains that the evolution of eukaryotic transducing and metazoic cellular organisms from prokaryotic transducing stock (e.g., various bacteria, cyanobacteria, algae etc.) took place around 2-1.5 b.y.a. (Schopf, 2001).

According to scientist Marguillis (1970), the symbiosis of various prokaryotic living beings gave rise to eukaryotic forms of life. However, scientists Mehler and Raff (1972) advocated that the prokaryotic cellular organelles went through a phase of mutation and evolution to form the "Eukaryotic cellular form." It has further been postulated that a low exposure to sunlight, the presence of moisture, availability of minerals, and free oxygen, helped diversify the physiological structures of primitive organisms. These organisms developed the traits of assimilation and metabolism (like photosynthesis, respiration, carbon assimilation etc.) as a result of the aforementioned conditions. The predecessor form of bacteria, "Archaea," has been found to be the living representative of microorganisms known as "extremophiles" that existed in the pre-Cambrian Period. Epigraphic records of living organisms have supported the 2nd transitional shift, i.e., the appearance and diversification of multicellular eukaryotic and metazoic cellular organisms, supposedly took place around 740-640 million years before the present form of the Earth. The primary shifts of paradigm in the evolution of plants on Earth were witnessed in the Proterozoic Era when the "age of prokaryotes" switched to the "age of eukaryotes" to continue the journey of transducing thallophytes on Earth.

According to paleontologists, the period between 750 to 550 million years (during the Cambrian Period of the Paleozoic Era) before the present form of the Earth, the evolutionary process of diversification of terrestrial life was noticeably slow (in lag phase). This was followed by a sudden explosion (in log phase) of diversification of marine life in the Upper Cambrian Period, (around 560-550 m.y.a.). The first "mass extinction incident" took place around 500 million years ago (during the Ordovician Period of the Paleozoic Era) and pushed most of the living organisms to the brink of extinction. Only some members of the aquatic algal plants, ancient thallophytes, and some lower group of animals belonging to insects continued the journey of evolution. While evolutionary biologists consider such events of mass extinction to be a negative occurrence, some groups of phylogenetic scientists claim that such calamities are essential to flush out the weaker forms of living organisms to allow the survival of only the fittest. Epigraphic evidence suggests that the terrestrial group of plants and bony fishes evolved and diversified during the Devonian Period of the Paleozoic Era and that the second mass extinction took place during this period (between 360-350 million years ago). This lead to the extinction of almost all of the living organisms, leaving only a select few (Kolbert, 2014). Paleontological evidence of the Carboniferous Period of the Palaeozoic Era recognizes this period as the age of diversification of amphibians and primitive reptiles.

The conventional idea about the evolution of advanced thallophytes and early tracheophytes is related to the migration process of diverse aquatic, oogamaous, photosynthesizing algal communities, to the land. This gave rise to a sexually distinguishable group of plants, bryophytes, and features such as the "alternation of generation," supposedly co-evolved during that time. However, the structural simplicity of bryophytes compared to the complex pteridophytes, forced the plant morphologists to be biased towards the earlier evolution of bryophytes. On the other hand, an independent mode of evolution was noticed in the chlorophyte algal groups of Cladophorales and Chaetophorales, suggesting the existence of isogamous algae in a primitive form. Whatever may be the ancestral form of chlorophyte algae that existed back then, the palaeobotanic evidence indicates that immediately after the first "mass extinction," a new transitional form of the mother-stock appeared in the primitive chlorophyte–thalloid members and the relatively advanced tracheophyte members. Eventually, the sexually reproducing, primitive transducers, bryophytes, evolved in the Carboniferous Period of the Paleozoic Era from that stock. The second paradigm shift in pant evolution was witnessed in the Ordovician and Silurian Periods of the Paleozoic Era. The "era of thallophytes" switched to the "era of tracheophytes" to continue the journey of the greens. Strategic movements, such as the migration of aquatic life to terrestrial life, played an important role in this.

On the basis of distinguished structural characteristics, the members of pteridophytes have been categorized into 3 groups: Lycoposids, Sphenopsids, and Filicopsids. According to the critical analysis of paleontological evidence, the evolution of tracheophyte plants, particularly the "evolution of pteridophytes," have been found to be a little complicated as the evolutionary trail of this group of plants is distinctively polyphyletic rather than monophyletic.

Palaeobotanic investigations ascertained that the transitional form between the primitive tracheophyte and the diversified tracheophyte epigraphical, known as "*Rhynia*" or "*Cooksonia,*" gave rise to the progenitor of early pteridophytes known as a hypothetical, paraphyletic "Protracyophyte" and "non-vascular polysporangiophytes" of the Palaeozoic Era. While the arborescent and herbaceous forms of pteridophytes evolved together from the primary mother-stock, the "Protrachyphytes" (the progenitor during the Lower Silurian Period) separated into two distinguished forms during the Carboniferous Period and the arborescent form of Pteridophytes finally disappeared during the later Permian of the Paleozoic Era. The herbaceous members of lycopods survived until the Tertiary Period of the Cenozoic Era, until they confronted challenges of habitat destruction and climatic changes in the Quaternary Period of the Cenozoic Era. The available fossil evidence helped the paleobotanists find a descendant of the complex structural identity, "*Psilophyton,*" in the later Silurian Period of the Paleozoic Era. This was the secondary mother-stock of the diversified pteridophytes, known as "Sphenopsids," popularly known as Horsetails. The members of Sphenopsida flourished during the Carboniferous Period of the Paleozoic Era and reached their phase of decline in the Permian Period, during the great geological incident of fragmentation of the super-continent of "Pangaea." Substantial palaeobotanic evidence further points to the origin of a most advanced group of pteridophytes, "Filicopsids," evolved from the same mother-stock: *Psilophyton*.

The landmark incidents regarding the evolution and diversification of life on Earth took place during Permian Period of the Paleozoic Era, on the super-continent of Pangaea, afloat on the super-ocean, "Panthalassa." At the same time, the third "mass extinction" incident occurred during this Permian Period (around 250 million years ago) to put an end to the flourishing primitive floral and faunal spectrum and with only a few members surviving. Tectonic movements induced the fragmentation of the supercontinent to their contemporary form as multiple continents (Oreskes, 2008). This was initiated during Lower Triassic Period of the Mesozoic Era, which initiated another paradigm shift. This entire process can be classified under four phases: the segregation of similar living forms; followed by the movement of these segregated living forms to distant locations; followed by the morphological and physiological acclimatization in different environmental, ecological, and edaphic conditions; this lead to the phylogenetic changes of the segregated living forms, distinguishing them from each other, allowing the speciation and evolution of life on Earth to continue.

The paleo-botanical and phylogenetic evidence helped the scientists consider gymnosperms as polyphyletic groups. They believed that the members of the early groups of gymnosperms: Ginkgoales (structurally primitive), Coniferales (structurally advanced) and Gnetales (structurally advanced), evolved from the common, secondary mother-stock "*Psilophyton*" of the later Silurian Period. The members of Ginkgoales and Coniferales flourished during the Triassic and Jurassic Periods, whereas the members of Gnetales flourished during the Cretaceous Period of the Mesozoic Era. Studies of the fossil remnants of early spermatophytes helped the plant phylogeneticists to identify 'Progymnosperms', a tertiary mother-stock that supposedly existed during thw later Devonian Period. On the other hand, the ascent of early gymnosperms, the seed ferns or "Pteridospermales," during Carboniferous Period of Palaeozoic Era from the tertiary mother-stock "Progymnosperms" placed the evolutionary progress of plants at the crossroads of two different hypothetical pathways (Cleal and Thomas, 2009).

According to a group of phylogeneticists, seed-ferns flourished during the Jurassic Period and disappeared during the Cretaceous Period of the Mesozoic Era. Other groups claim that the "Pteridospermales" themselves acted as the transitional "Quaternary stock" from which the members of Cycadales evolved during the later Triassic Period. They flourished during the Jurassic Period and a few members of the living fossils survived until the Quaternary Period of the Cenozoic Era.

During the Triassic Period, the diversification of primitive members of reptiles made them evolve into a massive form of reptiles. In the middle of the Mesozoic Era, i.e., the Jurassic Period, which is also known as the 'age of dinosaurs', these massive reptiles dominated the Earth and the forest habitat was dominated by diverse gymnosperms (the naked seeded plants or early spermatophytes).

The separation of the early continents continued with the progression of the evolution of life concomitantly during the Jurassic Period. During the fourth "mass extinction" (around 70 million years ago) that took place in the Cretaceous Period of the Mesozoic Era, the diversified groups of gymnosperms and dinosaurs were on their way to extinction and paved the way for modern smaller reptiles; turtles, snakes, crocodiles, and etc. (Kolbert, 2014).

The final shift of paradigm, which is recognized as the evolutionary change of regime of primitive spermatophytes to advanced spermatophytes, took place during the early Cretaceous Period of the Mesozoic Era. This directly or indirectly was influenced by the disintegration of the continental mass to create the contemporary seven continents. Studies of fossils have helped scientists decipher the evolution of various groups of animals like mammals, birds from reptiles, and the evolution of dicotyledonous angiosperms, the modern group of spermatophytes. They supposedly evolved from the "Quaternary mother-stock" progenitor "seed ferns" of "Pteridospermales" during the late Jurassic Period or the early Cretaceous Period of the Mesozoic Era (Cleal and Thomas, 2009). The diversification of this group of plants continued in the Tertiary Period of the Cenozoic Era. The evolutionary footprints of different plant groups on the Earth have been presented in Figure 1.

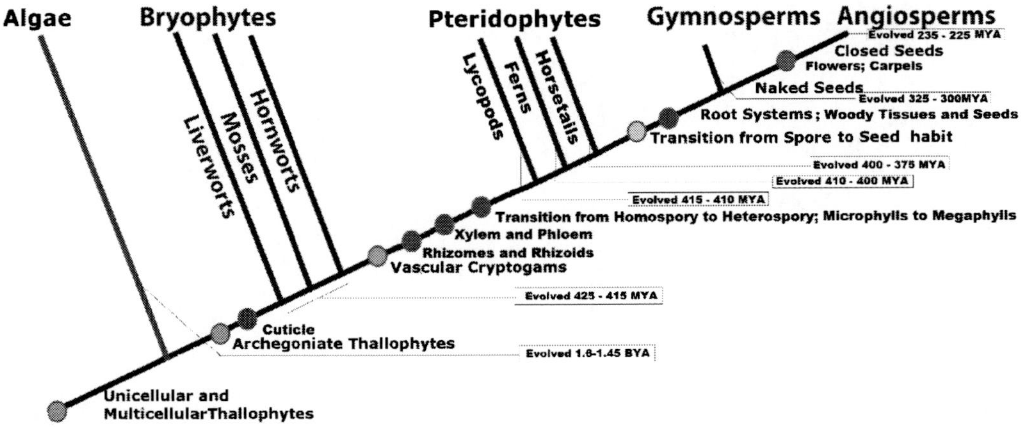

Figure 1. The evolution of different plant groups on the Earth.

The progress of evolution of life was noticed in the Tertiary Period of the Cenozoic Era when monocotyledonous spermatophytes were noticed to have evolved from dicotyledonous spermatophyte. The diversification of dicotyledonous plants, mammals, and the evolution of human ancestors took place during this period. The progress of life on Earth has been marked with incidents such as the diversification of dicotyledonous and monocotyledonous angiosperms, the evolution of modern humans (in the Quaternary Period); the progress of evolution of life has gone through a labyrinth of different life forms. The Green Revolution, Industrial Revolution, the age of deforestation, the phase of the Ozone Hole formation, the critical regime of global warming (still continuing through the era of Cyber Revolution of the Quaternary Period of the Cenozoic Era) are all chapters in this curious labyrinth!

Chapter 2

THE COLLECTIVE GEOLOGICAL HISTORIES OF LIFE AND TIME THROUGH THE LENSE OF PALEOBOTANY

GEOLOGICAL HISTORY THROUGH THE TRANSITION OF THE COSMIC CALENDAR IN THE ERA OF UNWINDING DNA (TEMPORAL REFERENCE)

"The biological change, as is with geological change, can occur in fits and bolts; there may be periods of rapid evolution followed by prolonged periods of constancy" - Stephen Jay Gould

The magnitude and extent of origin, evolution, and diversification of life on Earth is yet to be considered as a matter of controversy. The interpretation of any evolutionary incident has always been analyzed by different groups of scientists like paleobotanists, molecular systematics, evolutionary biologists, and geobiologists, with everyone having a different perspective in analyzing the same issue in their own time-scales. A prototypic "International Time Zone" format for interpreting a common evolutionary issue in the background of the "Cosmic Clock" does not exist. However, the contemporary hypothesis of evolution states that constant gradual accumulation of genetic variations would help in the evolution of a new species, although there is no strict methodology to determine the timing or rate of evolutionary progress. Hence the perception of time to judge evolutionary progress is still a qualitative process.

As we experience the thrill of the count-down of these incidents in the descending clocking system of Days – Hours – Minutes – Seconds - Nanoseconds …and so on, the evolution of all plant species has been concerned with a paleobotanical history of millions of years.

Even in our daily time frame, we find it inconvenient to sit in the same place for 10-12 hours. However, the tantamount features of life, particularly of the plant species, have remained on Earth without any change for thousands and millions of years. This make us thinking about the "conceptual time clock" of biological evolution. In order to understand these evolutionary incidents in a temporal capacity, the use of a "cosmic calendar" seems like the best way to customize 4.8 billion years of the history of the formation of the Universe as a calendar year. The formation of the universe (as per the Big Bang Theory) can be recognized as the maiden incident, taking place on 1st January, in relation to which the evolution of a living cell on Earth would take place on 8th April; the photosynthetic organism appeared on 1st July; the evolution of early land plants happened on 1st December; and the evolution of flowering plants took place between 21st and 23rd December. For this cosmic clock timeline to be conceivable, all these incidents must be considered to have transpired over a span of one second (Banks, 1970). This cosmic clock helps us understand the dynamics of the ancient plant species in any particular location in due course of time, as well as allow us to extrapolate the dynamics of a given species if the environmental factors, and atmospheric conditions remained static.

Eminent scientist Dr. H. P. Banks (1970) referred to the paleobotanical evidence and stated that in the Eocene Period (60-50 million years ago), the floristic elements of Oregon, (US) were predominantly sub-tropical, and were gradually succeeded by Arctic floristic elements as the climatic regime of this region changed to Temperate and Tundra type over the course of 1.8-2 million years of time. He found that the original subtropical floristic elements have shifted around 2000 miles southward since.

According to Dr. Banks (1970), at the migration rate of plants towards the south at 5 feet/year, the subtropical floristic elements could only move 1 mile south in a 1000-year period. The same floral elements would move 2000 miles southward in search of the sub-tropical climatic regime over the next 2 million years.

Thus, the analysis of the paleobotanical history with reference to the "cosmic clock" helps us understand the evolutionary past of any particular group of plants. It could also be treated as a futuristic tool to forecast the dispersal and migration of floristic elements and plant groups in the near future.

Geological History, Paleobotanical Processes and Evidence to Establish Relationships of the Past (Extinct) with the Present (Extant)

Literally, the term "fossil" means "extracted from the Earth," which is the epigraphic identity of a living organism in nature, formed as a combined effect of physicochemical reactions on the residue of living organisms after death. Paleobotany is the branch of science, dealing with the fossilized plants of the Earth. The geobotanical history of settlement of plant life on Earth has mainly been concerned with intensive analysis of ecological and floristic patterns of the living plants (i.e., extant) in the background of the dynamic variables of the atmospheric, climatic, edaphic, topographic, and biological changes. Furthermore, the fossilized predecessors (mostly extinct) of the living descendants provides us with a background of the paleoclimatic changes that the Earth went through.

There are a few impediments to create and secure enough fossilized representatives of the past:

i. The soft body of tissues of the dead organisms are found to be decaying rapidly;
ii. The chance of formation as well as availability of the fossil is remote;
iii. The paleoclimatic conditions and atmospheric temperatures might not be congenial to the forming of fossils;
iv. The sedimentation process might not be available to the living organism intended to be analyzed as a paleobotanical object.

Despite all the odd situations, we still could not wipe out the epigraphic footprints of life on Earth that have existed for a long time. The casts, impressions, petrified and other forms of fossil remnants that were excavated from the sedimentary rocks (e.g., shales, limestones etc.) of sea beds, buried under the suspended clay and minerals of water bodies or under solidified lava-streams of igneous rocks stand as testaments. These diverse forms of fossils are as follows:

i. Mold and Cast: When a stem, root or solid part of the plant gets buried and eventually decays in the sedimented rock, the surrounding sedimentary materials get hardened through the passage of time to form a template or mold. This is the external template of the bark or wood surface of the stem or root systems of a plant. When the mold or template of the plant part gets filled by the new sedimentary materials and solidifies to form a replica of the original plant part, they are known as casts, where no original plant materials or tissues exists anymore. Molds and casts may be found from the same fossil site.

The primitive members of arborescent pteridophytes, giant horsetails, and tree-ferns have been found largely in the form of molds and casts from the Jurassic Period.

ii. Impression and Compression: This process is also recognized as carbonization or coalification. When the plant materials get buried deep under the fine sedimentary deposits and sustain extreme pressure exerted by the upper strata, the water and air content of the tissues come out from the buried material to form a carbonized residue. This is known as compression or impression.

The flat, two-dimensional forms of these carbonized materials (also known as a positive) are known as compressions. The images of the carbonized materials or positives have been called impressions (also known as a negative) of the two-dimensional carbon film.

These diversified fossil remnants of the Carboniferous Period have created the black coal deposits (e.g., carbon deposit of giant tree ferns, gymnosperms etc.) to form diversified fossil fuel (e.g., lignite, anthracite, bituminous coal, diamond, graphite etc.) reserves on Earth.

The structural deformation noticed in such forms of fossils is due to the exertion of tremendous pressure combined with the physicochemical changes that the plant materials went through. However, the cuticular impression of the fossilized leaves and structural patterns of stems or trunks can be studied to reconstruct their extinct or primitive representation of their living descendants.

iii. Compaction: The plant materials of the Tertiary Period or post-Tertiary Period (of recent origin), were buried and went through a process of carbonization. These have been found buried in the lignified brown coal fossil forms, also known as compacted fossils. This process of fossil formation has been recognized as compaction.

As the petrified fossils of leaves, fructifications etc. are apparently soft and easy to examine after preparing their anatomical sections, they give a true representation of the fossil predecessors of extant members, helping conduct their phylogenetic analyses accurately.

iv. Petrification: When the plant parts got submerged in a waterbody that was enriched with calcium carbonate, silica, manganese, iron and other mineral salts, the organic chemicals of buried plant materials were gradually replaced by these mineral salts. Thus, the fossilized plant parts have morphologically and anatomically remained almost alike to their original plant parts for a long time.

Moreover, the cellular structure of petrified fossils have been found to be suitable for microscopic experiments. For instance, the petrified woods of Yellowstone National Park have provided us with an idea of the palaeoflora (mainly the Coniferous forest type) of North America in the Triassic Period.

GEOLOGICAL HISTORY, PALEOBOTANY, AND THE EVOLUTION OF LIFE: IN THE TEMPLATE OF A GEOLOGICAL TIME-SCALE

The 4.8 billion years of the dynamic history on Earth have gone through diversified transitional stages of geographical, geological, paleoclimatic, and atmospheric regimes. Paleobotany has become an integral part of documenting and narrating the glorified tale of the evolution of the contemporary diversified angiosperms on the land, which originated from unicellular prokaryotic thalloid life in the waterbody. The morphological changes of the Earth have been phylogenetically correlated to the origin, evolution, and diversification of these green transducers. The evolutionary changes of the Earth (as a symbolic abiotic component) and the green transducers (the biotic component) have been dynamic and ecological in their interaction. This has further been influenced by a number of universal, evolutionary variables that make the Earth suitable for the existence of a diverse group of plants in the different hours of our cosmic clock. These include the evolutionary changes of the atmosphere around the dark, cooled, proto-earth; to the formation of contemporary blue planet; gradual receding of the oceanic basins, expansion of land area and formation of a small range of mountains; origin of the mega ocean and mega-continent to give birth to more oceans and seven continents. These formed in the background of-

i. continental drifting,
ii. plate tectonics,
iii. paleomagnetic reversals,
iv. spreading of the sea floor,
v. formation of gigantic fissures,
vi. change of the paleoclimatic regime,
vii. seasonal patterns,
viii. glaciation,
ix. volcanism,
x. accumulation of suspended particles (chemicals, minerals and sediments in ocean and freshwater bodies) to form sedimentation.

While the evolutionary progress of life, particularly of the green transducers, has not been very well defined, the age-dating process of the paleobotanical evidence has been frequently used with the contemporary back-up of graphical reconstructions of these fossil plants. These include transmission electron microscopic analysis, molecular botanical analyses like cladistics analysis, DNA sequencing etc. However, the discrete, fragmentary fossil evidences have been chronologically documented throughout the determined geological time-scale.

With the process of age-dating it has been determined that sedimentary rocks mostly formed around 550-600 million years ago and therefore act as an epigraphic retrieval of the evolution of the transducers of the Paleozoic, Mesozoic and Cenozoic Eras. Geologically, the PreCambrian Period which represents the cosmic period of time prior to the Paleozoic Era unfortunately holds 90% of the history of the evolution of living plants and transducers on Earth with poor documentation due to a lack in formation of sedimentary rocks (Cleal and Thomas, 2009).

To get closer to the evolutionary history of the extinct transducer more accurately, these three eras have been sub-divided into the 12 geological periods and each geological period has subsequently been divided into a few epochs. As a snapshot review, if we look at the recent geological regime (Table 1A and Table 1B) of universal geological time-scale (ICS[3], 2014), the temporal span of the Cenozoic Era was 69 million years and was further subdivided into two periods; the Tertiary Period (63.2 million years of time span) and the Quaternary Period (starting around 1.79 million years ago and still continuing).

To review the evolutionary history of plant life over the past 70 million years, the geological "Tertiary Period" has been subsequently divided into five epochs-

i. Paleocene,
ii. Eocene,
iii. Oligocene,
iv. Miocene,
v. Pliocene.

According to paleo-climatologists, the most important event in this period was initiated with the PETM (Paleocene-Eocene Thermal Maximum), a sudden "thermal flux," evidenced by a 5-9^{0}C reduction in temperature (Jardine, 2011). This was induced by a negative surge of CO_2 levels in the atmosphere and the paleoclimatic conditions of the Earth in the Tertiary Period abruptly changed with this sudden cooling and dryness. The waterbody and landmass of the Earth in the Cenozoic Era went through changes like the receding of ocean bodies, gradual exposition of more landmass and rising of a series of low range mountains. All these paleoclimatic and geological changes highly influenced the evolution of diversified form of life (both plants and animals) on Earth.

The Quaternary Period (approx. 1 million years of lifespan) has also been divided into two geological regimes or epochs: the "Pleistocene Epoch" and the recent "Anthropocene Epoch" (Kolbert, 2011). In the Pleistocene Epoch, a series of four recurrent glaciations (with a short span of interglacial regimes) made most of the continental climatic conditions cooler and dryer. As this super-cooled environment was not suitable for the survival of a diversified group of plants and animals, they are thought to have existed before this period.

---

[3] ICS: International Commission of Stratigraphy.

This is further supported with the fact that most of the living organisms are found buried under ice-shields.

Table 1A. Biological evolution through geological time-scale

ERA	PERIOD	EPOCH	DURATION (Million years)	IMPORTANT EVENTS
CAINOZOIC	QUATERNARY	HOLOCENE	Contd. Starting from 0.01 million years before	Herbs and hardy tree species survived and diversified through evolution; modern birds, mammals, and man *(Homo sapiens sapiens)* emerged.
		PLEISTOCENE	0.79 (0.01 – 1.8)	Mass extinction event, many giant mammals extinct; genus Homo evolved; large tree extinct.
	TERTIARY	PLIOCENE	3.4 (1.8- 5.2)	Herbs diversified; large carnivores, pongids, hominids appeared.
		MIOCENE	18.1 (5.2- 23.3)	North and South America joined; honeycreepers emerged; hominids emerged; grazing animals proliferated; herbs invaded; trees dwindled.
		OLIGOCENE	12.1 (23.3 – 35.4)	Evoluation of grasslands and grazing animals; large running mammals proliferated; India collides with Laurasia; Angiosperms diversified.
		PALAEOCENE	10 (55 – 65)	Birds diversified; Mammals diversifed.
MESOZOIC	CRETACEOUS		76 (65-141)	Break-up of Gondwanaland; Dwindling of reptiles, gymnosperm; increasing of mammals and angiosperms.
	JURASSIC		54 (141-195)	Break-up of Pangea; age of gymnosperms and reptiles; birds appeared; origin of angiosperms.
	TRIASSIC		35 (195-230)	Mostly Cycadophytic gymnosperms; reptiles like Dinosaurs; mammals appeared.
PALEOZOIC	PERMIAN		50 (230 – 280)	Mass extinction event, Dwindling of pteridophytes, gymnosperms; reptiles emerged.
	CARBONIFEROUS		65 (280-345)	Dense rainforests of tree ferns, club mosses, gymnosperms; early reptiles, amphibians (early) appeared.
	DEVONIAN		50 (345 – 395)	Early vascular plants, pteridophytes, early bryophytes, fishes, amphibians appeared.
	SILURIAN		40 (395-435)	Migration of life from water to land; Early pteridophytes appeared, marine animals dominates.
	ORDOVICIAN		65 (435 – 500)	Marine algae and animals diversified.
	CAMBRIAN		70 (500-570)	Diversification of algae; invertebrate phyla appeared.
PROTEROZOIC			930 (570 – 1500)	Early sedimentary rocks with traces of blue-green algae; a few multi-cellular.
ARCHAEOZOIC			3000 (1500-4500)	Origin of life; evidence of bacterial and algal life on metamorphosed and Igneous rocks.

Table 1B. Biological evolution in the Mesozoic Era on Earth

ERA	Duration (in million years)	Series/Epoch	European Epoch	ICS (International Commission of Stratigraphy) Stage/Age
	6.5			Maastrichtian
				Campanian
M		Late (Upper)	Senonian	Santonian
		Cretaceous		Coniacian
				Turonian
E	97.5			Cenomanian
				Albian
				Aptian
S		Early (Lower)		Barremian
		Cretaceous		Hauterivian
			Neocomian	Valanginian
O	144			Berriasian
				Tithonian
		Upper Jurassic		Kimmeridgian
Z	163			Oxfordian
				Callovian
				Bathonian
O		Middle Jurassic		Bajocian
	188			Aalenian
				Toarcian
I				Pliensbachian
		Lower Jurassic		Sinemurian
	201			Hettangian
C				Rhaetian
		Upper Triassic		Norian
	231			Carnian
				Ladinian
	247	Middle Triassic		Anisian

The origin, extinction, re-emergence, diversification, migration, and dispersal of plant life on Earth has further been influenced by various other factors like the climatic, atmospheric, geological, topographic, edaphic, biological and anthropogenic changes on Earth with the changing dimensions of time. Epigraphical evidences retrieved by the sedimentary stones further support this. The geological time-scale acts as a dynamic template of reference to time, to help review the evolutionary progress of living organisms (plant and animal) on Earth. Furthermore, the cumulative changes of said factors helped evolve the diverse life spectrum of the recent "Anthropocene Epoch" (Kolbert, 2011). To visualize the evolutionary changes of living organisms as well as the geological changes of the Earth, the geological time-scale is a unique reference for analyzing epigraphic objects in nature.

Geological History and Hypothesis, Paleobotany, and Evolution of Plants on the Blue Planet

All the cosmogenic evidences suggest that in the primitive or proto-earth stage, due to an uninterrupted exposure to the solar vortex of nuclear explosions, the severity of volcanic eruptions, presence of ionized gaseous clouds in a reduced atmosphere, exposure to the heat-induced solar winds of the ancient Sun, and the impact of a series of meteoroids on its surface led to the ancient Earth's present solidified crust was in melted state in different strata.

The central core of the spherical proto-earth was consituted of dense metallic elements such as Iron, Nickel, etc. The outermost, apparently cold, solidified crusted layers of the surface of the sphere were made up of low-density stony patches, thus forming the lithospheric molten lava, rock level of the Earth, found in a floating state. The lava rocks of medium density were formed by the intercalated mantle layer in between the outermost crusted layer and the innermost central core. Studies of the Earth's magnetic field by the European satellite "Swarm" secured X-Ray images of the Earth's core, which is buried under 3000 km of solid rock. It has also been found that there is a molten jet stream of iron in the outer core of the Earth which is responsible for changing the Earth's magnetic field (Livermore et al., 2017). Critical analysis of data from the "Swarm" has helped us understand that the molten jet stream of the outer core is supposedly moving at the velocity of 40km per year towards the west, which is a thousand times faster than the tectonic plate movements. The "Swarm" mission is ongoing, to collect magnetic signals from the Earth's core, mantle, crust, oceans and ionosphere for better understanding of the Earth's magnetosphere. This could help solve the mysteries of organic and biological evolution on our planet. Through observations of the level of diversity on Earth, we are reminded of Newton's 3rd Law of Force; "every action has an equal and opposite reaction." The formation of a series of trenches supposedly took place as a result of some pressure gradients. This resulted in the formation of the island systems near the volcanoes as a result of upward geological pressure. This geological pressure or equilibrium has been maintained by "Geostasy[4]."

The important incidents of geological discovery, ideas, hypotheses, and their evolution are all accepted figments of the "Cyber Age" which are also known as the geological outcomes of the "Anthropocene Epoch." Some of these are paleomagnetic reversals, major fault lines, continental drifting, plate tectonics, sea-floor-spreading, and etc.

These played an important role in the evolution and diversification of plant species on Earth. The major fault lines of the geological matrix of the Earth are of various types that

[4] Geostasy: according to flood geological hypothesis, when continent rise, ocean basin sinks and geostasy is the load stabilizing natural phenomenon, encompassing a vast area where one part goes down, another part goes up to equalize it.

range from the fissure lines of earthquakes, the lava erupting mouths of volcanoes, epicenters of tsunami waves, etc.

According to the "Flood Geological Hypothesis," the formation of major fault lines was the result of a stress releasing mechanism in the process of formation of the Earth's crust which occurred as a result of homeostasis action, resulting in gradual rising of continents and sinking of the ocean floors. This hypothesis has helped determine that the rivers of the proto-earth were supposedly wider than the narrow oceans, before the "First flood" turned it around for the first time (Senter, 2011). The primitive continents were presumably close to each other. A prehistoric River "X" supposedly existed between South America and Africa. The onset of this maiden flood led to the filling-up of the River basin of "X," resulting in it sinking further which caused the continental masses to be relatively uplifted. This is referred to as the feasible explanation of the reciprocal match of the East coast of South American continents with the West coast of the African continent. On the other hand, the "Plate Tectonic Hypothesis" (Wegener, 1924) explains that these fault lines were created by the sliding of continental plates below one another. It has been accepted that the "Plate Tectonic Hypothesis" is the more likely geological process responsible for the shifting of continents.

1) Plate Tectonic Hypothesis: According to this hypothesis, the Earth's surface (either the top level of the continental shield or the bottom level of the ocean floor) geologically known as plates are considered to consist of layers that are a few miles thick. They have always moved sideways, while gently scrubbing each other, with one layer ultimately giving away its structural integrity and gradually bending to be overlapped by the other plate or layer. The bent, inner part would be trapped inside the molten lava while the overlying plate came out emerged from the ground at a right angle, solidifying instantly as a hard rock.

Though, as a hypothesis, "Plate Tectonics" is unique, the critics of this hypothesis have stated that it does not justify all the evolutionary expectations. The following aspects can be considered as counter-arguments to this theory:

i. If one end of the plate bends down straight underneath, the other end should go upward to form a sky-high two-dimensional plate on the ground. This however wasn't the case.
ii. Numerous cracks have been found on the plates of the ocean floor, through which sea-water could come in contact with the molten lava rock level, through capillary action, initiating massive volcanic eruptions and the resultant cascade of inundations would result in the flooding of the entire continent, but that did not happen either.

On the other hand, the supporters of this hypothesis, have referred to the "Mid-Atlantic Ridge" (extended from the North to the South of the Atlantic Ocean), a crack on the ocean floor from which lava streams gush, spreading out on the ocean floor. However, the downside of this supposition is that instead of the sideways movement of lava (necessary to build up the ocean floor), the lava debris have heaped-up in the same location to form submerged rocks on the ocean floor.

According to the geological history of the ancient Earth, all of our present continents originated from the ancient, single, mega-continent called "Pangaea," that was surrounded by the primitive ocean "Panthalassa." Geologists & evolutionists alike have considered that the ancient continental plates disintegrated, and drifted sideways to reach their current stage, a phenomenon popularly known as the "Continental drift" (Wegner, 1912).

2) Continental Drift Hypothesis: According to the proponents of this hypothesis, the floating continental plates did not recycle the sea-bed formations, rather these plates came in contact with each other, collided, fragmented, and drifted sideways to reach their current positions to form the landscape of all contemporary continents with the major oceans.

The collision of the continental plates has supposedly been evidenced in the form of earthquakes and the formation of large mountain ranges like the Himalayas, and the Alps. They have geologically been constructed as a result of the severe collisions of two continental plates in pre-historic times. On the basis of geological analysis, remote sensing, age-dating processes, phylogenetic observations, ultra-microscopy, and the advanced application of digital and crystallographic imaging, etc., the present shapes of the continents were determined and the contemporary movement of continents 5-10cms/year sideways has been ascertained. Unlike "Plate Tectonics", "Continental Drift" has three major supporting evidences to be considered as a real geological incident and plays an important role in the evolution of living organisms, particularly the transducers in pre-historic times. They are:

i. Paleobotanical/Fossil evidence in favor of the Continental Drift: Paleobotanical experiments have proved that the species diversity of plants and animals of different continents in the same atmospheric and climatic conditions without any physiographic barriers like high mountains or seas, remained identical for almost a thousand years. The availability of similar fossil evidences in Antarctica, the Arctic and Europe have helped predict the existence of identical living organisms in a common, broader matrix within the same climatic regime in a particular geological regime.

ii. Reciprocal match of continental masses in favor of the Continental Drift: The reciprocal match of inter-continental masses of the East coast of South America with West Coast of Africa makes this theory geologically evident.

iii. The vegetational match in favor of the Continental Drift: The identical atmospheric, topographic, climatic and edaphic conditions have played a key role in the formation of identical vegetational patterns. The West coast of Africa and the East coast of South American continental masses have been lithophytically enriched with diamonds and have an identical vegetational profile. This furthermore confirmed that these two are apparently separate continents formed from the same mass.

GEOLOGICAL HISTORY, PALEOBOTANY, AND DIVERSITY OF PLANTS ON THE BLUE PLANET

The diversified forms of extants, transducers, as well as the species diversity in plants, their gene diversity, and ecological diversity on the Earth's surface, are not a result of any sudden evolutionary explosions of life.

The current form of multicellular, eukaryotic, diversified, angiosperms have evolved through the interaction of various abiotic factors. Both abiotic and biotic factors contributed to the co-evolution of-

i. the Earth (metamorphosis of a dark, smaller, hot planet with a reduced environment, to the larger sized, cooler, blue planet with an oxidizing atmosphere)

ii. the Protocell (which has gone through evolutionary transitions of the geological timescale to form prokaryotic unicellular to eukaryotic unicellular and multicellular forms of life followed by the formation of multicellular, diversified forms of transducers in different ecological regions). The difference between prokaryotic and eukaryotic cells is the presence of a nuclear membrane in the latter to help to maintain the cytogenetic configuration of a particular species for the long-run. However, the "relief features of the first order" i.e., the continents and oceans of the globe in the past and present have inspired scientists to unveil the natural history of the Earth's origin and the evolution of its continents and oceans. Also, the abode of living organisms or primitive, simple, transducers on Earth, identified as footprints of evolution of the earliest forms of life, have been epigraphically restored on rocks (fossil remnants).

The present conventional phylogenetic studies draw phylogenetic inferences of extant plants to their lines of biotic evolution as well as discern their extinct predecessors. The eminent molecular phylogenetist, E.S. Barghoorn (1971), has narrated about 3.5–2 billion years of evolutionary journey of life on the Earth's surface in the last two decades. The ancient, simple, prokaryotic cellular forms of life, presumably evolved from the protocells that are identified have been "bacteria," "blue-green algae," or "green algae," which used to live in an aquatic habitat. They evolved around 3.5 b.y.a. and allow us to conceptualize that biological evolutions that were initiated 1 billion years after the formation of the Earth. During phylogenetic studies of some extant "stromatolites," which are the progenitors of cellular thalloid, and the earliest existence of simple forms of living organisms that are transducing, green algae were witnessed in the spherical, microfossil representations of *Eubacterium*, discovered from a 3.2 billion years old rock deposit in Transvaal, South Africa. The earliest microfossil of filamentous blue-green algae has been found in the 2 billion-year-old rock deposits of Western Ontario, Canada. Given the absence of a nuclear membrane, eukaryotes are not susceptible to instant changes in response to external stimuli and hence lengthen the outcome of biological evolution for a period of time. Eukaryotic forms are considered as advanced forms of the prokaryotic forms, based on the age-dating of 0.5 billion years old eukaryotic, green, algal, microfossil remnants. They were collected from the rock deposits of Northern Territory, Australia. From these paleobotanical evidences we have been able to infer that around 550-500 million years ago, the simple eukaryotic forms of the photosynthesizing living organisms went through an evolutionary explosion to diversify in the form of multicellular, thalloid green organisms on Earth.

Around 450-400 million years ago, transmigration of these thalloid forms of photosynthesizing living organisms on Earth led to the formation of living organisms which are highly diversified in terms of their structure, function, ecology, and genetics. Hence, they are advanced enough to face any challenges to their survival. Since this stage, plant organisms have gone through various different stages of evolution, diversification, colonization, and extinction, to appear, exist and disappear and then re-appear in a new form or group of plants in the labyrinth of the biological evolutions of life. So, paleobotany ascertains that biological evolution is not a single continuous process to initiate, originate, and conduce evolution and speciation on Earth, but is rather a cumulative outcome of diverse processes that happened over a due course of time to direct the advancement of different groups of transducers from simple to advanced forms, to cope up with the challenges to survival presented to Earth with the progression of the geological time-scale.

These diverse groups of transducers, as well as the transitional living groups and the outcomes of biological evolutions, have all been retrieved from the paleobotanical repository and broadly categorized as:

1) Thallophytes: This is a group of living, transducing organisms that originated around 3.2 billion years ago, in an aquatic habitat as simple, prokaryotic,

unicellular forms of algae, devoid of roots, stems and leaves. These algal communities have primarily been categorized in green, blue-green, brown, red and golden-brown types, based on their pigments. Physiologically, algae have little or no tolerance to survive in a terrestrial habitat. Hence, these niches were gradually occupied by the diversified, physiologically advanced, multicellular, eukaryotic algae. The structurally diversified, big, aquatic members of Phaeophyceae (popularly known as sea-grasses) appeared with root-like (holdfast), stem-like (stipe) and leaf-like (lamina) structures. And terrestrial algal members appeared in the due course of biological evolution, around 700-500 million years ago.

2) Archegoniate Thallophytes or Bryophytes: This group of plants is known as plant amphibians, as they transmigrated from waterbodies to terrestrial habitats. The internal tissue system in this group of plants is remarkably diversified. There is a complete absence of true leaves and stems in these bryophytes. However the formation of a leaf-like lamina has been noticed in some. The presence of conspicuous alternation of generations and reproductive bodies like antheridia (male reproductive organ) and archegonia (female reproductive organ) have been found in the thallus of some bryophytes. They are also recognized as archegoniate type thallophytes. Paleobotanical evidences are scanty for this group of transducers.
3) Early Tracheophytes or Pteridophytes: The early tracheophytes or pteridophytes have been subdivided into the following categories-

i. Clubmosses: Paleobotanical evidences reveal that the small, herbaceous members of the early tracheophyte groups (e.g., *Lycopodium* and *Selaginella*) are the descendants of arborescent members of the ancient lycopsids (e.g., *Lepidodendron*, *Sigillaria*). They evolved during the Carboniferous Period but disappeared during the Permian Ice Age.
ii. Horsetails: The ancient horsetails or sphenopsids in the Carboniferous Period have existed in water-logged swamps in two forms:
   a) arborescent form– *Calamities*
   b) herbaceous form – *Sphenophyllum.*

   The herbaceous forms are smaller like *Equisetums* which are the only living successors of the ancient horsetails. They disappeared during the severe glaciations of the Permian Period (also known as the Permian Ice Age).
iii. Pteropsida or ferns: The ancient members of pteropsids or ferns evolved during the Carboniferous Period in small herbaceous forms as under-shrubs of the forest floor and as arborescent forms of tree-ferns (e.g., *Psaronius*, a 23 feet long tree fern of the Carboniferous Period) later. The true ferns appeared during the transmigration of early land plants and were found in diversified forms up to the Tertiary Period.

"Seed ferns," another transitional plant group that forms an evolutionary link between the early tracheophytes (pteridophytes) and early spermatophytes. Gymnosperms evolved during the Carboniferous Era and have morphological resemblances to tree ferns but carry the seeds of the gymnosperm group of plants. This missing link of the group of tree ferns (e.g., *Medullosa*) is known to have flourished until the Devonian Period and disappeared during the Permian extinction.

4) Gymnosperms or early spermatophytes: The advanced tracheophytes or early spermatophytes are characterized by the presence of roots, stems, leaves, and seeds without protection and comprise of one extinct group (Cycadeoids) and three living groups of gymnosperms (Cycads, Ginkgos, and Conifers) (Sporne, 1959).

i. Cycadeoids: The paleobotanical evidences of the early Triassic Period showed that Cycadeoids are characterized with massive frond-like leaves, grown out of barrel-shaped trunks, and the trunks conspicuously possess two scar marks, one from the leaf base and another from the reproductive organs. This group of plants with arborescent habits most likely disappeared during the Cretaceous Period.

ii. Cycads: The structural resemblance of the leaf anatomy of Cycadeoids and Fossil Cycads make it difficult to separate these two groups of plants from one another, however, their reproductive structures differentiate them. The living members of Cycads have also been considered as living fossils which posess the primitive, relict characteristic features of primitive spermatophytes, which remained unchanged for millions of years.

iii. Ginkgos: These arborescent forms of gymnosperms have been represented by a solitary extant species characterized by fan-shaped leaves, with characteristic radiating parallel venation. Found in the remote corners of China, the Ginkgos have been considered as living fossils due to possessing relict features and remaining structurally unchanged for millions of years. It supposedly existed in its diversified state until the Permian Period.

iv. Conifers: These arborescent members have been characterized by evergreen trees with needle-like leaves. This group of plants comprise of Spruce, Pine, Fir, Hemlock, Juniper, Cypress, and Bald-cypress, which evolved in the pre-Carboniferous Period and diversified and existed in the temperate parts of the Northern hemisphere of the Earth.

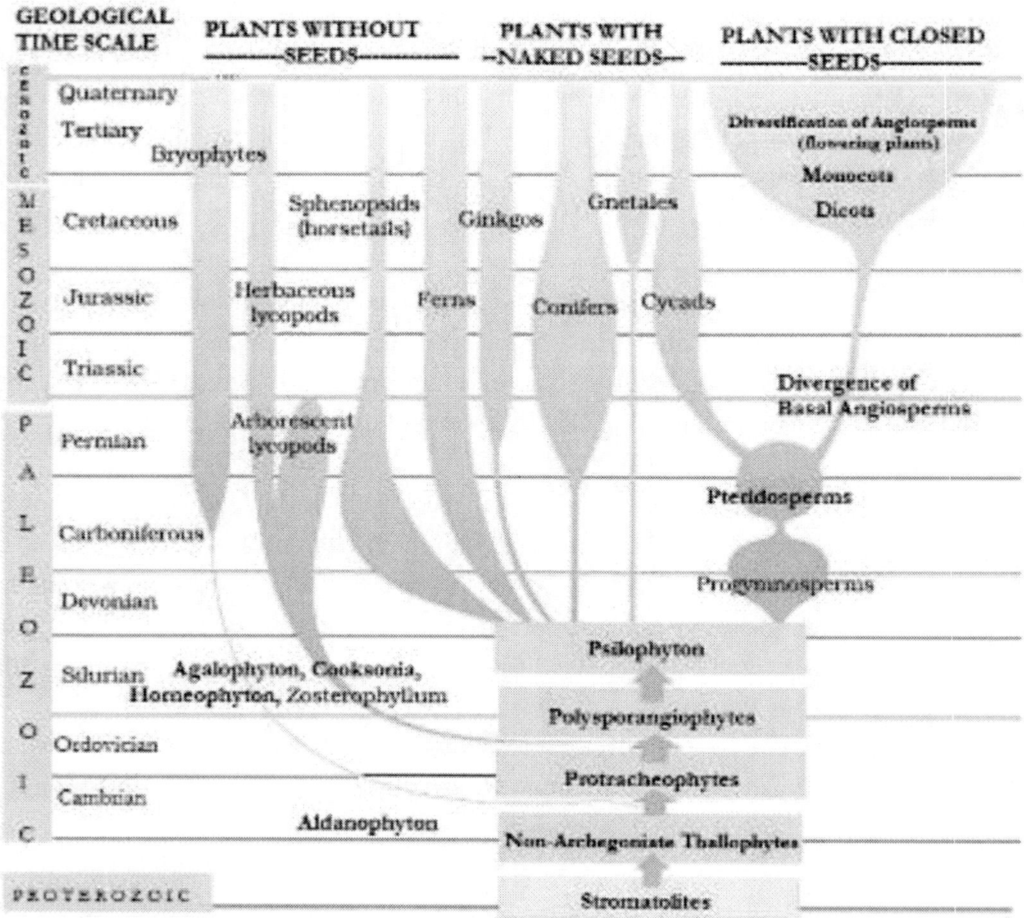

Figure 2. The divergence and diversity of different plant groups on Earth in different period of time.

5) Angiosperms: This advanced group of spermatophytes has been characterized by flower and seed-bearing capacity with closed seeds (ovule, encased in the ovary). They are found in diversified forms, represented by some 200,000 species (Heywood, 1978; Takhtajan, 1974, 1986; Turrill, 1953, 1964). Though a number of new species of angiosperms have been discovered in different corners of the Earth, the paleobotanical evidences revealed that this group of the most advanced spermatophytic plants evolved around 130-135 million years ago. The fossil evidences ascertained that around 70-100 million years ago this group of plants went through the glorified phase of diversification.

Two distinct groups of angiosperms, monocotyledons (having one embryonic leaf) and dicotyledons (having two embryonic leaves), have been broadly categorized on the basis of the number of embryonic leaves. The presence of vascular bundles in roots, stem, and leaves, and floral architectures are also taken into consideration. The group of angiosperm families of flowering plants have been categorized with a primitive, artificial system of classification on the basis of ascending sub-groups from simple to complex modes of attainment (Takhtajan, 1969; Turrill, 1964; Tutin et al., 1968). However, it was gradually regarded by the phylogenetic mode of classification on the basis of contemporary cladistic analyses, and the gradual advancement of molecular systematics, as well as on the basis of evolutionary diversification of the plants in due course of time. The modified sketch of divergence and diversification of the different plant groups (a modified diagram of NCERT[5] publication, 2006) in different periods of time has been presented in the background of a geological time scale (Figure 2).

Geological History and Paleobotany, and the Settlement of 'Green' on the Blue Planet: An Epigraphic Retrospect

The contemporary forms of plants should be considered as the result of short-term developmental processes of hundreds of years of organic evolution. This along with long-term gradual, evolutionary changes for hundreds of millions of years of inorganic and organic evolutions have also worked in its favor. Morphology plays a key role in establishing the evolutionary relationships of living groups of plants, and paleobotany plays a cardinal role in establishing the relationships between the contemporary living plants or extants with the plants that have become extinct long ago.

The morphologists and phylogenetists have made a radical comment that the evolutionary history of vascular plants could be traced back when a complete fossil record of the plants was available (Sporne, 1959). This implies that living plants on Earth provide us with an idea of their transitional phases or points of origin in prehistoric times. Sporne (1959) and Lam (1959) took it one step further and suggested that "Phylogeny is paleobotany and all else is at best pseudo-phylogeny." The real problem however, is that the available fossil evidences of the vegetative parts of the plants (e.g., leaf, stem, fronds, woods, rhizomes etc.) are majorly preserved in rocks, except the few number of floral (e.g., the petrified flower from the Oligocene deposit of the Baltic coast, Dancing; the petrified predecessor of the *Decodon* from the Deccan inter-trappean flora) and fruit-like (*Lepidocarpon, Miadesmia* etc.), reproductive organs (spores and pollen are exceptions), due to poor preservation of delicate tissues with a short life span. The study of spores and pollen grains has made immense progress due to the technical advancements of electron microscopy. Thus, palynology could solve phylogenetic challenges of evolution to some extent. The unique evidences of existence of angiosperms in the pre-Cretaceous Period have been represented by pollen grains obtained in the Jurassic coal deposits of Brora, Scotland (Turrill, 1953; 1964). These pollen-grains have been identified further as

---

[5] NCERT: National Council of Educational Research and Training, India, publication on Biology in the year of 2006.

monocolpate water-lilies and tricolpate lotuses of the aquatic angiosperm family-Nymphaeaceae by Takhtajan (1958).

Without reproductive structures, the vegetative fossil evidence has little value itself to unfurl the evolutionary mysteries of plant life. However, a phylogenetic relationship, has been established between the extant genus, *Stipa,* and the extinct genus, *Stipidium,* based on the comparative studies of their morphological features (Elias, 1942). The phylogenetic classification of a section of Cyperaceae also needs to be mentioned here, which was done on the basis of fossiliferous seeds of Miocene deposits of the Prairie region in Northern-Central America (Elias and Moore, 1965).

While paleobotany plays a supportive role in helping morphologists solve phylogenetic mysteries, there are some problematic aspects as well:

1) The deformed shape of vegetative parts, incomplete preservation (of reproductive parts) and fragmented parts of the plant body due to unfavorable stratification mechanisms
2) Incomplete casts, impressions or petrification of softer tissues, which otherwise could play a key role in considering the morphological and anatomical features for establishing phylogenetic relationships.
3) No deterministic paleobotanical evidence to establish the transitional state of plants to narrate the evolutionary processes-
   i. thallophytes to tracheophytes; tracheophytes to spermatophytes etc.
   ii. homospory to heterospory and heterospory to seed habit etc. are not also supported by evidences.
4) As paleobotanical evidences have fragmented outlines of the extinct plant forms, information gaps, as well as incomplete descriptions, they create judgment errors in bridging the evolutionary lines between the extant and extinct groups of species.

Furthermore, due to the incomplete and fragmented nature of morphological and anatomical information of extinct plants, the paleobotanists don't have the same advantage as botanists (until the beginning of the 21st century, the journey of cladistics analyses took over phylogenetic analyses, driven by molecular biologists), who enjoyed the benefit of diversified forms of extant plants with the feasibility of studying all the detailed structures to draw a phylogenetic relationship among plausible groups of plants. Hence, phylogenetic analyses of botanists on extant plants have left the rest to paleobotanists, who are engaged in determining the distribution of plant groups of a particular phylogenetic group through space and time. It could be defined as mix of combinations of observations and hypotheses.

Naturally, different opinions, concepts, and hypotheses regarding the origin and evolutionary history of the Earth have emerged from eminent scientists. Du Toit (1937), Fisher (1866), Muller (1968), Ollier (1981), Oxburgh (1979), Singh (1998), Taylor (1940), Wilson (1965), Wooldridge and Morgan (1960), have all thrown significant light on these aspects. Each scientist has a strong advocation of their hypotheses to clarify the natural history of the Earth, the origin of continents and oceans, the transition phase of evolution of continents and oceans from their origin to the present state, and the origin and evolution of life forms through the geological timescale.

Chapter 3

EVOLUTION OF PLANT LIFE ON PLANET EARTH THROUGH THE GEOLOGICAL TIME SCALE: A TALE OF THE SPATIOTEMPORAL DISINTEGRATION OF "PANGAEA" IN THE MIDDLE OF "PANTHALASSA"

While paleobotanical reviews did not help in producing a detailed mirror image of the entire plant spectrum of the universe since it was formed and went through several transitional phases of temporal dimensions, they still provided some important, fragmented footprints of the origin and evolution of plants (mostly extinct predecessors) of a particular time to conjecture the origin and evolution of the plant kingdom and correlate it with the diversity of the contemporary, living (extant successors) plants. Despite the fact that the primitive, paleoclimatic atmosphere of the primitive Earth has not been considered congenially in the favor of evolution of life, green transducers have been found to have evolved by utilizing external energy resources, and went through anaerobic, heterotrophic phases to evolve into aerobic, autotrophic forms in due course of the slow but gradual evolution.

Therefore, it may be considered that the energy production and assimilation processes have undergone a shift from the "fermentation processes" of anaerobic organisms to the "photosynthetic processes" of aerobic organisms. It can be further stipulated that the evolutionary changes were witnessed through a peep-hole of the *"evolutionary kaleidoscope,"* popularly known as the "Geological time-scale." This can specifically be determined after traversing the following conceptual nodes of evolution of angiosperms on the blue planet:

1) Complexity on the cellular level: The evolutionary progress of the "protocell" (the descendants of the complex organic molecules, co-acervates) into "unicellular prokaryotic forms" of life (single-celled with unorganized nucleic material), which transformed into "unicellular eukaryotic forms" of life (single-celled but with organized nucleus) and moved further to "multicellular eukaryotic forms" (multicellular organization with organized nuclei) which were better equipped in the dynamic environment of the primitive Earth.
2) Ecological transmigration: As the water level receded and greater land area was exposed, the plants of the aquatic habitat (either freshwater or ocean), transmigrated to the land, and the survival challenges in a different habitat forced them to go through structural and physiological adaptations resulting in better cellular formations (with thicker cell wall formation of stomata, etc.), and better conducting systems (for transportation of minerals, water, etc.).
3) Structural differentiation: In the post-transmigration era, plants were found better organized with the evolution of roots, stems, leaves, flowers, and seeds that helped in optimizing survival opportunities on land. The formation of the first three organs (roots, leaves, stems) helped the plants meet the survival challenges in their new habitat and are related to maintaining the vegetative growth. However, the formation of flowers and seeds have helped the land plants colonize a wider region with their improved reproductive opportunity, enhanced natural outcrossing, and the resultant evolution and diversification of plants in this dynamic atmosphere on Earth.
4) Predominance of herbs over trees: Apparently, it has been presumed that a large size represents more an advanced, complex, and diversified form of life. Thus, the big trees (or the ones with arborescent habitats) could be considered as the preferred ones in comparison with the herbaceous, smaller plants, when considering the survival potential of plants or their ecological preferences (bigger trees could act as a larger carbon sink). Paleobotanical instances ensured that giant trees (mainly gymnosperms) gradually disappeared during the different cycles of the Ice Age, hence the arborescent plants can be considered as relicts of plants rather than the herbaceous advanced forms. However, evolutionary evidences always support the fast-growing, advanced herbaceous plants with shorter life cycles than the plants with arborescent habit and longer life cycles.
5) Evolution of sex: Sexual reproduction has been prioritized in nature over asexual reproduction for maintaining the genetic variability and viability of plants in the due course of evolution on the ancient Earth. In primitive land plants, a homosporous feature is found to be prevalent, which evolved to the next level by heterospory among the advanced members of land plants; these lead to the seed habit of angiosperms, which is considered the best, natural outcome of out-

crossing. The regenerative potential and viability of seeds is found to be better than that of spores, and more well performed in sexual reproduction.

Among all theories and hypotheses, the "Continental Drift Theory," promulgated by Wegner (1912, 1924) in his book titled 'Die Entstehung der Kontinente and Ozeane', has ample merit to solve the problem of the origin and evolution of continents and oceans. This continental drift theory is supported by findings of geologists, cosmologists, paleoclimatologists, paleontologists, geophysicists, geochemists and phytogeographers, etc.

Wegner (1924) has categorically, stratified the sub-spherical Earth into three strata, the outermost strata, 'Sial', the innermost one, 'Nife', and the intermediate strata, 'Sima'. Wegner (1924) believed that the 'Continental masses' represented by 'Sial' floated on the 'Sima', i.e., on the ocean crust.

Wegner (1924) has successfully proved that all the present-day continents were unified in the form of one landmass, Pangaea, around 360-340 m.y.a., during the Devonian and Carboniferous periods of the Palaeozoic Era. Wegner's hypothesis of continental drift begins from the Devonian-Carboniferous Period, when 'Pangaea' was broken into two continents of which the continent of the Southern hemisphere was known as 'Gondwanaland', consisting of the South America, Africa, Malagasy, Peninsular India, Australia and Antarctica, and the continent of the Northern hemisphere, the 'Lauratia', comprising of north-western continents and Angaraland continents of Europe, Asia, and the former ones of North America, Greenland and Norway (Ollier, 1981). The intervening water mass between 'Gondwanaland' and 'Lauratia' has been recognized as the 'Tethys Sea'.

The drifting of Gondwana and Lauratia was primarily caused by the equator-ward (due to gravitational force and buoyancy) and Westward movements (due to the force of Solar and lunar tides).

During the Triassic Period, Angaraland and the north-western continent of Lauratia brok away from each other, and North America drifted away 30^0 westwards from Lauratia and the north-western continents in this period. During the Cretaceous Period, Gondwanaland was broken-up and the isolated landmasses of the Indian peninsula, Malagasy, Australia, and Antarctica drifted away from Pangaea under the influence of solar and lunar tidal forces (DuToit, 1937).

In the period of Eocene, India collided with the broken part of Laurasia i.e., Asia and merged with it. It took a long period of time, nearly 323.6 million years from the Carboniferous-Devonian period to the Pliocene period, through disruption, drifting and displacement of continental blocks, to attain the present-day arrangement of the continents and oceans. It took about 4494.8 million years, to obtain the seven continents (which cover 29.2 per inch of the surface area of the globe) of present day Africa, North America, South America, Antarctica, Asia, Europe, and Oceania from the supercontinent 'Pangaea', and the three oceans, (which cover of 70.8 per inch of the surface area of the globe), Indian Ocean, Pacific Ocean, and the Atlantic Ocean. Due to the continental drift, polar and equatorial regions of the world were shifted from time to time which lead to climatic changes (Wooldridge and Morgan, 1960), as witnessed through the geological time-scale.

Major paleographic changes of the ancient Earth to its present form can be understood through the origin and evolution of plant life i.e., the origin of the simple plants on the ancient Earth and their diversification over the geological timescale.

> "Fossil Plants represent the debt that botany owes to geology"- Birbal Sahni

Plant fossils are the retrieval reservoir of natural history, that start with the origin of life and is passed through the evolution of the biosphere in the due course of time. Plant fossils provide encoded data on plant morphology, taxonomic profile, ecological diversity, distributional accesses, and evolutionary patterns through the passage of time. The above statement of professor Birbal Sahni, the doyen of Indian Paleobotany, has been quoted by Venkatachala (1991).

Timeless efforts have been made to decode the natural history of the biosphere and of the primitive Earth which is 3.5 billion years old (Venatachala, 1991). Paleobotanical studies were initiated in the 18th Century (Sonnerat, 1782) to find the solution to the problem of the 'Origin of life', particularly of plant life, and to decode the time of the advent of angiosperms, one of the most important areas of paleobotanical study.

However, the origin of plant life, particularly of angiosperms and their dissemination, is reviewed in Indian context through a geological timescale. Major events in the time scale (ICS, 2014) are presented in Table 1A and the stages (according to the European Commission of Stratigraphy) of the Mesozoic Era have been presented in Table 1B. A brief history of the ancient plant life in general has been traced through the geological timescale. This study is a pre-requisite and is essential for the better understanding of the evolution and diversity of angiosperms; logically in terms of their widespread migrations from one continent to another, modes of dispersal, colonization, insular confinement, extinction, etc. It should be mentioned here that the preliminary conditions like the formation of organic compounds from inorganic molecules followed by the formation of plasma membrane-bound protocell-like structures from organic compounds, initiated the evolutionary journey of life. The formation of the unicellular, prokaryotic transducers from protocells was followed by the formation of multicellular, eukaryotic thalloid organisms from the prokaryotic cell, and helped traverse the primary phase of evolutionary journey.

The formation of archegoniate thalloid transducers, from unicellular prokaryotes and multicellular thallophytic transducers, followed by the formation of spore-bearing tracheophytes from spore-bearing archegoniate thallophytes, helped the smooth transition of the second phase of the evolutionary journey. Fossil evidences ascertain that prior to 500-425 m.y.a., the aquatic living organisms (e.g., algae, fishes, fungi etc.) existed on this

Earth followed by the dominance of the first land settlers (e.g., pteridophytes, arthropods) post the 500-425 m.y.a. geological time slot (Randhawa et al. 1969). Factors like the migration of plants from water to land, alternation of generation, dominance of independent sporophytic generations over gametophytic generations, transitional progression from spore formation to seed formation, preference of the cross-pollination mode of fertilization over self-pollination, etc., directed the progress of evolution of green plants to a particular direction.

The paleoclimatic variations, atmospheric changes, co-evolution of pollinators such as insects and birds, migration, colonization and distribution of plants in the pre-human era on Earth, and the human-induced migration, colonization and distribution of plants played an important role in influencing the immediate developmental changes in plants and hence the long-term evolutionary process of life on Earth. This is further evidenced by the frequent changes in the position of the equator and the poles of the Earth due to the continuous change of climatic, atmospheric, edaphic, and topographic conditions, and the drifting of continental blocks, etc.

The major geological regimes and the floristic profile of the Earth in these geological regimes is discussed here.

The history of paleobotanical studies in South Asia date back to the latter part of the 18th century (Sonnerat, 1782; Brongniart, 1828-1837) and has been elucidated by the illustrated publication of *Glossopteris angusfifolia,* from the Raniganj coalfields of India. The earliest plant fossils from India were recorded in the Dharwar system of the Archaean Era, which was a sulfur bacterium, *Phormidium* (Venkatachala, 1991). The Angiosperm plants inhabited the Earth during the Triassic-Cretaceous period (Chapman and Reiss, 1955; Randhawa et al. 1969). A brief overview of plant life on Earth has been presented here, according to the geological time-scale (in ascending order).

i. *Archaeozoic Era (Lower Pre-Cambrian):* The earliest natural history of the Earth is witnessed in rocks, i.e., Archaean rocks, which are 3000 million years old. It seems that the Archaeozoic Era covers one-fourth of the age of the Earth. Comprising of igneous and metamorphic strata, Archaean rocks form all the great mountain chains of the world and are the foundation of ancient plateaus.

Being the earliest on Earth's crust, Archaean rocks should have traces of the earliest forms of life as the palaeoclimate of this era favored the existence of life. However, perhaps due to metamorphism, all life that existed perished without leaving behind any direct evidence.

Indirect evidences of life can be seen through the occurrence of beds of graphite, limestone, and iron ores. The enormous quantities of limestone with bedded graphite schists may be of algal origin, as algae alone are capable of precipitating lime carbonate.

The Eozoon[6] coated with lime carbonate and *Atikokania,* which is the crude, cone-shaped algal body, is best known as the limestones of the Rainy Lake region of South-West Ontario, Canada, and have been found in Archaean rocks. The occurrence of iron ore in the Hudson Bay in North America has been attributed to bacteria, which causes deposition of ferric-hydroxide. It seems that primeval oceans, which were devoid of lime, received it from the connecting rivers. The primitive algae, protozoa, coelenterate, e.g., *Foraminifera,* and *Radiolaria,* utilized their lime through the formation of shells. No fossil evidences of plants have been traced thus far in the Archaean rocks.

ii. *Proterozoic era (Upper Pre-Cambrian):* The natural history of the Proterozoic Era is presented by a barren landscape and jagged rocks which erupted through volcanic activity and lasted for 930 million years. The Proterozoic Era is marked with continental uplift and mountain building, followed by intense glaciation in North America, China, South Australia, South Africa, North-West India and West Pakistan.

The very existence of Proterozoic life is represented by Protozoa, Myxophyceae, a slime-algae (perhaps the first land plant). The iron-ore deposits i.e., hematite (Fe2O3), found at the bottom of the shallow seas of U.S, Brazil, and Sweden, seem to be organic in origin, deposited by bacterial action. Various unicellular calcareous flagellates, with filamentous forms, recognized as *Rivularia*, and *Oscillatoria,* were found in Southern Ontario, Canada. Impressions of Jellyfish, *Brooksella canyonensis*, were found in the the Proterozoic strata under the Grand Canyon, Arizona, U.S.

iii. *Paleozoic Era:* The Paleozoic rocks, (the period lasted for 340 million years) with fossiliferous strata are found in the mountain ranges of the British Isles, China, Pakistan, and etc. The Palaeozoic Era comprises of six Periods; Cambrian, Ordovician, Silurian, Devonian, Carboniferous, and Permian.

a. *Cambrian Period:* The marine calcareous algae found in Cambrian rocks (this period lasted for 70 million years), resemble the existing species of Warm seas. Some fragmentary shoots with microphyllous leaves were found in the Middle Cambrian strata of the Aldan Mountain Range in Siberia, referred to as *Aldamophyton antiquissimum* Kryshtofovich, which is the earliest land plant (Randhawa et al., 1969). This is further supported by the spores and tracheid with bordered pits on the Cambrian rocks of India, Sweden, and Russia (erstwhile USSR or C.I.S.).

[6] Eozoon: It is a complex, fossil or pseudo-fossil (related to foraminifera) laminated form of calcite and serpentine found in Precambrian marbles.

b. *Ordovician Period:* The ancient alagal representative, called *Solenospora* sp., a close relative of the *Lithothamnion* sp. and a member of Rhodophyceae, was first discovered from the Ordovician rocks (lasted for 65 million years) in Estonia; *Grivanella* sp., discovered from the Ordovician limestone of Scotland is related to the Codiaceae family. *Dimerphosiphon* sp., related to the genus *Halimeda* sp., the oldest members of the Condiaceae family, were discovered from the Ordovician rocks of Norway. The paleoclimatic environments of North-America & Eurasia were tropical, characterized by the tropical rainforest vegetation, but most of the floristic elements have disappeared due to the shift in the climatic regime. The existence of those primitive plants is evidenced by paleobotanical samples in the fossiliferous coal deposits of Europe. Some fern-like plants of very simple structure with profusely branched leafless stems have been recorded from the Lower Ordovician limestone in Wyoming, North America.
c. *Silurian Period*: The unique pteridophyte predecessor, *Hedeia corymbosa*, a dichotomously branched lycopod, *Yarravia* with a terminal synangium of five or six sporangia, and *Baragwanathia longifolla*, another lycopod found in the Silurian rocks (period lasted for 40 million years), an overlying inseparable stratum in the Ordovician and Silurian deposits of Australia. *Hedia* sp. *Prototaxites* sp., the vascular plant fossils of Silurian are extended up to the Devonian. These Silurian terrestrial plants seemed to be the progenitors of the land vegetation of the Devonian Period. The plant fossils of this period are poorly preserved in the southern hemisphere.
d. *Devonian Period:* The Devonian rocks are 50 million years old and comprised of three kinds of rocks (red sandstones, conglomerates, shales and sedimentary rocks) containing fragments of terrestrial plants with fossils of fishes and crustaceans. The desert-like conditions prevailed as the water-body gradually receded, changing the habitat and triggering the chain of transitions (from hydrophytic adaptations to "apparently xerophytic" adaptations) like the response of plants through developmental changes like acquiring mechanical strength, better conduction systems, dependence on sporophytic generations etc. A diagrammatic sketch of the areas of the ancient Earth in the Devonian Period is presented in Figure 3.

From the evolutionary viewpoint of origin of land plants or amphibian transmigration, the members of Psilophytes have been considered as the symbolic representative of ancient plants in the Devonian Period. *Psilophyton principes*, *Rhynia gwnne–vaughanii*, *R. major*, *Horneophyton* sp., *Asteroxylon* sp., *Taeniocrada decheniana* are all early vascular land plants having an identical resemblance with the members of Psilotales, the simple members of Cryptogams of the present day. *Archaeopteris latifolia* (the pre-ferns), *Archaeosigilliaria* (apparently considered as one of the oldest trees on Earth),

Eospermatopteris (tree-fern), *Prototaxites*, *Cooksonia*, and *Calamophytons* sp. (pre-gymnosperm), are predominant members of this period. From water to land, this transition of plants seems to have started with non-archegoniate thallophytes e.g., *Coleochaete* sp. or *Nematothallus* sp.

e. *Carboniferous Period:* The worldwide fossiliferous coal beds (period lasted for 65 million years) represent the Carboniferous Period. The Lower and Middle Carboniferous rocks are absent in most parts of the Eastern Hemisphere, except North-West Himalayas and parts of China. The Upper Carboniferous Period began with a Glacial Epoch. Plant life in the Carboniferous period is marked by the dominance of several plant groups such as lycopods, arthrophytes, ferns, seed-ferns, and diverse groups of gymnosperms. A diagrammatic sketch of the Earth in the Carboniferous Period is presented in Figure 4.

The Carboniferous lycopods are chiefly arborescent members of *Lepidodendron*, *Sigillaria*, *Lepidostrobus diversus*, *Lepidocarpon*, *Archaeosigillaria primaeva* etc.

The articulated predominant taxa of Carboniferous comprise of arborescent representatives: *Kalloostachys scotti*, *Calamostachys casheana*, *Arthropitys*, Calamodendron, *Cingularia typica* etc. as well as herbaceous representatives *Litostrobus iowensis*, *Bowmanites*, *Sphenophyllum* etc.

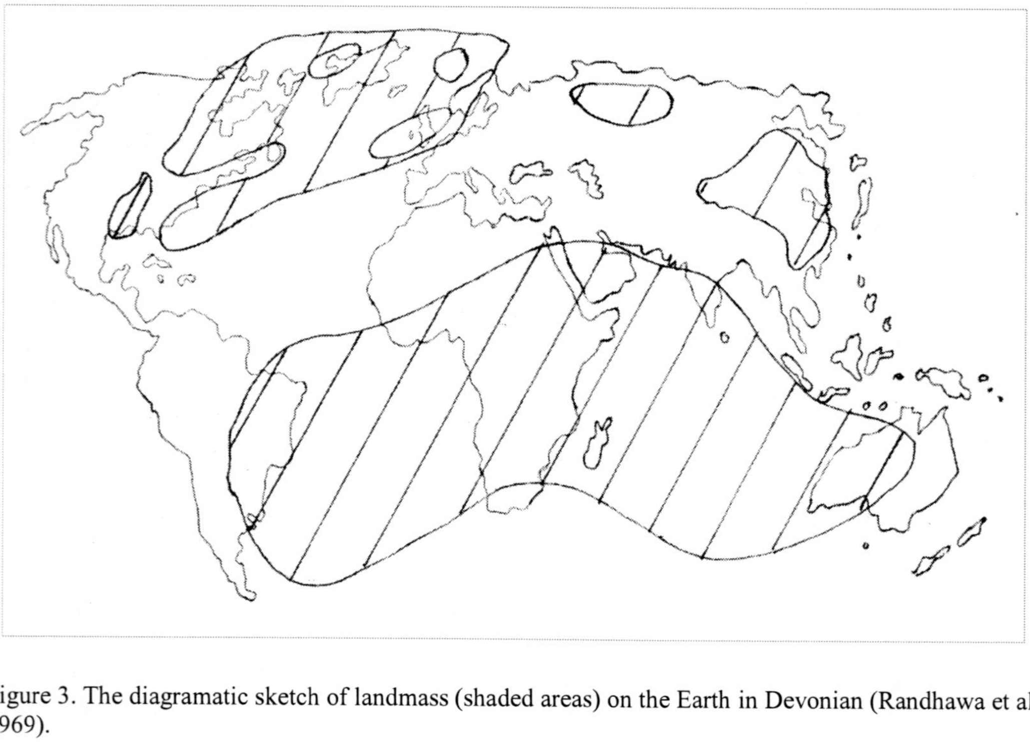

Figure 3. The diagramatic sketch of landmass (shaded areas) on the Earth in Devonian (Randhawa et al. 1969).

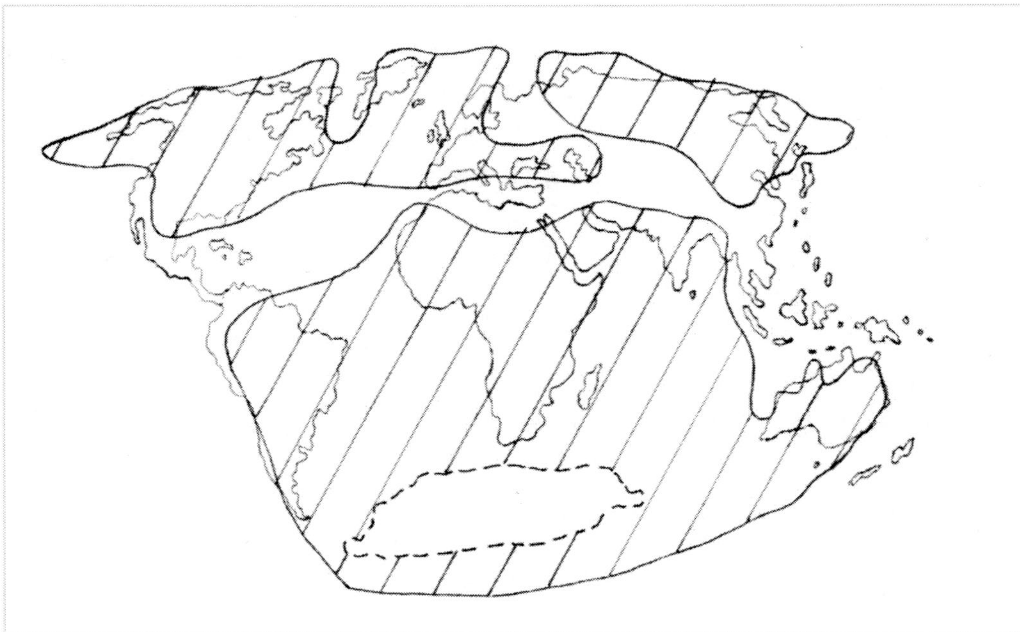

Figure 4. The diagramatic sketch of landmass (shaded areas) on the Earth in Carboniferous (Arlt, 1907).

The Carboniferous ferns are represented by *Anachoropteris elavata*, *Tubicaulis stewartii*, *Senftenbergia stri*, *Renaultia gracilis* etc.

The seed ferns of this period are - *Pecopteris phegopteroides*, *Glossopteris indicus*, *Medullosa* sp., *Gangamopteris* sp., *Gnetopsis elliptica*, *Whittleseya media* etc.

Cordaites, dominant tree elements of Carboniferous forests are represented by C*ordiates* sp. *Poroxylon* sp. *Cordaianthus zeilleri* etc, predominant in this period.

Conifers of the Carboniferous period are *Lebachia* sp., *Ernestiodendron* sp. etc.

Even, the predecessors of *Ginkgo* sp., *Dichonophyllum moorei,* are reported to be from this period.

f. *Permian Period:* This period lasted for 50 million years. Around 280 m.y.a., the African plate collided with the coastal Atlantic part of the north-American plate, resulting in the formation of the Appalachian mountain ranges.

Likewise, Ural mountain ranges were formed as a result of the collision of the European and Asian plates. The tectonic plate fragmentation linked with the land-mass movement, which initiated in the Carboniferous Period, attained its highest pace in the Permian Period after volcanic activity. The main coal formation period of Gondwanaland is Permian, seen in a salt range of Western and North-West Himalayas of Pakistan, Eastern Himalayas of Nepal, Bhutan, and India, to the Arakan-Yoma ranges of Myanmar (Burma). The prolonged glaciation of this period (Ice Age) made the climate more unfavorable to plant life.

However, some of the plant representatives have sustained through such adverse climatic conditions, e.g. *Sigillaria*, of Lycopod gr., *Equisetites*, *Calamites,* of Articulates gr., *Lebachia, Ernestio-dendron, Ullmannia* sp. and *Pseudevoltzia* of the Conifer group; *Sphenobaiera* sp., *Trichopitys heteromorpha, the* predecessors of *Ginkgo* sp. were found in this period. The three floral provinces of this period are known as:

- Gigantopterids Flora of the Far East (China and Korea): *Asterophyllites, Annularia, Calmites* etc.
- Glossopterids Flora of Gondwanaland: *Glossopteris*, *Pityosporites antarcticus. Gangamopteris*, *Gondwanidium* etc.
- Kusnetzk Flora of Angaraland: *Ginkgoites, Gangamopteris, Cladophlebis, Wallichia, Callipteris* etc.

iv. *Mesozoic Era:* The Mesozoic Era existed for 165 million years, with fossiliferous strata, deposited in the mountain and table-lands of Austria, rocky mountains of North-America, Spain, France, Italy, Germany, Balkans, Britain, Peninsular India, lower Myanmar (Burma), Malay Peninsula, Indonesian Islands, New Guinea, Australia, Alaska, Andes mountain of South America, and etc. The Mesozoic Era has been subdivided into three periods – (a) Triassic, (b) Jurassic and (c) Cretaceous Periods.

a. *Triassic Period:* The Triassic Period lasted for 35 million years. The red-bed with trias, or 3 stratified layers, with the top one being marine limestone and the bottom layer being sandstone, intercalated in between a layer of mud, were found in Germany and parts of Western and North-West Europe. The modernization of flora commenced during the Triassic Period. The plant life of the Triassic Period highly is represented in the Gymnosperm group of plants. The diagrammatic sketch of water and land masses of the Earth in the Triassic is presented in Figure 5.

The habitats of the plant life in the Triassic Period has prevailed through arid and semi-arid conditions as the continental masses are not far from each other to open up shorelines enough, for getting exposure to moist climatic conditions. However, by the end of the Triassic Period, a transition was witnessed from dry semi-arid types of vegetation to moist vegetation.

The leading plant community of the Triassic Period comprised of Liverworts and Moss. The liverwort is represented by *Ricciopsis florinii* and mosses are represented by *Naiadita lanceolate, Naiadita* sp.

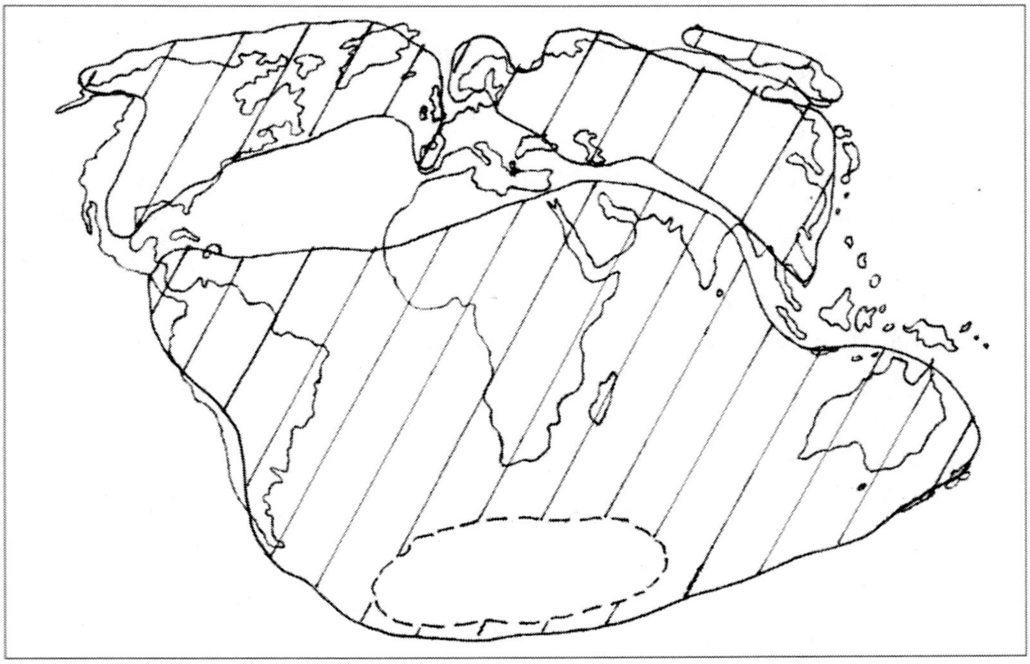

Figure 5. The diagramatic sketch of landmass (shaded areas) on the Earth in Triassic (Arlt, 1907).

The Triassic Lycopods are represented by herbaceous *Selaginellites polaries*, *Lycopodites* sp. etc. and arboreal *Pleuromias* sp. Articulates are represented by *Equisites platydon, Schizoneura gondwanesis, Phyllotheca etteridgei* etc. Ferns of the Triassic have been represented by *Marattiopsis hoerensis, Clathrropteris meniscoides, Gleichenites* etc. The preponderance of Pteridosperms in this period has been noticed by *Unkamasia macteani, Zuberia zuberi, Lepidopteris natalensis, Pilophorosperma granulatum, Thinnfeldia* etc. The cycadeoids group of the gymnospermous plants have been represented by *Wielandiella angustifolia, Anomozamites minor, Sturiella langeri* etc.

The extinct Cycad groups of plants have been evidenced by the genera *Cycadospadix, Palaeocycas* etc.

The extinct Cordaite of this period has been identified by the solitary member *Noeggerathiopsis hislopi*.

The fossiliferous Ginkgoales are evidenced with *Ginkgoites lunzensis*, *Baiera* sp., *Sphenobaiea furcata* etc.

The extinct Conifers of the Triassic have been found by *Voltziopsis* sp. *Voltzia* sp., *Cycadocarpidium* sp., *Stachyotaxus* sp., *Woodworthia arizona*, *Cedroxylon* etc.

The Taxads of this period are represented by *Rhexoxylon africanum, Harrisia marsillides, Ptilozamites, Palaeotaxus* etc.

The ancient angiosperm-like plants of the Triassic Period have been epigraphically existing, as evidenced by *Fraxinopsis major*, *F. minor* (from Argentina) *Furcula granulifer* (from Greenland), *Sanmiguelia lewisii* (North America) etc.

b) *Jurassic Period:* The Jurassic[7] Period had a span of 54 million years, popularly known as the age of reptiles or age of dinosaurs, and existed in the stratified limestone of the European Alps.

The diagrammatic sketch of water and landmasses on Earth in the Jurassic period is presented in Figure 6. Prior to 180 m.y.a., the super-continent Pangaea was surrounded by the super-ocean Panthalassa, however towards the end of the Jurassic period, Pangaea was subdivided into two sub-continents Gondwana (comprising of Africa, South Africa, Australia, Antarctica, and India) and Laurasia (comprising of North America, Greenland, Asia) as a result of initiation of the continental-shield fragmentation which began around 180-120 million years ago.

Due to further northward movement of Laurasia and southward movement of Gondwana, the shoreline opened up enough to be exposed to the humid climatic atmosphere, represented by the moist, rainforest-type of plant communities dominated by ferns, tree ferns, and giant-sized tree species (diversified group of gymnosperms and some pro-angiosperms) in place of the medium-sized plants with semi-arid adaptations.

The plant life during the Jurassic Period comprised of advanced groups of pteridophytes, ferns and dominant groups of gymnosperms: Pteridosperms, Bennettitales, Cycads, Cordaites, Conifers, Pentoxylales, Ginkgoales, and Taxads along with the progenitor of angiosperms.

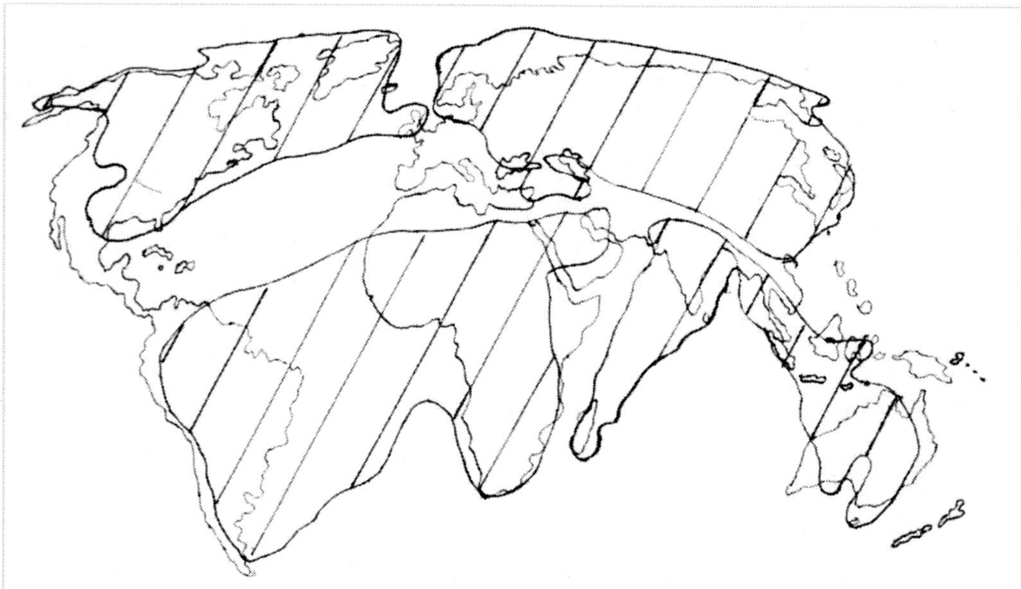

Figure 6. The diagramatic sketch of landmass (shaded areas) on the Earth in Jurassic (Arlt, 1907).

[7] The naming of the period has been ascribed in the name of "Jura" mountain of Alps mountain ranges of Europe.

Jurassic Lycopods are represented by the *Lycoxylon* sp. The Ariculates of the Jurassic period are represented by *Equisites, Phyllotheca, Neocalamites* etc.

Jurassic ferns are represented by *Marattiopsis, Cladophlebis denticulata, Osmundites, Cyathocaulis, Gleichenites glauca, Klukia* etc.

Jurassic pteridosperms are represented by *Thinnfelidia, Dicroidium, Cycadopteris, Czekanowskia* etc.

The Pentoxylales group of plants are represented by *Nipanioxylon raoi*, *Sahnia nipaniensis*, *Carnoconites compactum*, *Pentoxylon* etc.

Jurassic Cycadeoids are represented by *Williamsoniella coronata*, *Willamsonia sewardiana*, *Willamsonia santalensis*, *Cycadeodea sahnioxylon* etc.

The Cycads of the Jurassic Period are *Nilssonia compta*, *Androstrobus manis*, *Beania gracilis* etc.

The Conifers, dominant group of plants in the Jurassic Period, are represented by *Mehtaia nipaniensis*, *Taxodioxylon*, *Araucarioxylon*, *Cedroxylon*, *Cupressinoxylon*, *Brachyphyllum* etc.

The Taxad groups of plants of this period are represented by *Taxus jurassica*.

The predecessor of angiosperms of this period are *Sassendorfites benkeri*, *Phyllites*, *Propalmophyllum liasinum* etc.

The Jurassic Period is an evolutionary jumping board for the phylogenetic studies of the diversified group of plants. Due to the availability of the logistical advantage of "pre-disintegration of the Pangaea regime," widespread presence of all plants all over the continents, the ancestors of these primitive groups of plants are supposed to have existed in all continents and the availability of the fossil representatives in the pre-Jurassic regime in different continents far and wide with different climatic and atmospheric regimes, further support the continental drift hypothesis. The confinement of some fossil representatives in very localized and isolated states were noticed among some advanced floristic elements, which evolved after "post-disintegration of the Pangaea regime" as well as in the post-Jurassic Period of the Mesozoic Era.

c) *Cretaceous Period:* During this period, Gondwana further broke into six pieces: Africa/South America, India, Antarctica, and Australia, around 120 million years ago. The co-evolution of birds, small mammals, and angiosperms in this period made a huge impact on the diversification process of the angiosperms. The Cretaceous Period lasted for 76 million years; stratified rocks of calcium carbonate enriched the Paris basin of France, Chicxulub impacted the crater of the Neuquen basin of Argentina and Yucatan Peninsula, at the Gulf of Mexico. This period is characterized by the formation of different islands, the formation of inland seas with prevalence of humid and warm climatic conditions. The landmasses and waterbodies on Earth in this period are represented in Figure 7.

Intensive review of the fossil evidences of small mammals "Marsupials," helped discern that this group of animals evolved around 100 m.y.a., during the Cretaceous Period in Laurasia in the North, and moved to Gondwana in the South at the time of continental disintegration.

Once the marsupials reached South America they reached further down to Antarctica, going through Australia through the land bridge that existed at the time between South America and Australia, then from South America, they moved to the African continent until the geological disintegration of these two continents occurred around 100 m.y.a. (Nilsson et al., 2010). The shift of the climatic regime in the Northern hemisphere from warm to cool conditions, is considered to be the driving force for these warm-blooded mammals to migrate from the North to the South.

Considerable change in the plant life of the world took place at the beginning of the Cretaceous Period. Prior to the commencement of the Cretaceous Period the ferns and gymnosperms dominated the plant world, but the origin of the progenitor of angiosperms changed the scenario. Cretaceous plant communities comprised of Pteridophytes and Gymnosperms to a lesser extent, and the progenitors of angiosperms to a greater extent.

The pteridophytes of this period are *Lycopodites, Sellaginellites, Nathorstia alta, Matonidium indicum, Tempskya* sp. etc.

The Gymnosperms of this period have been represented by *Drepanolepis, Splenobaiera, Ginkgoites* etc.

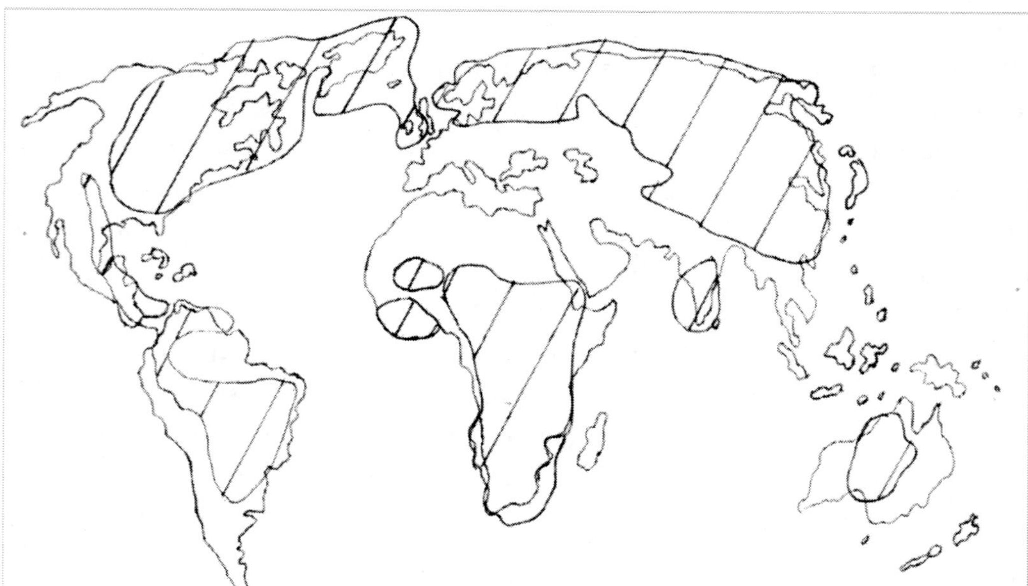

Figure 7. The diagramatic sketch of landmass (shaded areas) on the Earth in Cretaceous (Arlt, 1907).

The advent of pro-angiosperms or ancient groups of angiosperms have been marked in this period with the fossil remnants of: *Magnoliophyllum* (Magnoliaceae), *Menispermites* (Menispermaceae), *Populites* (Salicaceae), *Cinnemomoides* (Lauraceae), *Araliaephyllum* (Araliaceae), *Cretaegites* (Rosaceae), *Dalbergites* (Leguminosae), *Sapindopsis* (Sapindaceae), *Ficus* (Moraceae), *Sabalites* (Palmae) etc.

v. *The Cenozoic Era:* The Cenozoic Era lasted for 69 million years. This era has been subdivided into two periods – a. Tertiary Period (63.2 million years of time span) and b. Quaternary Period (starting around 1.79 million years ago till present time).

a. *Tertiary Period:* The plant life of the Tertiary Period was studied under five Epochs.

ai. Paleocene,
aii. Eocene,
aiii. Oligocene,
aiv. Miocene,
av. Pliocene.

According to paleo-climatologists, the most important event in this period was the PETM (Palaeocene-Eocene Thermal Maximum), a sudden "thermal flux," evidenced by a 5-9^{0}C temperature change induced by a surge in the CO_2 level in the atmosphere. It took supposedly 200,000 years to get rid of the extra CO_2 levels from Earth's atmosphere and is almost similar to the global warming of the contemporary Earth in the Quaternary period. The waterbody and landmasses of the Earth in the Cenozoic Era are presented in Figure 8.

ai. Paleocene Epoch: The Paleocene Epoch existed for a span of 10 million years. During this time, Laurasia was further fragmented into three plates: North America, Eurasia, and Greenland. Between 60-50 m.y.a., the Indian plate collided with the Asian plate to form the Himalayan mountainous ranges. The paleozoological evidence indicates that placental mammals evolved in Laurasia during this time, which helps conjecture that placental mammals moved from North American territory to South America through the corridor popularly known as the "Isthmus of Panama" and that these placental mammals outnumbered the marsupials in the course of the evolutionary struggle. Placental mammals failed to reach the Antarctic/Australian plate due to the absence of a land bridge.

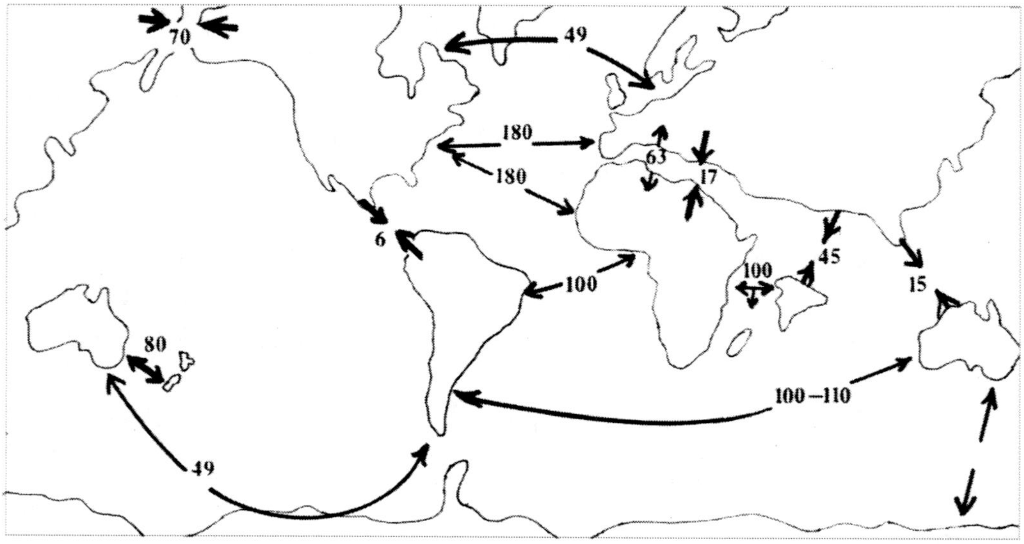

Figure 8. The diagramatic sketch of landmass on the Earth, isolated from/collided to each otherfor million years in Cenozoic (Raven and Axelrod, 1974).

During the Paleocene Epoch, Antarctica segregated from Australia and moved southward, subjected to a severely cold climatic regime, resulting in the disappearance of almost all the mammals. Contrastingly, the Australian plate moved northward, close to the equator, which helped the marsupials survive. The marsupials of Australia have occupied most of the ecological niches and it's most surprising feature is that those identical ecological niches in other continents have ecologically been shared by placental mammals.

This was also the age of marine transgression. Due to intense crustal movements, compressions, invaginations of newly formed lava rocks in the Tethyan region, the mountain ranges of European Alps and the Himalayan Mountain ranges of Asia were formed. The stratified Paleocene rocks were found in the Yucatan Peninsula. In Greek "Paleocene" literally refers to the combination of "Old" and "New," a unique situation induced by the extreme (thermal flux) weather conditions. This epoch is known for the atmospheric and oceanic carbon circulation, disappearance of foraminifers to the benthic level of the ocean, avian dinosaurs on land, and the "Gymnospermous Rainforest" habitat.

aii. Eocene Epoch: The Eocene Epoch lasted for 19.6 million years. The Himalayas were uplifted and formations of widespread delta established in this epoch. The ^{13}C (constitute 1.1% of global carbon sink) level lowered in comparison to ^{12}C (constitute 98.9% of global carbon sink) levels in the atmosphere.

It has been proven that modernization of flora was initiated and continued as 23% of Hordle beds flora of North-West Europe show resemblance to the present flora. Literally, Eocene is the epoch marked as the dawn of new fauna on Earth. The plant life of the Eocene period is marked by diversification of extant marine algae, the dominance of angiosperms and a limited number of gymnosperms with primitive characteristic features. Mainly

Conifers, Cycads, Pinus etc. were newly recognized as "living fossils," surviving without diversifications.

The predominant algae of the Eocene Epoch are represented by *Neomeris, Acicularia, Chara, Dissocladella, Chara* etc. The marine algae of the Eocene epoch are represented by *Lithophyllum, Mesophyllum, Triploporella, Oligoporella, Neomeris, Acicularia* etc.

The pteridophytes of the Eocene epoch are represented by *Salvinia intertrappea, Azolla intertrappea* etc.

The gymnosperms of the Eocene Period are represented by *Indostrobus, bifidolepis, Mohgaostrobus sahnii, Gymnoovulites, Dadoxylon, Spiroxylon Pinus* sp. *Abies*, *Sequoia, Picea* etc.

In the Eocene Epoch, the floristic elements comprised of predominantly angiosperms: *Trema garoensis, Grewia foxii, Nelumbium* sp. *Eriodendron* sp. (Leaf) *Nypa sahnii, Neolitsea sahnii, Cynometroxylon indicum, Kayea, Terminalia* sp. *Dipterocarpus* sp., *Cocos sahnii, Mesua tertiaram, Magnolia, Ficus, Columbia, Betula, Laurus, Potamogeton, Rubus, Corydalis* etc.

aiii. Oligocene Epoch: The Oligocene Epoch existed for 12.1 million years. The topographical unrest like the uplifting of mountains in North America, and the formation of islands concomitantly, paved the way for the formation of lagoons, as the sea advanced and retreated along the margins of continents.

The emergence of Alps in the Euro-Mediterranean region happened as a result of the northward movement of the Eurasian plate by a push from the African plate and the rising and folding of Himalayas from Tethys resulting in the disappearance of the Tethys Ocean. The identical floral and faunal remnants ascertain that there is supposed to have been a land bridge between Europe and North America in the early Oligocene Epoch. Due to the continuation of tectonic plate movements, South America supposedly detached from Antarctica and moved towards North America, and Antarctica moved towards the south pole and consequently became an ice-capped continent during the Oligocene Epoch. Volcanisation was frequently witnessed in Eastern and South-Eastern Asia and Africa.

The evolution of flora in the Oligocene Epoch continued, as is evidenced from the Reuver flora of north-west Europe which has 34% resemblance to the present flora. The floristic elements were considered as a relict form of a transitional state between the tropical broadleaf representatives and the modern representatives of the grassland ecosystems of the Miocene Epoch. Seemingly, the grassland ecosystems of the Oligocene Epoch, as well as the savannah forest ecosystems, widened, consequently resulting in the broadleaf tropical forest ecosystems to be confined to the equatorial regions.

Oligocene flora is represented by the predominant angiosperms: *Ailanthus* sp., *Albizzia* sp., *Celtis, Carpinus, Fagus, Dalbergia lalifolia, Garcinia borooahii, Mangiferaxylon scleroticum, Sonnersioxylon dakshinense, Zelkova* sp. *Plantanus* sp., *Juglans* sp., *Ulmus sp.* etc.

The gymnosperms of the Oligocene Epoch are represented by the *Mesembrioxylon* sp.

aiv. Miocene Epoch: The Miocene Epoch lasted for 18 million years. In this epoch, almost all the great mountains of the present day evolved and attained their height during this period. The Tethys Sea gradually receded and the submerged areas were exposed, and the shallow sea floor areas expanded further. The sediments of Tethys were found in the sedimentary rocks of the rising Himalayas, known as Siwalik, in the Miocene Epoch. They retained various fossil forms of the animal and plant life of the time. According to most of the paleoclimatologists, the rising of Himalayas stopped the glaciation induced cold waves of the Northern Hemisphere from reaching the Asian continent. However, the seasonal wind circulations and precipitations changed in Asia, which induced a change in the floristic patterns.

Thus, remarkable changes in the floristic pattern were witnessed. It was further deciphered that flora of the foothills of Siwalik, that were predominantly tropical (C_3) types of plants, gave way to tall grass (Savannah forest ecosystems), shrubs, scattered trees, and grasses of C_4 vegetation type (Valdiya, 1999). The disappearance of a land bridge between South and North America paved the evolution of two separate, distinguished modern flora of the two continents.

Miocene plant life is represented by the Arctic-Tertiary flora, comprising of temperate hardwood, and deciduous and coniferous forests, and the Metro-Tertiary, comprising of oak woodlands and thorn forests. This epoch is considered as the transitional, geological, temporal, cold atmospheric phase of the Earth (due to the initiation of the ice ages), gradually moving towards the cold Pliocene Epoch from the hot Oligocene Epoch. This was also known as the "MMCT" or the Middle Miocene Climate Transition. The appearance of 'Kelp Forest Ecosystems', dominated by Laminariales groups of algae were observed in the temperate and polar benthic ocean zones, tantamount to the regression of tropical forest ecosystems[8], furthered by the gradual progression of the grassland ecosystems in this epoch. The appearance of "*Homo sapiens*" occurred in the Miocene Epoch; the dominance of terrestrial and oceanic mammals were also witnessed during this time. The Miocene Tehachapi flora in California has shown 65% resemblance to the present flora of Mexico.

The Miocene flora is represented by some of the dominant angiosperms: *Betula, Engelhrdtia, Corylus*, *Carpinus* etc.

av. Pliocene Epoch: The Pliocene Epoch lasted for 3.4 million years. According to the palaeoclimatologists, the atmospheric temperature in the Pliocene Epoch is predicted to have been 2-3^{0}C higher; the CO_2 level was found to be the same as the present day levels, and the sea level was supposedly 25m higher than the current sea level. Glaciations induced sudden ice-cap formation in Greenland as proven by the analysis of the oxygen isotope ratios and ice pebbles of Atlantic and Pacific Ocean beds. The elevation of land from the seafloor continued at the expense of the sea. All the continents of the present day seem to

[8] Regression of tropical forest ecosystems are due to less moisture retention capacity of the cold palaeoatmosphere in Miocene Epoch.

be more or less the same as in the Pliocene Epoch. The Pliocene seas receded to get smaller than those of the Miocene Epoch. Further upliftment of the Himalayas continued and became snow covered in high latitudinal regions.

The Mediterranean ocean got cut off from the Tethys Sea due to the collision of the African and European plates in the Pliocene Epoch. The descending levels of seas exposed the submerged marine-rocks as well as the earlier land-bridge between (Alaska) Europe and Asia, supposedly used as a migratory route of earlier plants in the pre-Pliocene Epoch, which were well adapted in aquatic habitats. Due to tectonic plates inducing the drifting of South America, it moved closer to North America and got connected by the Isthmus of Panama, which helped the migration of modern animals and plants across the continents in the post-Pliocene Epoch. The formation of the "Isthmus of Panama" stopped the inundation of the equatorial warm ocean currents and the onset of the redundant glaciations and the Ice age-induced Atlantic cooling cycles, which led to the gradual cooling of the Arctic and Antarctic Oceans.

The commencement of the Ice Age took place around 2 m.y.a., the glaciers and ice-shields induced a cold climate which is responsible for the disappearance of a number of transducers, but phylogenetically, the super-cold atmospheric conditions acted as a fine genetic screening process and induced a selective pressure on certain alleles to evolve into stronger, modern species having better survival potentials.

The global cooling as well as modernization of flora, initiated during the Triassic Period, was finally almost complete during the Pliocene as 78% of the Reuver flora of North-West Europe showed resemblance to present day flora. The Pliocene plant life was greatly influenced by the physiological, xerophytic conditions (due to the low moisture content in the cold atmospheric conditions). The gradual receding of the tree-dominated tropical forest ecosystems, and its confinement to the equatorial zone, has been inversely proportional to the global expansion of the deciduous and savannah-grassland type ecosystems in Asia and Africa. Coniferous and Tundra forests of the Northern hemisphere shaped a congruent floristic composition of the Earth in the Quaternary Period. The Pliocene floristic elements are represented by the Coniferous pollens and spores of the grass family, collected from the stratified rocks of Andes and Patagonia, Mediterranean seabeds, and Eastern Himalayan mountain ranges.

The climatic change from warm to cold, and the changing patterns of vegetation from tropical to temperate during the Cenozoic Era might have given rise to the herbaceous invasion during the late Cenozoic Era (the diversification of C-1 level herbivorous animals was witnessed during this time, supporting the prey-predator equilibrium in energy tropics). This era has been characterized by the floristic migrations in response to climatic changes.

b. *Quaternary Period:* The formation of rocks of the Quaternary Period started 1.79 million years ago. It is subdivided into the Pleistocene (glacial age) Epoch and the

Holocene (recent) Epoch. A diagrammatic sketch of the landmass and water bodies in the Quaternary Period of the Earth is presented in Figure 9.

bi. Pleistocene Epoch: The Pleistocene Epoch (popularly known as the "Ice Age") was initiated 1.79 million years ago. During the Pleistocene, the occurrence of four major glaciations were witnessed, popularly known as Gunz, Mindel, Riss, and Wirm. Among all major glaciations in the Himalayan region, the second is considered the most severe. The duration of interglacial periods is noticed to be much longer than the glacial periods. During the glacial phase, the water bodies of oceans were locked up in the glacial ice-shields and the sea-level receded, leading to the exposition of more land-mass (Kellman, 1975).

During the interglacial period, the water from melting ice-shields raised the seashore lines up to 30m above the present sea level. Glaciations occurred in the high mountain tops of Scandinavia, the British Isles, Alps, California, Alaska, Aleutian Is., Andes, and Patagonia. As a cumulative result of the slow pace of tectonic movements and consecutive cycles of glaciation in the Pleistocene Epoch, the Himalayan mountain ranges ascended to their maximum height at this time.

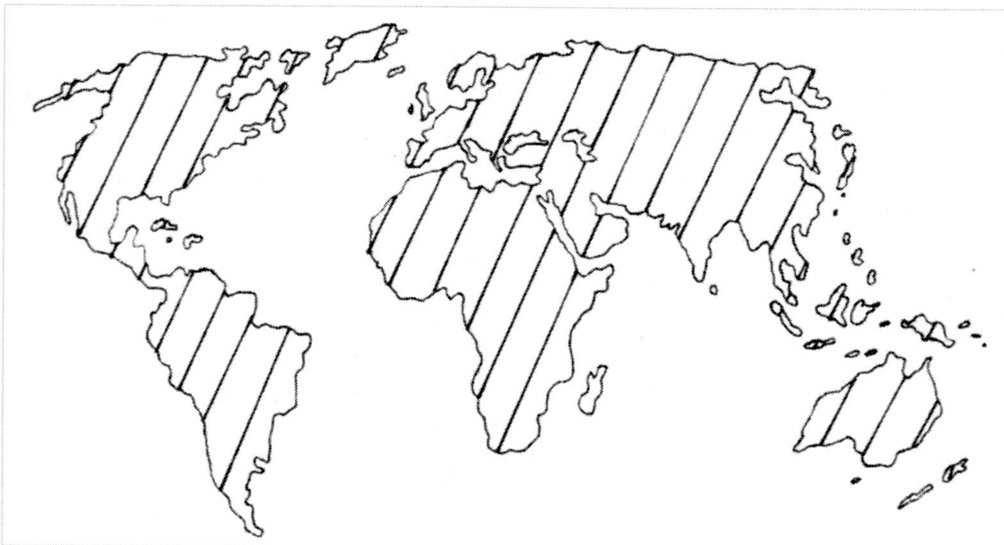

Figure 9. The diagramatic sketch of landmass (shaded areas) on the Earth in Quaternary (Arlt, 1907).

With successive climatic and vegetational changes in the various glacial and interglacial phases, some of the plant species: *Corema intermedia*, *Trapa natans*, *Naja minor*, *Azolla filiculoides* etc. became extinct, however a few of them were resurrected with the onset of favorable climatic conditions. The big, arborescent, relict type stable form of chromosomal configurations (mostly diploid form) of plants have faced more challenges in survival due to extremely cold weather conditions than the smaller herbs and shrubs, modern type polyploidy (having a varied number of chromosomal configurations) that strived to acclimatize in the varied climatic conditions. During the Ice age, the conifers migrated far south from the north; e.g., *Picea glauca,* which grew further north around Huron, and flourished in Texas during the last glaciation in North America. In North-Western Himalayas, the number of Oaks and Alders declined in the Kashmir Valley (further North-West of Kumaon) during this glaciation. Paleobotanists observed that the species of *Engelhardtia, Pterocarya* have not extended beyond the Western Himalayas. These used to grow in the North-Western Himalayas in the pre-glaciation phase and got wiped-out from its earlier habitat during the second glaciation. "It seems reasonable to conclude that the influence of glacial periods (of the Pleistocene Epoch) has enhanced the number of polyploids and decreased diploids, as the latter could not survive in competition" - (Sahni, 1991).

bii. Holocene Epoch: The Holocene Epoch began around 0.01 million years ago and has been considered as an interglacial transition of the concurrent cycles of glaciations. The formation and diversification of forest ecosystems, mode of seral successions, rate of turnover of plant species on diversified habitats (in the form of species diversity, ecological diversity, and gene diversity), and the formation of congruent flora on Earth in the Holocene epoch have been highly influenced by anthropogenic activities and various deterministic factors. These include the elevated level of Greenhouse gases, albedo effect and the meltdown of polar ice caps, Ozone Hole in the stratospheric level, changing patterns of monsoonal wind, disrupted oceanic currents of El-Nino, fall-out of nuclear explosions, and etc. The formation of ice caps in the Arctic and Antarctic ice shields in the Holocene Epoch have resulted in the changing depth of the sea-level, deposition of sediments to form undulating shapes of landscapes on Earth, formation of plateaus, changing patterns of landscape near shorelines, and the formation of riverine networks and pluvial lakes immediately after the last glaciation of the Pleistocene Epoch, locally known as: "Antarctica"; "Otiran" in the Pacific region (New Zealand), "Llanquihue," "Merida" in South America (Chile, Venezuela), "Weichselian," "Wurm" and "Devensian" in Europe (Germany, Russia, UK) and "Greenland," "Pinedale," "Tahoe, Tenaya & Tioga," and "Wisconsin" in North America. During the Holocene Epoch, the dominance of subarctic flora represented in the open Tundra-like vegetation comprising of *Salix herbacea, Dryas octopetala, Alnus* sp., *Quercus* sp. *Abies* sp. and etc. has been observed. The late glaciation of this epoch was followed by the formation of birch, *Betula pubescens,* forests and later by pine or a composite form of birch-pine mixed forest ecosystems in the early phase, which was subsequently replaced by pine and mixed oak wood or beech-oak-hornbeam forest communities (Sharma, 1993).

While floral migration commenced due to the climatic changes of the Earth, the influence of the Neolithic man giving up nomadic life in the post glacial phase and the consequent domestication of wild flora and fauna, and utilization and then over-utilization of wild germplasm of flora and fauna was an entirely new determinant in the evolution of flora on Earth. Anthropogenic civilizations undertook the conservation of plant species, manipulation of the formation of forest ecosystems, and control of the logistics of wild germplasm of floristic components which further made floral migration more dynamic due to the exploitation of plant resources and their distributional pattern in various forms (community, flora or biome), which continued during this period across the globe (Chapman and Reiss, 1995).

Apparently, it has been presumed that the present continental shape of the Earth is static, but the Greenhouse effect and global warming, Ozone Hole in atmosphere, melting of polar ice, changing monsoonal wind cycles, precipitation levels, nuclear fall-out, high exposures related to the mutagenic effects of U.V., gamma rays etc. on plants, and other natural catastrophes (like increasing occurrence of typhoons, tsunamis, tornados) would bring some change in the present plant life on Earth. These incidents are highly impacted by anthropocentric actions which has pushed the evolution and diversification of modern plant groups, particularly angiosperms, to an entirely new direction in the late nineties. Moreover, a group of geological scientists in 1960, tried to introduce a new name to designate the contemporary geological epoch as "Anthropocene" (yet to be accepted globally), which has further been supported by a group of geoscientists in the International Geological Congress, 2016, at Cape Town, South Africa (Dawson, 2016; Zalasiewicz et al., 2008).

Chapter 4

Evolutionary Stepping Stone: The Alternation of Generations, Origin and Evolution of Sex in Algae

Since the beginning till the end, every living organism undergoes the cycle of life from the juvenile phase, maturity, reproduction, old age, and finally death; plants are no exceptions. All these stages go in a cyclical order from the beginning to the end for all plants, referred to as the lifecycle. Whether it's thallophytes, tracheophytes or spermatophytes there are two alternative phases in the lifecycle of all types of plants.

The uniqueness of these two generations might be either morphologically distinct or may be indistinguishable, there may be one generation dependent on the other for nutritional support, or they may independent of one another. Among these two distinct generations, the particular phase, which bears the sex organ producing male gametes, or antherozoids, and female gametes, or eggs, have been designated as the gametophytic generation or the sexual generation, and the particular phase or generation that bears the asexual bodies, responsible for producing spores, has been designated as the sporophytic or asexual generation.

By the process of gametogenesis (by means of meiotic division), the gametophytes yield male and female gametes. The sexual fusion (either by conjugation in lower groups of plants or by fertilization in higher group of plants) of male (antherozoids) and female gametes (eggs) with a haploid set of the chromosomes (n) lead to the formation of the diploid zygote (2n). The beginning of sporophytic generation is when the mature sporophyte (2n) with diploid set of chromosomes produces spores through the process of sporogenesis, and the germination of mature spores form either monoecious (bearing male and female sex organs in the same plant) or dioecious (where male and female sex organs are produced in different plants) gametophytes.

A unique phenomenon of the integral, consecutive, and rotational cycle of the "gametophytic generation" or the "haploid state," and the "sporophytic generation," i.e., the "diploid state" in a plant's life is known as the "alternation of generations." Most of the time, the alternation of generation occurs in plant life in nature but does not necessarily happen in the life-cycle of all plants. In few occasions, the clonal propagation or vegetative reproduction of any gametophytic generation leads to the formation of successive sub-cycles of generative generations before the sporophytic generation of a plant. In some situations, the egg, the end-product of the gametophytic (female gametophyte) generation gives rise to the sporophytic generation, bypassing sexual fusion of the male gamete, known as parthenogenesis. Naturally, the stereotypic interpretations of the cyclic rotation of the "haploid" and "diploid" states don't reconcile with the phenomenon of "alternation of generations."

In some occasions, the consecutive alternation of "sporophytic" and "gametophytic" generations have been defined as the "alternation of generations," but that does not necessarily mean a cyclic alternation of the "diploid" and "haploid" states. Sometimes, the sporophytes yield gametophytes without forming spores, known as "apospory," and gametophytes yield sporophytes by vegetative budding rather than sexual fusion of gametes, known as "apogamy."

Thus, from an evolutionary viewpoint, "alternation of generations" plays a key role in maintaining the progress of plant life on Earth by stabilizing the phenomenon of "proliferation of the homozygous state" through far and wide dispersal. This is done either by the formation of numerous spores of identical genetic configurations and phenotypic characteristics, or the "recombination of unique characteristics to maintain the heterozygous state" by merging the characteristic features of different individuals of a particular type of plant through gametic fusion. Basically, the "alternation of generation" is a balancing podium between a homozygous and heterozygous state in nature and acts as a natural safety valve to control genetic erosion as it maintains the equilibrium state of any gene-pool. In some archegoniate thallophytes (e.g., mosses), the sexual fusion of gametes has stunted the formation of sporophytes, so the alternation of gametophytes and sporophytes is visibly absent, leading to the continuation of homozygous generations. The continuation of the homozygous state of any particular plant makes it genetically weaker and unable to sustain in the dynamic conditions of the Earth's biosphere, with fewer chances of new species formation in the long run of evolution and diversification of plant life on the blue planet.

Algae, the ancient thalloid form of life reproduces literally by making a copy or copies of its prototype or multiplying by means of vegetative, asexual and sexual modes of reproduction. From an evolutionary viewpoint, reproduction is the most important phenomenon in the life of any living organism as it helps continue the journey of life by the formation of (offspring) a number of copies of the mother stock when the lifeline of the mother-stock is about to lapse. Basically, "reproduction" is an integral part of evolution and diversification of living organism on Earth.

As some algal members multiply or reproduce in a vegetative state, they are devoid of the rejuvenation of protoplasm and form offspring that are almost identical to the mother plant body (due to a lack of recombination of diversified characteristics). Thus, the formation of a homozygous colony is most likely; moreover, homozygous living forms cannot play a positive role in the evolution and diversification of life on Earth.

The asexual mode of reproduction involves the formation of typical homozygous communities due to mitotic nuclear division (equal division), such that the offspring are supposed to be absolutely identical to their mother plants with little chance of genetic variation and are highly susceptible to inbreeding depression. However, sexual reproduction, which involves the fusion of gametes of the same or different individuals under the same group of plants. The chances of sustainability among the heterozygous (as genetic recombination induced infusion has offset the genetic erosion of the existing gene pool) type of plants in the changed scenario of the Earth's biosphere is greater than that of homozygous (as those are plagued by genetic erosion) types of plants due to the better acclimatizing capability to survival challenges. It is noteworthy that by means of asexual reproduction, algae produce some specialized reproductive cells (either motile or non-motile), known as spores, which on maturity give rise to new individuals without taking part in sexual fusion. However, the gametes and zoospores, the specialized sexual cells in algae, are morphologically alike size-wise, zoospores are a little bigger than gametes. When zoospores, the unit of asexual reproduction, give rise to new plants, which the gametes could not do without sexual fusion, and the formation of the zygote (2n). So, the appearance of a phenomenon like a fusion of gametes is the base of sexual reproduction in algae, which could also be considered as the origin of sex in primitive plants.

Hence, a transitional state between asexual reproductive units e.g., zoospores and sexual reproductive units, gametes, must be there. In some of the members of a primitive group of a green algal family of Chlorophyceae (e.g., *Chlamydomonas* and *Ulothrix*), the formation of structurally alike but different sized asexual units: zoospores (bigger size) and microzoospores (smaller size), have been noticed. It has been noticed by scientists that a number of divisions of the protoplasmic mass in the mother cell gives rise to numerous smaller zoospores (having little protoplasmic content not enough to give rise to new plants) and bigger sized zoospores or macro zoospores (having enough protoplasmic content to give rise to new plants) that go through a small number of divisions of protoplasmic mass. During unfavorable situations (e.g., absence of moisture, excessive cold, physical drought, physiological dryness etc.) the microzoospores cannot come out of the sporangia, so they stay inside the sporangia and functionally act like gametes; the pair of perennating micro-zoospores, and the functional gametes fuse with each other to form zygotes, which come out of the sporangia in favorable conditions and germinate to form new plants. This sequential phenomenon of formation of microzoospores, perennation in unfavorable conditions and fusion of paired microzoospores, has been defined by phycologists as the attraction of two nutritionally compatible cells in the "Sarvation Hypothesis," which has laid the foundation of the origin of algae as well as primitive simple form of a thalloid group of plants on Earth.

Therefore, the origin of sex as well as sexual reproduction is not only an extended process of formation of haploid (n) generations from diploid mother plants (2n) and fusion of a pair of haploids (n) to form diploid (2n) offspring, instead of the formation of diploid (2n) offsprings = from diploid (2n) mother plants by vegetative and asexual reproduction rather than sexual reproduction, is the most important natural phenomena of living organisms to survive generations after generations on Earth, keeping the journey of evolution on Earth uninterrupted and going through the diversification processes to offset the survival challenges in spatiotemporal variations and evolve as new species after coming a long way. This supposedly originated in algae as well as a lower group of ancient plants as an accidental fusion of microzoospores (structural zoospores but functional gametes) during unfavorable spatiotemporal conditions.

It has been notes that the external factors of changing of temperatures (e.g., *Collinsiella, Urospora*), photoperiodism (e.g., members of Phaeophyceae), light intensity (e.g., members of Rhodophyceae), moisture (e.g., *Prasiola*), salinity (e.g., filamentous algal members belong to Chlorophyceae), tidal waves (e.g., marine algal members belong to Chlorophyceae) and the internal stimuli (e.g., the "pseudo-hormonal" chemical "Ectocarpen" secretion from female gametophyte of the algae *Ectocarpus siliculosus* attracts the male gametes for sexual fusion), play the key role in initiating sexual reproduction, instead of the conventional vegetative and asexual reproductive cycles in algae (Sambamurty, 2005).

Once, the origin of sex in algae has been established on Earth, the next phase is its evolution and diversification to optimize diversified survival challenges to stay in the survival race as well as acquire better adaptability to achieve sustainability in the dynamic biosphere to evolve as better, superior and higher forms of plants to acclimatize to the Earth's continuous changes to reach its current form. The evolution of sex in thalloid forms of primitive organisms (e.g., algae) has been initiated with "isogamy," passed through the primary transitional phase of "physiological anisogamy," followed by the secondary transitional phase of "morphological anisogamy," subsequently followed by the tertiary transitional phase of "homothallic oogamy," to reach the quaternary evolutionary stage of sexual fusion known as "heterothallic oogamy" (Bold and Wynne, 1978).

i. Isogamy: It has been found among some algal members (e.g., *Chlamydomonas, Ulothrix* etc.) where sexual fusion takes place between two structurally and functionally identical gametes, which is a preliminary approach towards genetic recombination of thalloid plants.

ii. Physiological anisogamy: In some algal members, a sexual fusion of gametes has taken place between the pair of gametes which are structurally identical but functionally different (e.g., *Spirogyra, Ulothrix* etc.), known as functional anisogamy or physiological anisogamy.

These types of apparently identical gametes either originated from a different type of filament (not to be identified as microsporangium or megasporangium) so one is functionally more proactive than the other during sexual fusion or are structurally identical and originated from the same location but one is physiologically more active than the other during sexual fusion. Physiologically, it is easier to distinguish the proactive ones as male or (+) gamete and other one as female or (-) gamete and this phenomenon could be treated as a preliminary approach to have a distinct sexual identity, the obligatory condition of genetic recombination.

i. Morphological anisogamy: In some algal members (e.g., *Clodium, Ectocarpus* etc.), the fusion of gametes took place between the pairs which are structurally different (morphologically identical but of different sizes) and is designated as morphological anisogamy and these participatory gametes are known as anisogametes. The larger gamete is known a macrogamete (functionally female gamete), and is produced from the microsporangium whereas the smaller one designated as microgamete (functionally male gamete), is produced from the microsporangium and this stage may be considered as a well-defined evolutionary journey of primitive plants to reap the harvest of genetic recombination to achieve the ultimate heterozygous status.
ii. Homothallic oogamy: In the members of some advanced groups of algae (e.g., *Chara, Fucus, Oedogonium* etc.), sexual fusion has taken place between the gametes of structurally and physiologically distinct types, known as a relict type of oogamy. The smaller, more active gamete has been designated as the male gamete or antherozoids (motile with the help of flagellated movements), produced in antheridia, and the larger size, non-motile female gamete has been designated as an egg, produced in oogonia. As this relict type of oogamous sexual fusion of gametes has been noticed among most of the homothallic algal members, this phenomenon has also been designated as "homothallic oogamy," a penultimate effort of thalloid forms of plants to express affinity towards heterozygous formation through genetic recombination with a relict homothallic (same thallus bearing antheridia and oogonia) body. Most of the algal species are homothallic entities rather than heterothallic structures.
iii. Heterothallic oogamy: Though it is a rare phenomenon, the heterothallic algal species (e.g., Spermatia) were also discovered by phycologists that could produce the morphologically and physiologically identified male gamete (without

motility), known as spermatium, produced inside the spermatangium (produce singly instead of producing a number of male gametes), driven by water currents to the female gametophyte, carpogonium (egg), to conduct the sexual fusion of the male and female gametes, designated as heterothallic oogamy.

Thus, if we consider the evolutionary journey of the primitive, thalloid, simple form plants to structurally diversified ones, an advanced form of plants is a shift of paradigm, origin, and evolution of sex in the quest for the formation of the heterozygous form by utilizing the advantage of genetic recombination, which has been considered as a basic condition (Foyn, 1929). Hence the evolutionary journey of plants in terms of better structural and physiological developments and the diversification of primitive plants should be considered in the background of its effort to go for better reproductive practices. The earnest effort to switch from a vegetative and asexual form of reproduction of the primitive thalloid algal plants on Earth, which was started with the isogamous type of sexual fusion of gametes, went through different transitional stages of development to reach the heterothallic oogamy, monophyletically. We can visualize the occurrence of sexual reproduction of male and female plants, an earlier but futuristic form of sexual reproduction found among structurally and physiologically diversified, modern, dioecious plants, evolved in the late Tertiary or Quaternary Period.

Chapter 5

CONSECUTIVE EVOLUTION OF THALLOPHYTES-1 (ALGAE): (TRANSMIGRATION FROM WATER TO LAND ALONG WITH METAMORPHOSIS PROKARYOTIC CELLS TO EUKARYOTIC FORM)

Literally, the meaning of "Algae" (in Latin, Seaweed) is Hepataceae, which was used by Linnaeus (1753) for the first time. A number of diversified forms of composite features, like- habit (unicellular prokaryotic to multicellular eukaryotic), habitat (aquatic, epiphytic, lithophytic etc.), size (microscopic unicellular to 5 feet long sea grasses), ecology (autophytic, halophytic, symbiotic, saprophytic, thermophilic etc.), biochemistry (chlorophyll-a, b, c, d and e, xanthophyll, phycocyanin, phycoerythrin etc.), reproduction (vegetative, isogamy, anisogamy, oogamy etc.), and etc., have been found among these group of ancient, transducing, simple forms of life as well as the plants on Earth.

i. Though algae and bryophyte plant groups are categorized as Thallophyta (in Greek Sprout like plants) and the common trends like the initial evolutionary effort of structural diversification and
ii. transmigration from water to land was witnessed, but there are two distinguishing features that help distinguish algae from bryophytes-
    a) the female sex organs in algae are normally unicellular (the members of Chlorophyceae are an exception), in bryophytes they are multicellular, known as archegonia (hence bryophytes are referred as "archegoniate Thallophytes") and
    b) in post-fertilization phase, during the formation and development of the zygote, no embryo formation was found in algae.

Until the middle of the 20th century, this group of ancient, simple form thalloid, ever-present photosynthesizing plants have been categorized either based on their habitat (aquatic vs. terrestrial, etc.) or their thallus orientation (unicellular vs. multicellular, etc.). In his preliminary approach, "The structure and reproduction of algae," Fritsch (1935), considered the characteristic features of pigmentation, assimilatory food reserves, nature of flagella, etc. and categorized it in 11 classes.

Based on the earlier phylogenetic principles, Papenfuss (1946; 1955) studied all algal members under 8 phyla and 12 classes. Following these earlier approaches of the artificial system of classification of algae by Fritsch (1935; 1945), Chapman (1962) and Prescott (1968) developed the earlier format by adding up the two important aspects of the nature of nuclear materials, and the life and reproduction cycle. Staying away from the classical approaches to classification, Round (1973), Chapman and Chapman (1973) solely considered the nature of nucleic material, i.e.,

i. absence or presence of a nuclear membrane,
ii. nucleolus attached to nuclear chromatin and
iii. absence or presence of membrane-bound organelles, through which algae have been broadly categorized into two groups: i. Prokaryota and ii. Eukaryota and are subsequently classified under class divisions. Following the phylogenetic approach of Papenfuss (1946, 1955), Bold and Wynne (1978) classified algae into 9 divisions. In the initial approach, in a contemporary phylogenetic system of classification, in "Phycology," 2nd edition, 1989, Lee (1989) classified algal members into 4 evolutionary groups and 15 divisions and classes. Finally, in his phylogenetic revisionary work on algae, "Phycology," Lee (1999, 2008), revised his earlier work and critically reviewed the works of Jordan and Green, 1994; Preisig, 1995; Clay et al., 1999; and Harper and Saunders (2001, 2002), to reformat all algal members into four phylogenetic groups, nine phyla; with each phylum under a class; each class under order and so on, under contemporary, systematic classification of botany-

Taxonomic Rank	Suffix
Phylum	+phyta (e.g., Chlorophyta)
Class	+phyceae (e.g., Charophyceae)
Order	+ales (e.g., Coleochaetales)
Genus	(e.g., *Coleochaete*)
Species	(e.g., *scutata*)

An outline of Lee's (1999; 2008) phylogenetic classification of algae has been given herewith:

Group A: Prokaryotic algae

1) Phylum- Cyanophyta (Cyanobacteria or blue-green algae): The members of this group have been characterized by- cells with membrane-less nucleus; cell walls made up of muramic acid and mucopeptide; chromatophores absent, presence of pigments in thylakoids- chlorophyll a, c-phycocyanin, c-phycoerythrin, etc., absence of phycobiliprotein; presence of food storage like- cyanophycean starch, cyanophycean protein; reproduction by asexual or vegetative methods, devoid of sexual reproduction. e.g., *Anabaena*, *Oscillatoria* etc.

Group B: *Eukaryotic algae* having chloroplasts surrounded by a chloroplastic envelope (double membrane) –

1) Phylum- Glaucophyta: This algal group possesses Chlorophyll A and phycobiliproteins; photosynthesis of this group of algae has been carried out by endosymbiotic cyanophycean algae (cyanelles) as the evolution of its chloroplast is in transition and the host body known as cyanom, proteinaceous pyrenoid inherited from cyanelle, while the metabolic food storage takes place in the cyanome cytoplasm. e.g., *Cyanophora, Geosiphon* etc.
2) Phylum- Rhodophyta (red algae): This group of algae possesses pigments like – Chlorophylls A and D, phycocyanin, phycoerythrin, and phycobiliproteins; they contain Floridian starch as food storage and have no flagella. The class Rhodophyta has been assigned under the class- Rhodophyceae.

a. Class- Rhodophyceae- It has been further assigned in 10 orders:

i. Order- Porphyridiales. e.g., *Prophyridium, Rhodosorus* etc.
ii. Order- Rhodochaetiales- e.g., *Rhodochaete.*
iii. Order- Acrochaetiales- e.g., *Audouinella, Rhodochorton* etc.
iv. Order- Bangiales- e.g., *Bangia, Porphyra* etc.
v. Order- Batrachospermales. e.g., *Batrachospermum.*
vi. Order- Nemaliales. e.g., *Galaxaura, Nemalion* etc.
vii. Order- Corallinales. e.g., *Amphiora, Corallina* etc.
viii. Order-Gelidiales. e.g., *Gelidiella, Gelidium* etc.
ix. Order-Gracilariales. e.g., *Gracilaria, Melanthalia* etc.
x. Order- Ceramiales. e.g., *Ceramium, Chondria* etc.

3. Phylum- Chlorophyta (green algae): This algal group possesses chlorophylls A and B and starch is stored inside the chloroplast. It has further been categorized under 4 classes-

a. Class- Micromonadophyceae (micromonads): Phylogenetically, this unicellular, ancient group, gave rise to the young members of green algae having a number of flagella attached to the depressed bases (instead of attaching to the papilloid location of the cell wall). It has a scale wall and flagella covered by scales, produced by Golgi bodies, and an interzonal mitotic spindle which formed during cytokinesis; e.g., *Pyramimonas*.
b. Class- Charophyceae: The flagella are attached laterally to the motile cells of this algal group. The flagellar root is comprised of a broad microtubule with concentric smaller microtubules, eyespots are absent, it has a multilayered wall, an interzonal mitotic spindle found in its telophase, and a zygote formed through sexual fusion. It has been further characterized under 4 orders-

i. Order- Klebsormidiales- e.g., *Klebsormidium, Raphidonema* etc.
ii. Order- Zygnematales- e.g., *Cosmerium, Spirogyra* etc.
iii. Order- Coleochaetales- e.g., *Coleochaete*.
iv. Order- Charales- e.g., *Chara, Nitella*.

c. Class- Ulvophyceae: This is a marine algal group with the flagella attached to the anterior side of the motile cell. Its flagellar roots are comprised of 4 micro-tubules (sometimes with rhizoplasts); eyespots are present; multilayered cell wall is absent; the interzonal mitotic spindle is found in the telophase; dormancy of zygote is absent; and alternation of generation is present in its lifecycle. It has been characterized under 6 orders:

i. Order- Ulotrichales- e.g., *Geminella, Ulothrix* etc.
ii. Order- Ulvales- e.g., *Monostroma, Ulva* etc.
iii. Order- Cladophorales- e.g., *Chaetomorpha, Cladophora* etc.
iv. Order- Dasycladales- e.g., *Acetabularia, Neomeris* etc.
v. Order- Caulerpales- e.g., *Caulerpa, Codium* etc.
vi. Order- Siphonocladales- e.g., *Siphonocladus, Valonia* etc.

a. Class- Chlorophyceae: This algal group exists in fresh water; the flagella are attached to the anterior side of the cell; flagellar roots are comprised of 4 micro-tubules with rhizoplasts; eye spots are present; multilayered cell wall is absent; its interzonal spindle collapses in the telophase; and dormancy of zygote has been noticed. It has been further characterized under 8 orders:

i. Order- Volvocales - e.g., *Chlamydomonas, Volvox* etc.
ii. Order-Tetrasporales - e.g., *Palmella, Tetraspora* etc.
iii. Order- Parsiolales - e.g., *Parsiola, Hormidium* etc.

iv. Order- Chlorococcales - e.g., *Chlorella, Hydrodictyon* etc.
v. Order- Sphaeropleales - e.g., *Microspora, Sphaeroplea* etc.
vi. Order-Chlorosarcinales - e.g., *Chlorosarcina, Desmochloris* etc.
vii. Order- Chaetophorales - e.g., *Chaetophora, Stigeoclonium* etc.
viii. Order- Oedogoniales - e.g., *Bulbochaete, Oedogonium* etc.

Group C: Eukaryotic algae having chloroplasts, encased within a single membrane of endoplasmic reticulum-

1. Phylum - Euglenophyta (euglenoids): This algal group possesses chlorophylls A and chlorophylls B; proteinaceous pellicle strips are found under its plasma membrane; storage product is paramylon; it has a single flagellum of the spiral; and its mode of cell division is not conspicuous. It has only one class- Euglenophyceae.

a. Class - Euglenophyceae: This unicellular, algal group possesses chlorophylls A and B; have a single membraned chloroplast and endoplasmic reticulum; have flagella with fibrillary hair; their metabolic food, 'paramylon', is stored in the cell; sexual reproduction is absent in its life cycle; and its nucleus is mesokaryotic. It has been categorized under 3 orders:

i. Order- Heteronematales - e.g., *Entosiphon, Heteronema* etc.
ii. Order- Eutreptiales - e.g., *Eutreptia, Eutreptiella* etc.
iii. Order- Euglenales - e.g., *Euglena, Phacus* etc.

2. Phylum- Dinophyta (dinoflagellates): This algal group possesses chlorophylls A and chlorophylls C; thecal pellicle is found under its plasma membrane; a mesokaryotic nucleus is present; cell division along the girdle of its cingulum lead to the formation of an epicone and a hypocone; it is helically coiled; and a transverse flagellum has been found in this group. It has been assigned under the class of Dinophyceae.

a. Class- Dinophyceae: This group possesses chlorophylls a and chlorophylls c_2 along with peridinin (carotenoids) if photosynthetic, unless it is a colorless heterotroph; the chloroplast is encased within a single membrane of chloroplast of endoplasmic reticulum; dinoflagellates are comprised of an epicone and a hypocone; eye spots are present; and its nucleus is either dinokaryotic or mesokaryotic. It has been classified under 4 orders:

i. Order- Porocentrales - e.g., *Porocentrum*.
ii. Order-Dinophysales - e.g., *Dinophysis, Ornithocerus* etc.
iii. Order-Peridiniales - e.g., *Ceratium, Peridinium* etc.
iv. order- Gymnodiales - e.g., *Amphidium, Gymnodinium* etc.

Group D: Eukaryotic algae with a chloroplast encased within a double membrane of endoplasmic reticulum-

1. Phylum- Cryptophyta (cryptophytes): This group of algae possess chlorophylls A and chlorophylls C, phycobiliproteins; nucleomorph is present between the two membranes of the endoplasmic reticulum; and storage starch has been found inside the chloroplast of endoplasmic reticulum. It has been assigned under the class- Cryptophyceae.

a. Class- Cryptophyceae: This group of algae grows in freshwater and marine habitat; it has a dorsiventral body; possess chlorophylls A and chlorophylls C_2; its chloroplast is pyrenoid or heterotrophic occasionally; and phycobiliproteins have been found inside the thylakoids of chloroplasts. e.g., *Chilomonas, Cyathomonas* etc.

2. Phylum- Heterokontophyta (heterokonts): this group of algae possesses chlorophylls A, chlophylls C_1, and chlorophylls C_2, fucoxanthin; the metabolic food, chrysolaminarin, is stored in the vesicles of the cytoplasm; dimorphic flagella have been noticed (the anterior one is tinsel type and the posterior one is whiplash type) in this algae; eyespots are present; and has an endoplasmic reticulum. This phylum has further been classified into 9 classes-

a. Class- Chrysophyceae (golden-brown algae): Identical features of phylum- Heterokontophyta; has a bilaterally symmetrical body; one of the flagella is perpendicular to the another; contractile vacuole is present in the front part of the body. e.g., *Anthophysa, Paraphysomonas* etc.

b. Class- Synurophyceae: This algal group possesses chlorophyll C_1; its body is bilaterally symmetrical; chloroplast endoplasmic reticulum is absent; the flagella is impregnated to the cell, poised parallel to each other; photoreceptors are present at the base of flagella; eyespots are absent; chloroplast and endoplasmic reticulum are absent; and it has a contractile vacuole on the back side of the body. It has been assigned under 3 orders:

i. Order- Goniomonadales- e.g., *Goniomonas.*
ii. Order- Cryptomonadales- e.g., *Cryptomonas.*
iii. Order- Chroomonadales- e.g., *Chroomonas.*

c. Class- Dictyochophyceae (silicoflagellates): This group of algae has identical pigments to the golden brown algal group; its body cell is amoeboid; it has locomotory organs such as tentacles or rhizopods that are a part of the vegetative amoeboid cell. This group has further been categorized under 3 orders:

i. Order- Rhizochromulinales- e.g., *Chrysoamoeba, Phaeaster* etc.
ii. Order- Pedinellales- e.g., *Apedinella, Pedinella* etc.
iii. Order- Dictyocales- e.g., *Dityocha, Naviculopsis* etc.

d. Class- Pelagophyceae: It has a unicellular algal body, very small (3-4.5μ), and spherical with an inconspicuous protoplasm. e.g., *Pelagomonas*.

e. Class- Bacillariophyceae (diatoms): This group is either unicellular or survived in a colony; cells are radially or bilaterally symmetrical; this group of algae possesses chlorophylls A, chlorophylls C_1 and chlorophylls C_2, fucoxanthin; their box type body consists of a top lid, epitheca, and the bottom part, hypotheca; silica granules are found in its cell wall; metabolic food, chrysolaminarin, volutin, and fat is stored in the cytoplasmic vesicles; and they have a posterior flagellum on spermatozoa. It has been categorized under two orders-

i. Order- Biddulphiales- e.g., *Chaetoceros, Melosira* etc.
ii. Order- Bacillariales- e.g., *Navicula, Pinnularia* etc.

f. Class- Raphidophyceae (chloromonads): This group of algae lives in a marine habitat; this algal group possesses chlorophylls A and chlorophylls C; they have two membranes of endoplasmic reticulum; with a tinsel flagellum in the anterior part and a whiplash flagellum in the posterior part of the body. e.g., *Chattonella, Heterosigma* etc.

g. Class- Xanthophyceae (yellow-green algae): This group of algae possesses chlorophylls A and chlorophylls E, β-carotene, xanthophyll; their chloroplast is not pyrenoid; they have a tinsel flagellum in the anterior part and a whiplash flagellum in the posterior part of the body; storage food is oil; and vegetative, asexual, and sexual reproduction has been found in this group. It has been subdivided into 3 orders-

i. Order- Tribonematales- e.g., *Tribonema*.
ii. Order- Botrydiales- e.g., *Botrydium*.
iii. Order- Vaucheriales- e.g., *Vaucheria*.

h. Class- Eustigmatophyceae: This group of algae is unicellular, living in an aquatic-terrestrial habitat; they possess chlorophyll A, and β-carotene; conspicuous eyespots have been found on the chloroplast; and a tinsel flagellum is present on the anterior part of the body and the smaller whiplash flagellum is present on the posterior part of the body, e.g., *Pleurochloris, Polyedriella* etc.

i. Class- Phaeophyceae (brown algae): This algal group lives in a marine habitat; possess chlorophylls A, C_1, C_2, and fucoxanthin; their metabolic food, chrysolaminarin, mannitol, and fat, is stored in the cytoplasmic vesicles; tinsel flagellum has been found in the anterior part and whiplash flagellum has been found in the posterior part of the body; and its sexual reproduction ranges from isogamy to oogamy. This group has been divided into 7 orders-

i. Order- Ectocarpales- e.g., *Ectocarpus, Pilinia* etc.
ii. Order-Desmarestiales- e.g., *Desmarestia, Phaeurus* etc.
iii. Order- Cutleriales- e.g., *Cutleria, Zanardiana* etc.
iv. Order- Laminariales- e.g., *Chorda, Laminaria* etc.
v. Order- Sphacelariales- e.g., *Battersia, Sphacella* etc.
vi. Order- Dictyotales- e.g., *Dictyota, Pachydityon* etc.
vii. Order- Fucales- e.g., *Fucus, Sargassum* etc.

3. Phylum- Prymnesophyta (haptophytes): This algal group possesses chlorophylls A, C_1 and C_2, fucoxanthin; having two membranes of chloroplast endoplasmic reticulum; its metabolic food, chrysolaminarin, is stored in a cytoplasmic vesicle; the outside of the body is covered with scales; and two whiplash flagella are connected to a peg like base, "haptonema." It has been assigned under the class – Pyrmnesiophyceae.

a. Class- Prymnesiophyceae: These family members have been sub-divided under two orders:

i. Order-Prymnesiales- e.g., *Hymenomonas, Prymnesium* etc.
ii. Order- Pavlovales- e.g., *Pavlova, Rebecca* etc.

It was mentioned earlier, that 2.9 billion years old paleobotanical evidence at the Birbal Sahni Institute of Paleobotany in India, has ascertained that single-celled, prokaryotic, transducing cyanobacteria (popularly known as blue-green algae), appeared around 3 billion years ago (Sergeev et al., 2012). In support of this view, the 3 billion years old fossilized eobiont from the African continent, "*Eobacterium isolatum,*" is referred to here (Barghoorn and Schopf, 1966). Similarly the existence of 3.5 and 3.7 billion years old Stromatolites (a chemo-rganotrophic bacteria, interspersed with anoxygenic photosynthesizing bacteria) from two locations in Australia, and Greenland respectively further affirms the earlier view regarding the evolution of the simple form of life on Earth (Barghoorn and Schopf, 1965; Nutman et al., 2016; Van Kranendonk et al., 2008).

A number of evolutionary biologists like Marguillis (1970), Mahler and Raff (1972) postulate that either complex symbiotic relationships among different prokaryotic living beings, or a combination of symbiosis and mutation among prokaryotic organelles, lead to the latter evolving to the "eukaryotic cellular form." Even the epigraphic evidence indicates that evolution of the derived form of the eukaryotic, metazoic, transducing organisms have supposedly originated from prokaryotic root-stock (Fritsch, 1916; Archibald, 2015) of ancient algae, bacteria, and cyanobacteria around 2-1.5 billion years ago.

The unique features of cyanobacteria, like possession of unorganized nucleus, lack of chromatophores, flagella and sexual reproduction have attributed this group of unicellular microorganisms as an ancient form of life. From an evolutionary aspect, it maintains a transition between the ancient form of prokaryotic bacteria and the advanced form of algae Rhodophyta. The common features of cyanophycean members shared with bacteria are as follows-

1) The presence of an unorganized nucleus and organelles without a membrane.
2) The cell wall being composed of mucopeptide (murein) instead of cellulose.
3) Being capable of sulfur and nitrogen metabolism. For utilizing the N_2 in the air (high content of atmospheric Nitrogen was most likely in the ancient Earth), the presence of a special functional organ, "heterocyst," supposedly evolved in the group of Cyanophyceae.
4) Sexual reproduction is absent, though genetic recombination has been noticed (e.g., *Anacystis nidulans*), same as bacteria.
5) The cell wall is sensitive to antibiotics like bacteria.

The common features of Cyanophycean members, shared with Rhodophyta are as follows-

1) The absence of motile cells in both groups.
2) The cyanophycean algal pigments- like c-phycoerythrin (red) and c-phycocyanin (blue), have a structural resemblance to the pigments of Rhodophyta (r-phycoerythrin and r-phycocyanin).
3) The presence of intra-cellular pit-connections in Cyanophyceae make them phylogenetically closer to Rhodophyceae.

Gaidukov (1903), an eminent plant physiologist, noticed a unique feature among the cyanophycean members of algae, known as the "complementary chromatin adaptation." Regarding the evolutionary phase of photosynthesizing organisms, the paleo-atmospheric conditions, as well as the light intensity and color spectrum, were supposedly more diversified than the contemporary stable form of the wavelength of life and standard color spectrum, in order to optimize the rate of photosynthesis.

The cyanophycean members most likely evolved with a diverse range of pigments to cope with different wavelengths of light, the intensity in different atmospheric conditions, and habitats to energize the bacteriochlorophyll for maximization of conversion of light energy to biochemical energy required for metabolic activities, known as the "Gaidukov Phenomenon."

According to the contemporary algal classifications of Lee (1999, 2008), the members of the phylum: Glaucophyta (under Group 2 of Eukaryotic algae) have been observed without chloroplasts (i.e., a lack of energy production and storage system). It is quite amazing that this group of apparently heterophytic, eukaryotic algae has offset this impediment by establishing an endosymbioticrelationship with a cyanophycean group of algae, formally known as "cyanome-cyanelle" (Keeling, 2010). In this endosymbiosis, the body of glaucophyte algae acts as a host (cyanome) and the cyanophycean algal member in the cytoplasm of cyanome is known as "cyanelle."

The ancient paleobotanical evidence along with molecular phylogenetic studies ascertained the "inclusive engulfing of plastids" in eukaryotes or "endobiotic surrogacy of photosynthesizing organelle" and "chloroplast genomes" from prokaryotes (e.g., Cyanobacteria) to have taken place around the late Proterozoic Era (i.e., around 3,500 m.y.a.) (Green, 2011; Yoon et al., 2004). The cyanelle is devoid of a cell wall and is encased either within an old membrane of the food vesicle of the cyanome or, the plasma membrane of the cyanome. In the course of evolution of transducing capabilities in eukaryotes, the transitional divergence of the "cyanome-cyanelle" unit is the half-way stage before chloroplasts (Gould et al., 2008) and endoplasmic reticulum appear in eukaryotes.

Prior to the transmigration of plants from water to land, all green plants, comprising of green algae (including chlorophyte and streptophyte) and embryophytes were recognized as "Viridiplantae" (Becker and Marin, 2009; Leliaert et al., 2011) and this monophyletic group with its diverse forms, depends upon the structural, physiological, ecological, biochemical, and reproductive adaptations. While algae occupy freshwater and ocean ecosystems, the land plants are considered as key components of the terrestrial ecosystems (Lewis and McCourt, 2004). According to recent phylogenetic investigations, the embryophyte and streptophyte groups of algae conform with the Streptophyta division, whereas the rest of the green algae form the Chlorophyta division (Leliaert et al., 2011; Sanderson et al., 2004). According to Leliaert et al. (2011), the "origin of oxygenic photosynthesis" went through 3 evolutionary lineages: the green plants, the red algae, and the Glaucophytes. Hence, the event of photosynthesis initially diversified due to the successful occurrence of "secondary symbiosis," the transfer of plastids in green or red algae from the cyanophycean stock to the non-photosynthesizing, larger eukaryotes (Gould et al., 2008). Furthermore, "plastid acquisition" become an evolutionary trend as secondary symbiosis was taken over by a phylogenetically distant group of eukaryotes, known as "tertiary endosymbiosis" (Archibald, 2009; Keeling, 2010; Leliaert et al., 2011). According to the recent phylogenetic evidence (Leliaert et al., 2011), the surrogacy of plastids in those groups of eukaryotes, was done by a common algal donor endosymbiont, recognized as "Chlorarachniophyte," comprised of euglenids and dinoflagellates, dwelling in tropical oceans. It has also been considered that the diverse derived groups of eukaryotic, transducing algal members, like- charophytes, ulvophytes, euglenoids, dinoflagellates, bacillariophytes, xanthophytes, phaeophytes, and haptophytes emerged as a result of "plastid acquisition" from the red algal mother-stock as a result of a recurrent cycle of "endosymbiosis" (Archibald, 2009; Keeling, 2010; Leliaert, 2011).

From an evolutionary perspective, the divergence of Streptophyta and Chlorophyta is related ecological adaptations like the fresh-water adaptation of Streptophyta versus the affinity towards marine habitat of Chlorophyta. Hence, being part of the Streptophyta, the primitive embryophytes (ancient land plants) have an evolutionary advantage in freshwater habitat (even before the appearance of alternation of generations set in the life cycle of embryophytes), from where it's much easier to migrate to a terrestrial habitat. The evolution and diversification of land plants (like terrestrial algae, bryophytes, pteridophytes, early seed plants and flowering plants) on terrestrial ecosystems have an ancestral inheritance (Beraldi-Campesi, 2013), to the earliest effort of transmigration of aquatic plants to terrestrial ecosystems as the ancient groups of aquatic progenitors of land plants overcame challenging factors like the changing O_2 concentration in the atmosphere, better morphological, ecological, and physiological adaptations to colonize the terrestrial habitats, etc. (Scott and Glasspool, 2006).

According to contemporary phylogenetic observations of Kenrick and Crane, (1997); Sanderson et al., (2004), and Becker and Marin (2009), the terrestrial successors of Streptophyta migrated and colonized the terrestrial habitats successfully in the Ordovician Period of the Paleozoic Era (between 470-450 m.y.a.). The divergence of primitive land plants or embryophytes was initiated by the simple forms of unicellular, flagellated, scaly, freshwater algae of Mesostigma of the division- Streptophyta, and went through a complex mode of evolution (like structural diversification and modifications such as the terrestrial plants) in the derived group of Streptophyta, e.g., Charales, which has also been considered paraphyletic to early land plants (Becker and Marin, 2009). According to Becker and Marin (2009), the Streptophyta group of algae diverged with the evolutionary toolbox equipped with- phragmoplast, cellulose synthetase, photorespiration, flagellated cells, sexual reproduction, zygote viability, and etc., necessary to continue transmigration from water to land.

Chapter 6

ALTERNATION OF GENERATIONS AND THE EVOLUTION OF SPOROPHYTES AND GAMETOPHYTES IN EM-BRYO-PHYTES

Heteromorphic alternation of generations has been witnessed in the life-cycle of bryophytes with cyclic rotations of the following-

a. independent, autotrophic, haploid (n) and gamete producing "gametophytic phase," started by the germination of spores and followed by-
b. the "sporophytic phase," also known as "Sporogonium," which is a diploid (2n), engaged in sexual fusion of haploid (n) gametes by means of syngamy, dependent upon the gametophytic phase.

Perceived from phylogenetic aspects of the plant kingdom, evolution of sex is related to the alternation of generations on one hand, and the origin of seed habit on the other as a natural avoidance of self-fertilization in the natural habitat acts as a safeguard against inbreeding depression. Genetic drifting and genetic bottleneck provide a better chance of cross-fertilization with better opportunities of diversification and better acclimatization in dynamic environments. The origin of new species would also be facilitated in long run and can ensure the optimization of evolutionary progress in the spatiotemporal dimension. There are two hypotheses to elucidate the alternation of generations:

a) Homologous or Transformation Hypothesis:

This hypothesis was postulated by Pringsheim (1876) and subsequently supported by Scott (1895, 1908), Church (1919), Zimmermann (1930), Bold (1948), Taylor, et al., (2005). According to this, the gametophytic and sporophytic generations evolved independently during the transmigration of plants from water to land. When gametophytic generations were able to support themselves independently, sporophytic generations became partially or fully dependent on the gametophytic generations. A few biological incidents can justify this hypothesis:

ai. Apogamy[9] and apospory[10] have been noticed in the life-cycle of a number of bryophytes and pteridophytes.

aii. Autotrophic nutrition of sporophytes has been effectuated by the presence of green photosynthetic tissues in the sporophytic generations of some bryophytes, which helps recognize the sporophytic generation as autotrophic instead of dependent on gametophytic generation for nutrition.

aiii. Identical structures between gametophytes (features like the cylindrical shape, dichotomous branching and continuous apical growth etc.) and sporophytes of primitive pteridophytes (e.g., *Psilotum* of Rhyniopsida) have been cited to refer to the homologous hypothesis of alternation of generations.

b) Intercalation or Interpolation Hypothesis:

This hypothesis was promulgated by Celakovsky (1874), and subsequently supported by Bower (1890); Strasburger (1894), Cavers (1910), Campbell (1903) and Blackwell (2003). According to this, the gametophytic generation is the original generation in the life-cycle of simple thalloid plants like algae, whereas the sporophytic generation is the result of the gradual elaboration of the zygotes (also recognized as transient sporophyte) of ancient green algae.

The sporophytic generation could be interpreted as an intercalated or interpolated phase (which is distinct from the gametophytic generation with respect to its identity and functionality) in the life cycle (predominantly gametophytic) that is initiated during dynamic fusion. It lasts until the beginning of meiotic division during the formation of gametes. This validates the importance of sex in the due course of the evolution of plant life.

Furthermore, the sporophyte is an advanced, diversified, functionally active, multicellular stage of the unicellular, primitive, dormant zygotic form, evolved in a course of time that became functionally more distinct than the gametophytic phase.

---

[9] Apogamy: When haploid (n) sporophyte has directly been formed from the gametophyte without sexual fusion of gametes or syngamy.

[10] Apospory: When diploid (2n) gametophyte has directly been formed from sporophyte without formation of spores.

EVOLUTION OF "SPOROGONIUM" AS WELL AS SPOROPHYTES IN BRYOPHYTES

The sporophyte of the bryophyte has been structurally built with the aerial sporangia connected by a stem-like stipe with a bulbous foot (root-like) attached to the gametophytic thallus for procuring nourishment (functionally identical to the quasi-parasitic mode of nutrition).

In the post-fertilization period, the sporophytic generation of bryophytes the formation of the diploid zygote inside the archegonium was initiated.

The simple form of sporophyte has been characterized by the zygote made up of central sporogenous tissues and surrounded by outer layers of sterile tissues whereas in the complex form of a zygote, sporogenous tissues have been found well differentiated. Basically, the evolution of the sporophytic generation among ancient plant groups should be considered as a step closer to the evolution of seed formation in angiosperms.

The formation of sporophyte has been explained by two groups of scientists, discussed under two opposing hypotheses:

a) The hypothesis of progressive evolution or evolution of sporophytes by the progressive sterilization of sporogenous tissues:

The hypothesis of progressive evolution was proposed by Bower (1908), and later supported by Cavers (1910) and Campbell (1940). This hypothesis states that the ancient sporophyte was considered to be a structurally simpler form, where most of the sporogenous tissue was found to be fertile (e.g., Liverwort: *Riccia*). This gave rise to the complex form of sporophytes (e.g., Moss: *Funaria*), where parts of the sporogenous tissues gradually transformed into sterile tissues instead of being fertile, in the due course of evolution.

Eventually, step-wise evolution as well as the elevated level of sterilization of sporogenous tissues, would help transform the single-layered jacket of sterile cells around the fertile sporogenous tissues with a low degree of sterilization. Formed from a multi-layered jacket of the foliose sporophyte with a distinct part of the foot, seta, capsule and elaters with intermediate level of sterilization of sporogenous tissues have formed simple type of capsules. With a few-layered jacket, columella, endothecium, amphithecium and pseudo-elaters to the complex capsule with a multi-layered jacket, columella, apophysis, peristome teeth etc. with highest level of sterilization of sporogenous tissues formed inner structure of capsule. Functionally, those sterile tissues engaged in structural diversification of sporophytes. This builds up a multi-level protection system to protect the spores to sustain the challenges of climatic, edaphic, topographic changes, potential threat of destruction by animals, and perennation in unfavorable conditions and attain maturity for optimizing the dispersal and germination potential of spores to initiate gametophytic generations.

b) The hypothesis of regressive evolution or evolution of sporophytes due to progressive reduction:

The hypothesis of regressive evolution has been proposed by Kashyap (1919), Church (1919), Goebel (1930) and Evans (1939). This hypothesis states that the complex type of foliose sporophytes with distinct sterile tissues in capsular formations of the diverse moss group of bryophytes have been considered as a relict type of sporophyte, which gave rise to an advanced type of sporophyte of fertile sporogenous tissues. This is surrounded by a jacket of a few layered, sterile cells, by the process of progressive reduction of the complex form of sporophytes to yield a simple form in the due course of evolution. In support of this hypothesis, the evolutionary biologists have referred to the following incidents:

bi. The presence of a multilayered capsule (primitive sporophyte) found in the members of Moss (e.g., *Sphagnum*), Hornwort (e.g., *Anthoceros*) and Jungermanniales (e.g., *Porella*), that went through progressive reduction to yield a single-layered sporophyte (advanced) of the members of Liverworts (e.g., *Riccia, Marchantia*).

bii. The quasi-parasitic, foliose sporophytes of the members of bryophytes have gone through progressive reduction to get rid of leaves, followed by the seta and the bulbous foot in the course of evolution, to become a mass of sporogenous tissues (advanced) embedded in the gametophytes of liverworts.

biii. The high content of sterile tissues and low content of sporogenous tissues was noticed in primitive type foliose sporophyte of the members belonging to the moss group of bryophytes. The progressive reduction of sterile tissues in these primitive sporophytes has been found to be inversely proportional to the gradual increase of the content of sporogenous tissues, found among the apparently advanced types. Simple forms of sporophytes of hornworts (comprised of foot and capsule) in the due course of evolution formed the most advanced type of sporophytes (comprised of capsule only), found in the members belonging to the Liverwort group of bryophytes.

Undoubtedly, the formation of sporophytes in lower group of plants is a reality which is an integral part of the evolution of sex in plant life and plays an important role in the evolution of seed habit among vascular plants. The origin and evolution of angiosperms and post-Cretaceous diversification and dominance of angiosperms on this blue planet was also affected by the same, but it's not clear whether the evolution of sporophytes is the result of

I. progressive sterilization of sporogenous tissues of simple sporophytes (advanced form) or

II. progressive reduction of sterile tissues of complex sporophytes (primitive form).

EVOLUTION OF GAMETOPHYTES IN BRYOPHYTES

There is a lack of paleobotanical evidence (poor preservation of non-cutinised epidermis, non-lignified and non-vascular tissues of bryophytes in the process of fossilization) nthe evolution of gametophyte. In Bryophytes, "evolutionary development"[11] has been substantially conjectured, so most of these issues have tried to be resolved by the reviewing of two opposing hypothesis, as follows:

a) The progressive evolution or Upgrade Hypothesis:

The upgrade or the progressive evolution hypothesis has been promulgated by Campbell (1940) and Cavers (1910). According to this, the ancient gametophyte has been characterized by a simple internal structure and external formations, prostrate and dorsiventrally symmetrical in the form of liverworts, that went through gradual ascending nodes of evolution to reach the complex form of upgraded structural forms and reproductive content of the gametophyte, found among members of moss group. According to the hypothesis of Cavers (1910), the gradual upgradation of gametophytes started with the hypothetical Sphaero-Riccia, the most simple gametophytic form, diversified into *Sphaerocarpus* in one evolutionary line which went through the transitional, upgraded nodes of evolution of *Riccia, Tessellina, Boschia, Exormotheca, Wiesnerella, Pressia* to reach the complex form of gametophyte of *Marchantia*.

The hypothesis of Campbell (1940), suggests the simple thalloid gametophytes of *Riccardia* and *Metzgeria* form the primitive, simplest form of gametophyte of bryophytes from which the complex form of gametophytes of bryophytes have evolved parallelly through two different lineages:

a. According to the first conjectured thought of evolution, the gametophytes of some members of bryophytes have retained the outer simple form but went through internal changes at a cellular level to get their complex structural forms (e.g., found in members belonging to Marchantiales).

In *Marchantia*, the formation of air chambers, organization of reproductive organs, and wide arrays of sexual receptacles have been noticed.

In *Riccia*, the simple thallus formed the sex organs that have been found in the median part of the thallus whereas, in *Marchantia,* the stalked receptacle (gametophore) bears the

[11] Evolutionary development: evo devo.

sex organs and the transitional form of organization[12] of gametophyte between *Riccia* and *Marchantia* has also been observed.

b. According to the second idea of evolution, the gametophytes of some members of bryophytes have retained their internal simple form of cellular organizations but went through external changes like the addition and alteration of appendages.

The simple, thalloid forms of gametophytes have gone through complex forms of modifications to get the shape of leafy structural formations (e.g., members belonging to Calobryales and Jungermanniales), visually. Further evolutionary upgradation influenced the internal differentiation of cellular configurations of the advanced gametophytes of some mosses (e.g., the members of Bryopsida).

b) The regressive evolution or Downgrade Hypothesis:

The downgrade or regressive evolution hypothesis has been postulated by Wettstein (1903-1908), Church (1919), Kashyap (1919), Goebel (1930) and Evans (1939). This hypothesis states that conspicuous generations characterized by leafy, gametophytic shoot of radial symmetry, belonging to the moss group of bryophytes and members of Calobryales have been considered as the primitive gametophytes. The regressive evolution, as well as downgrading or gradual simplification of structures gave rise to advanced forms of gametophytes, characterized by the thallus with dorsiventral symmetry (found among the members of Hornworts and Liverworts). In support of this hypothesis, the work of Kashyap (1919), Mehra (1969) on Marchantiales group of bryophytes has been referred herein:

bi. In complex gametophytic structures of Marchantiales (e.g., *Preslia*), barrel shaped pores are found on the thallus and the disc of gametophores; further downgraded in the similar gametophytes of the members of *Reboulia*, where complex, barrel shaped pores are found on the disc of gametophores but simple forms of pores are also found on the thallus; going further lower in the simple gametophytes of the members of *Exormotheca*, the simple form of pores were noticed on the disc of the gametophore and on its thallus and pores are completely absent in the simplest forms of gametophyte of *Riccia*.

bii. The gradual degeneration of assimilatory filaments has also been seen among the members of Marchantiales. Plenty of assimilatory filaments are found in the air cavities of the complex type of gametophytic thallus of *Marchantia*; going lower in the thallus of *Concocephallum* and *Wiesnerella* is marked by the presence of a small number of short,

---

[12] where sex organs have been borne by cushion type of receptacle, arranged on the thallus terminally or dorsally

rudimentary assimilatory filaments; further in the gametophytes of *Dumortiera*, the assimilatory filaments are totally absent in their thallus.

biii. Mehra (1969) proposed that the shifting of antheridiophore (male reproductive organ bearing stalk) and archegoniophore (female reproductive organ bearing stalk) was noticed from the terminal position to the dorsal position of gametophytes, noticed among the member belonging to Marchantiales group of bryophytes. The antheridia and archegonia are borne terminally on the stalked gametophores in *Marchantia*; noticed remarkably in the lower gametophyte of *Preslia*, where archegoniophores and antheridiophores have been found in terminal positions initially, and gradually shifted to dorsal positions on attaining maturity; the ultimate lowered feature in gametophytes were observed in the member of *Corsinia,* where archegoniophore is found in sessile form.

In the absence of determinant paleobotanical evidence, it is difficult to be biased towards any particular hypothesis to draw the evolutionary foot-prints of gametophytes of bryophytes. It could be ascertained that the evolution of gametophytes, sporophytes, and alternation of generations in the life-cycle of bryophytes, play an important role to inch closer to the enigmatic issues of the "evolution of seed habit of angiosperms."

Chapter 7

EVOLUTION OF THALLOPHYTES-2 (EM-BRYO-PHYTES): AN ENIGMATIC STORY OF AMPHIBIAN MIGRATION AND TERRESTRIAL COLONIZATION

The simplest form of herbaceous, non-vascular, ancient embryophytes[13], more advanced and diversified than thallophytes and physiologically well adapted to the complete transmigration from water to land through the completion of their reproductive cycle is partially dependent upon the availability of water.

The contemporary bryological investigations show that this unique group of plants, popularly known as 'Plant Amphibians' are represented by the 1,237 genera and 18,000 species (Groombridge, 1992), distributed over moist, shady regions in the Southern and Northern hemispheres, conspicuously identified by persistent gametophytic generations with short-term sporophytic generations dependent on the gametophytic generation for nourishment. They are also recognized as "sporogonium" with the earliest form of structural trinity (comprising of leaves, stem, and roots) of plants, which comprise of an aerial sporangium and a bulbous foot, attached by an aerial stalk or seta.

On the basis of the existence of primitive characteristics in bryophytes, the thalloid members of algae, fungi, and lichens have collectively been recognized as Bryophyta along with liverworts, hornworts and mosses in the 18th century (Braun, 1864). After a decade, Schimper (1860) segregated the members belonging to liverworts, hornworts, and mosses from algae, fungi, and lichens and assigned them in the division of "Bryophyta." The members of algae, fungi, and lichens were assigned to a new division, "Thallophyta." Based on distinctive features, Eichler (1883) segregated the members of Bryophyta into two distinctive classes: i. Liverworts or Hepaticae and ii. Mosses or Musci.

---

[13] Embryophytes: The group of plants where zyogote has matured to form embryo.

On the basis of distinguishing feature of the members of hornworts (as the members stand-out from the rest of the members of Musci) they have been regrouped under a new class: "Anthocerotales" (Howe, 1899).

Since the end of 18th Century till the middle of the 19th Century, most of the bryologists and taxonomists traditionally followed Eichler's (1883) template of assortment (Bessey, 1907; Bower, 1935; Melchoir and Werdermann, 1954). A different approach of an artificial system of classification has been found in the work of Cavers (1910), who subdivided Bryophyta into ten groups (of equal status), deviating from Eichler's (1883) bi-fold classification of Bryophyta. This is as follows:

1. Sphaerocarpales,
2. Marchantiales,
3. Jungermanniales,
4. Anthocerotales,
5. Sphagnales.
6. Andraeales,
7. Tetraphidales,
8. Polytrichales,
9. Buxbaumiales and
10. Eubryales.

A paradigm shift was witnessed in the field of taxonomic classification of the plant kingdom in the middle of the 19th Century, as most of the systematic botanists tried to switch from the established "artificial system of classifications of plants" to "phylogenetic system of classification of plants." This could be considered as the virtual continuation of the "Voyage of Beagle," initiated by the great naturalist "Charles Robert Darwin," to provide satisfactory answers about the evo-devo of life on the planet in his "An Abominable Mystery." Influenced by phylogenetic approaches of classification of bryophytes as well as the entire plant kingdom, which is known to narrate the origin, evolution and diversification of plants on the basis of true phylogenetic relationships among themselves, a group of scientists (Campbell, 1940; Rothmaler, 1951; Takhtajan, 1953; Smith, 1955; Proskauer, 1957; Stotler and Crandall-Stotler, 1977; Crandall-Stotler, 1980; Schuster, 1966; Kenrick and Crane, 1997 and Crandall-Stotler and Stotler, 2000) from the mid-19th Century till now, prefer to follow the trifold template of Howe (1898) rather than one bi-fold template of Eichler (1883). For classification of Bryophytes, Rothmaler (1951) has amended the erstwhile "classes" (viz. Hepaticae, Anthocerotae, and Musci) as per the code of ICBN[14] to use the suffix "opsida" to reintroduce three classes:

---

[14] ICBN: International Code of Botanical Nomenclature.

1. Hepaticopsida,
2. Anthoceropsida and
3. Bryopsida.

Proskauer (1957) changed the name of the class Anthoceropsida to Anthocertopsida in his classification. In the quest of finding the origin and evolution of transducers and evolution of seed habits of angiosperms on the Earth, our botanical journey has traversed a long way through the labyrinths of diversified modes of classifications of plants (either artificial or phylogenetic). Particularly the "amphibian transducing migrators," bryophytes and contemporary classification (born from molecular systematics) is pro-cladistic analysis oriented, rather than the comparative structural analysis of plants.

An outline of contemporary, phylogenetic classification of around 18,000 members of the division: Bryophyta, under three coordinated phyla (i.e., Marchantiophyta, Anthocerophyta, and Bryophyta) which are evolutionarily correlated, have been presented here (Crandall-Stotler and Stotler, 2000):

A. Phylum- Marchantiophyta or Hepatophyta or Liverworts: It has been divided into two subclasses:

A1. Class- Marchantiopsida: This group has been divided into two subclasses-

a. Subclass- Sphaerocarpidae: It has been studied under the single suborder:

1. Order- Sphaerocarpales: It has been classified under two suborders:

1a. Suborder- Sphaerocarpineae: further been studied under the single family-

i. Family: Sphaerocarpaceae (e.g., *Geothallus tuberosus, Sphaerocarpos texanus*)

2a. Suborder- Riellineae: further been studied under the single family-

i. Family: Riellaceae (*Riella helicophylla*).

b. Subclass- Marchantiidae: This subclass has further been subdivided under three orders:

1. Order- Monocleales: further been studied under the single family –

i. Family: Monocleaceae (e.g., *Monoclea frosteri*).

2. Order- Marchantiales: It has been studied under four sub-orders:

1a. Suborder- Marchantiineae: It has been sub-divided under eight families-

i. Family: Aytoniaceae (e.g., *Reboulia hemispheraeica*).

ii. Family: Wiesnerellaceae (e.g., *Dumortiera hirsute*).

iii. Family: Conocephalaceae (e.g., *Conocephalum conicum*).

iv. Family: Lunulariaceae (e.g., *Lunularia cruciata*).

iv. Family: Marchantiaceae (e.g., *Marchantia polymorpha*).

v. Family: Monosoleniaceae (e.g., *Monosolenium tenerum*).

vi. Family: Cleveaceae (e.g., *Athalamia hyalina*).

viii. Family: Exormothecaceae (e.g., *Exormotheca bullosa*).

2a. Suborder- Corsiniineae: It has been categorized under two families:
i. Family: Cyathodiaceae (e.g., *Cyathodium tuberculatum*).
ii. Family: Corsiniaceae (e.g., *Corsinia coriandrina*).
3a. Suborder- Monocarpineae: It has been studied under a single family-
i. Family: Monocarpaceae (e.g., *Monocarpus sphaerocarpos*).
4a. Suborder- Targioniineae: It has been studied under a single family-
i. Family: Targionaceae (e.g., *Targiona hypophylla*).
3. Order- Ricciales: It has been subdivided into two families-
i. Family: Oxymitraceae (e.g., *Oxymitra paleacea*).
ii. Family: Ricciaceae (e.g., *Riccia fluitans*).
A2. Class- Jungermanniopsida: This group has been divided into two sub-classes:
a. Subclass- Metzgeriidae: It has been categorized under five orders-
1. Order- Haplomitriales: It has been studied under a single family-
i. Family: Haplomitriaceae (*Haplomitrium moicum*).
2. Order- Blasiales: It has been studied under the family-
i. Family: Blasiaceae (e.g., *Blasia pusilla*).
3. Order- Treubiales: This group has been studied under two suborders-
3a. Suborder- Treubiineae: It has been studied under the family-
i. Family: Treubiaceae (e.g., *Treubia scapinioides).*
3b. Suborder- Phyllothalliineae: It has been studied under the family-
i. Family: Phyllothalliaceae (e.g., *Phyllothalia nivicola*).
4. Order- Fossombroniales: This group has been studied under two suborders-
4a. Suborder- Fossombroniineae: It has been studied under two families-
i. Family: Fossombroniaceae (e.g., *Fossombronia foveolata*).
ii. Family: Allisoniaceae (e.g., *Allisonia cockaynei*).
4b. Suborder- Pelliineae: It has been categorized under two families-
i. Family: Pelliaceae (e.g., *Pellia crispa*).
ii. Family: Sandeothallaceae (e.g., *Sandiothallus radiculosus*).
5. Order-Metzgeriales: This group has been categorized under two suborders-
5a. Suborder- Pallaviciniineae: It has been categorized under three families-
i. Family: Pallaviciniaceae (e.g., *Pallavicinia spinosa*).
ii. Family: Makinoaceae (e.g., *Makinoa crispata*).
iii. Family: Hymenophytaceae (e.g., *Hymenophyton pedicellatum*).
5b. Suborder- Metzgeriineae: It has further been divided into four families-
i. Family: Aneuraceae (e.g., *Riccardia chamedryfolia*).
ii. Family: Mizutaniaceae (e.g., *Mizutania riccardioides*).
iii. Family: Vandiemeniaceae (e.g., *Vandiemenia ratkowskiana*).
iv. Family: Metzgeriaceae (e.g., *Metzgeria furcata*).
b. Subclass- Jungermanniidae: It has been categorized under five orders-
1. Order- Lepicoleales: This order has further been divided into four suborders-

1a. Suborder- Ptilidiineae: It has been categorized under three families-
- i. Family: Ptilidiaceae (e.g., *Ptilidium ciliare*).
- ii. Family: Mastigophoraceae (e.g., *Mastigophora diclados*).
- iii. Family: Chaetophyllopsidaceae (e.g., *Chaetophyllopsis whiteleggei*).

1b. Suborder- Lepicoleniellineae: It has been divided into two families-
- i. Family: Vetaformataceae (e.g., *Vetaforma dusenii*).
- ii. Family: Lepicoleaceae (e.g., *Lepicolea loriana*).

1c. Suborder- Perssoniellineae: It has been categorized under two families-
- i. Family: Schistochilaceae (e.g., *Schistochila pinnatifolia*).
- ii. Family: Perssoniellaceae (e.g., *Perssoniella vitreocincta*).

1d. Suborder- Lepidolaenineae: It has been sub-divided into four families-
- i. Family: Trichocoleaceae (e.g., *Tricholea tomentella*).
- ii. Family: Lepidolaenaceae (e.g., *Lepidolaena clavigera*).
- iii. Family: Neotrichocoleaceae (e.g., *Tricholeopsis sacculata*).
- iv. Family: Jubulopsidaceae (e.g., *Jubulopsis novae-zelandiae*).

2. Order- Jungermanniales: This group has been studied under eight suborders-

2a. Suborder- Herbertineae: This group has been divided into four families-
- i. Family: Herbertaceae (e.g., *Herbertus buchii*).
- ii. Family: Pseudolepicoleaceae (e.g., *Blepharostoma trichophyllum*).
- iii. Family: Trichotemnomataceae (e.g., *Trichotemnoma corrugatum*).
- iv. Family: Grolleaceae (This family has also been considered under Herbertaceae by some bryologists) (e.g., *Grollea antheliopsis*).

2b. Suborder- Balantiopsidineae: It has been studied under the single family-
- i. Family: Balantiopsidaceae (e.g., *Balantiopsis diplophylla*).

2c. Suborder- Lophocoleineae: It has been divided under five families-
- i. Family: Geocalycaceae (e.g., *Calypogenia muelleriana*).
- ii. Family: Gyrothyraceae (e.g., *Gyrothyra underwoodiana*).
- iii. Family: Plagiochiliaceae (e.g., *Plagiochila ptychanthoidea*).
- iv. Family: Acrobolbaceae (e.g., *Marsupidium epiphytum*).
- iv. Family: Arnelliaceae (e.g., *Arnellia fennica*).

2d. Suborder- Lepidoziineae: It has been categorized into three families-
- i. Family: Phycolepidoziaceae (e.g., *Phycolepidozia exigua*).
- ii. Family: Calypogeiaceae (e.g., *Calypogeia fissa*).
- iii. Family: Lepidoziaceae (e.g., *Lepiidozia reptans*).

2e. Suborder- Cephaloziineae: It has been divided under four families-
- i. Family: Cephaloziaceae (e.g., *Haesselia roraimensis*).
- ii. Family: Cephaloziellaceae (e.g., *Allisoniella nigra*).
- iii. Family: Jackiellaceae (e.g., *Jackiella sinensis*).
- iv. Family: Adelanthaceae (e.g., *Wettsteinia inversa*).

2f. Suborder- Antheliineae: It has been studied under the family-

i. Family: Antheliaceae (e.g., *Anthelia filum*).

2g. Suborder- Brevianthineae: This group has been divided into two families-

i. Family: Brevianthaceae (e.g., *Brevianthus flavus*).

ii. Family: Chonecoleaceae (e.g., *Chonecolea acutiloba*).

2h. Suborder- Jungermannineae: It has been categorized into four families-

i. Family: Jungermanniaceae (e.g., *Mylia aequata*).

ii. Family: Mesoptychiaceae (e.g., *Mesoptychia sahlbergii*).

iii. Family: Gymnomitriaceae (e.g., *Acolea alpina*).

iv. Family: Scapaniaceae (e.g., *Scapinia albicans*).

3. Order- Poreallales: This order has been sub-divided into two suborders-

3a. Suborder- Porellineae: It has been studied under the family-

i. Family: Porellaceae (e.g., *Porella apiculata*).

3b. Suborder- Jubulineae: It has been sub-divided into four families-

i. Family: Goebeliellaceae (e.g., *Goebeliella cornigera*).

ii. Family: Jubulaceae (e.g., *Jubula complanata*).

iii. Family: Bryopteridaceae (e.g., *Bryopteris filicina*).

iv. Family: Lejeuneaceae (e.g., *Leptolejeunea tridentata*).

4. Order- Radulales: It has been studied further under the family-

i. Family: Radulaceae (e.g., *Patarola uvifera*).

5. Order- Pleuroziales: It has been studied under the family-

i. Family: Pleuroziaceae (e.g., *Pleurozia acinosa*).

B. Phylum- Anthocerotophyta or Hornworts:

B1. Class- Anthocerotopsida: This group has been studied under the solitary order-

1. Order- Anthocerotales: This order has been divided into four families-

i. Family: Anthocerotaceae (e.g., *Folioceros assamicus*).

ii. Family: Phaeocerotaceae (e.g., *Phaeoceros himayalensis*).

iii. Family: Dendrocerotaceae (e.g., *Dendroceros crispus*).

iv. Family: Notothylaceae (e.g., *Notothylas orbicularis*).

B. Phylum- Bryophyta or Mosses: This group has been studied under six groups-

C1. Class- Takakiopsida:

1. Order- Takakiales: This group has been studied under the family-

i. Family: Takakiaceae (e.g., *Takakia ceratophylla*).

C2. Class- Sphagnopsida: This group has been subdivided under two orders-

1. Order- Sphagnales: It has been studied under the family-

i. Family: Sphagnaceae (e.g., *Sphagnum flexuosum*).

2. Order- Ambuchananiales: It has been studied under the family-

i. Family: Ambuchananiaceae (e.g., *Ambuchanania leucobryoides*).

C3. Class- Andreaeopsida: This group has been studied under the order-

1. Order- Andreales: It has further been studied under the family-

i. Family: Andreaceae (e.g., *Andreaea nivalis*).

C4. Class- Andreaeobryopsida: this group has been studied under the order-
  1. Order- Andreaeobryales: It has been studied under the family-
    i. Family: Andraeobryaceae (also recognized under Tetrahidaceae by bryologists) (e.g., *Tetrodontium repandum*).
C5. Class- Polytrychopsida: This group has been divided into two orders-
  1. Order- Tetraphidales: It has been divided into three families-
    i. Family: Tetraphidaceae (e.g., *Tetraphis pellucida*).
    ii. Family: Oedipodiaceae (e.g., *Oedipodium griffithianum*).
    iii. Family: Buxbaumiaceae (e.g., *Buxbaumia aphylla*).
  2. Order- Polytrichales: It has been studied under the family-
    i. Family: Polytrichaceae (e.g., *Polytrichum commune*).
C6. Class- Bryopsida: This group has further been categorized under four sub-classes-
  a. Subclass- Diphysciidae: This group has been studied under-
    1. Order- Diphysciales: It has been studied under the family-
      i. Family: Diphysciaceae (e.g., *Pseudotimmiella pocsii*).
  b. Subclass- Funariidae: This group has been categorizing under three orders-
    1. Order- Timmiales: It has further been studied under the family-
      i. Family: Timmiaceae (e.g., *Timmia sibirica*).
    2. Order- Encalyptales: It has been studied under the family-
      i. Family: Encalyptaceae (e.g., *Encalypta vulgaris*).
    3. Order- Funariales: This group has sub-divided into three families-
      i. Family: Funariaceae (e.g., *Funaria hygrometrica*).
      ii. Family: Disceliaceae (e.g., *Discelium nudum*).
      iii. Family: Gigasperemaceae (e.g., *Gigasperemum repens*).
  c. Subclass- Dicranidae: This group has been categorized under five orders-
    1. Order- Grimmiales: This order has further been divided into four families-
      i. Family: Grimmiaceae (e.g., *Grimmia anodon*).
      ii. Family: Ptychomitriaceae (e.g., *Ptychomitriopsis africana*).
      iii. Family: Scouleriaceae (e.g., *Scouleria siskiyouensis*).
      iv. Family: Drummondiaceae (e.g., *Drummondia prorepens*).
    2. Order- Archidiales: This order has been studied under the family-
      i. Family: Archidiaceae (e.g., *Archidium andersonianum*).
    3. Order- Seligerales: This order has been divided into two families-
      i. Family: Seligeriaceae (e.g., *Seligeria diminuta*).
      ii. Family: Wardiaceae (e.g., *Wardia hygrometrica*).
    4. Order- Dicranales: This order has been divided into thirteen families-
      i. Family: Bryoxiphiaceae (e.g., *Bryoxiphium mexicanum*).
      ii. Family: Fissidentaceae (e.g., *Fissidens acacioides*).
      iii. Family: Dicranaceae (e.g., *Dicranella alpina*).
      iv. Family: Bruchiaceae (e.g., *Bruchia aurea*).

v. Family: Dicnemonaceae (e.g., *Dicnemon pallidum*).
vi. Family: Leucobryaceae (e.g., *Leucobryum glaucum*).
vii. Family: Ditrichaceae (e.g., *Ditrichum aureum*).
viii. Family: Rhabdoweisiaceae (e.g., *Rhabdoweisia crispata*).
ix. Family: Rhachitheciaceae (e.g., *Rhachithecium papillosum*).
x. Family: Erpodiaceae (e.g., *Erpodium glaucum*).
xi. Family: Schistostegiaceae (e.g., *Schistostega pennata*).
xii. Family: Eustichiaceae (e.g., *Eustichia ulei*).
xiii. Family: Viridivelleraceae (e.g., *Viridivellus pulchellum*).

5. Order- Pottiales: This order has further been subdivided into six families-
i. Family: Pottiaceae (e.g., *Pottia antarctica*).
ii. Family: Ephemeraceae (e.g., *Ephemerum apiculatum*).
iii. Family: Serpotortellaceae (e.g., *Serpototortella crytophylla*).
iv. Family: Calymperaceae (e.g., *Calymperes afzelii*).
v. Family: Cinclidotaceae (e.g., *Cinclidotus aquaticus*).
vi. Family: Bryobartramiaceae (e.g., *Bryobartramia novae-valesiae*).

d. Subclass- Bryidae: This group has been categorized under three superorders-
d1. Superorder- Bryanae: This group has been sub-divided under four orders-

1. Order- Splachnales: This order has been categorized under three families-
i. Family: Splachnaceae (e.g., *Splachnum austriacum*).
ii. Family: Meesiaceae (e.g., *Meesia longiseta*).
iii. Family: Catoscopiaceae (e.g., *Catoscopium nigritum*).
2. Order- Orthotricales: It has been studied under the family-
i. Family: Orthotrichaceae (e.g., *Orthotrium tuberculatum*).
3. Order- Hedwigiales: It has been categorized into three families-
i. Family: Hedwigiaceae (e.g., *Hedwigia attenuata*).
ii. Family: Helicophyllaceae (e.g., *Helicophyllum torquatum*).
iii. Family: Rhacocarpaceae (e.g., *Rhacocarpus alpinus*).

4. Order- Bryales: This group has further been categorized under eight families-
i. Family: Aulacomniaceae (e.g., *Aulacomnium palustre*).
ii. Family: Bartramiaceae (e.g., *Bartramia affinis*).
iii. Family: Orthodontiaceae (e.g., *Orthodontium gracile*).
iv. Family: Bryaceae (e.g., *Bryum abyssinicum*).
v. Family: Phyllodrepaniaceae (e.g., *Phyllodrepanium falcifolium*).
vi. Family: Pseudoditrichaceae (e.g., *Pseudoditrichium mirabile*).
vii. Family: Mniaceae (e.g., *Mnium ambiguum*).
viii. Family: Leptostomataceae (e.g., *Lepstomum erectum*).

d2. Superorder- Rhizogonianae: This group has been studied under the order-

1. Order- Rhizogoniales: This order has been categorized under eight families-
i. Family: Hypnodendraceae (e.g., *Hypnodendron brevipes*).

ii. Family: Rhizogoniaceae (e.g., *Rhizogonium hattorii*).
iii. Family: Calomniaceae (e.g., *Calomnion milleri*).
iv. Family: Mitteniaceae (e.g., *Mittenia plumula*).
v. Family: Cyrtopodaceae (e.g., *Bescherellia cryphaeoides*).
vi. Family: Spiridentaceae (e.g., *Spiridens camusii*).
vii. Family: Pterobryaceae (e.g., *Pterobryon convolutum*).
viii. Family: Racopilaceae (e.g., *Racopilum africanum*).

d3. Superorder- Hypnanae: It has further been divided under two orders-

1. Order- Hookeriales: This group has been sub-divided under two suborders-

1a. Suborder- Hookeriineae: this group has been divided under six families-

i. Family: Hookeriaceae (e.g., *Hookeria albata*).
ii. Family: Leucomiaceae (e.g., *Leucomium debile*).
iii. Family: Daltoniaceae (e.g., *Daltonia apiculata*).
iv. Family: Adelotheciaceae (It has been considered under the Family: Daltoniaceae) (e.g., *Adelothecium bogotense*).
v. Family: Pilotrichaceae (e.g., *Pilotrichum affine*).
vi. Family: Hypopterygiaceae (e.g., *Hypopterygium balantii*).

1b. Suborder- Ptychomnineae: It has been sub-divided into two families-

i. Family: Ptychomniaceae (e.g., *Ptychomniella ptychocarpon*).
ii. Family: Garovagliaceae (It has been considered under family: Ptychomniaceae or Pterobryaceae) (e.g., *Garovaglia angustifolia*).

2. Order- Hypnales: This order has been divided into forty-three families-

i. Family: Rutenbergiaceae (e.g., *Rutenbergia cirrata*).
ii. Family: Trachylomataceae (e.g., *Trachyloma archeri*).
iii. Family: Amblystegiaceae (e.g., *Amblystegium aduncum*).
iv. Family: Cratoneuraceae (It has been considered under the family: Amblystegiaceae) (e.g., *Cartoneuron aommutatum*).
v. Family: Helodiaceae (It has been considered under the family: Thuidiaceae) (e.g., *Helodium blandowii*).
vi. Family: Hylocomiaceae (e.g., *Hylocomium dentatum*).
vii. Family: Rhytidiaceae (It has been studied under the family: Hylocomiaceae) (e.g., *Rhytidium rugosum*).
viii. Family: Leskeaceae (e.g., *Leskea adnata*).
ix. Family: Regmatodontaceae (e.g., *Regmatodon acidodon*).
x. Family: Pterigynandraceae (e.g., *Pterigynandrum boscii*).
xi. Family: Rigodiaceae (e.g., *Rigodium arborescens*).
xii. Family: Thuidiaceae (e.g., *Thuidium aciculum*).
xiii. Family: Campyliaceae (It has been considered under The family: Amblystegiaceae) (e.g., *Campylium arcticum*).
xiv. Family: Brachytheciaceae (e.g., *Brachythecium acutum*).

xv. Family: Stereophyllaceae (e.g., *Stereophyllum anceps*).
xvi. Family: Myriniaceae (e.g., *Nematocladia tesserata*).
xvii. Family: Fabroniaceae (e.g., *Fabronia assamica*).
xviii. Family: Meteoriaceae (e.g., *Meteoridium remotifolium*).
xix. Family: Plagiotheciaceae (e.g., *Plagiothecium alpinum*).
xx. Family: Fontinalaceae (e.g., *Fontinalis baltica*).
xxi. Family: Entodontaceae (e.g., *Entodon adyris*).
xxii. Family: Climaciaceae (e.g., *Climacium japonicum*).
xxiii. Family: Pleuroziopsidaceae (It has been considered under the family: Climaciaceae) (e.g., *Pleuroziopsis ruthenica*).
xxiv. Family: Hypnaceae (e.g., *Hypnum aciculatum*).
xxv. Family: Catagoniaceae (e.g., *Catagonium emrginatum*).
xxvi. Family: Symphyodontaceae (e.g., *Symphyodon asper*).
xxvii. Family: Sematophyllaceae (e.g., *Sematophyllum adnatum*).
xxviii. Family: Hydropogonaceae (It has been considered unthe the family: Sematophyllaceae) (e.g., *Hydropogon fontinaloides*).
xxix. Family: Myuriaceae (e.g., *Myurium alaris*).
xxx. Family: Cryphaeaceae (e.g., *Cryphaea nervosa*).
xxxi. Family: Prionodontaceae (e.g., *Prionodon densus*).
xxxii. Family: Leucodontaceae (e.g., *Leucodon assimilis*).
xxxiii. Family: Pterobryaceae (e.g., *Pterobryon densum*).
xxxiv. Family: Phyllogoniaceae (e.g., *Phyllogonium viride*).
xxxv. Family: Orthorrhynchiaceae (e.g., *Orthorrhynchium cylindricum*).
xxxvi. Family: Lepyrodontaceae (e.g., *Lepyrodon alaris*).
xxxvii. Family: Neckeraceae (e.g., *Neckera americana*).
xxxviii. Family: Leptodontaceae (e.g., *Leptodon irriguus*).
xxxix. Family: Lembophyllaceae (e.g., *Lembophyllum orbiculatum*).
xxxx. Family: Anomodontaceae (e.g., *Anomodon viticulosus*).
xxxxi. Family: Theliaceae (e.g., *Thelia hirtella*).
xxxxii. Family: Microtheciellaceae (e.g., *Microtheciella kerrii*).
xxxxiii. Family: Sorapillaceae (e.g., *Sorapilla sprucei*).

Bryophytes are primitive representatives of simple, archegoniate thallophytes, well adapted in moist, terrestrial habitats, and supposedly evolved from the specialized group of algae, belonging to Charophyceae. They transmigrated from water to land and are conventionally regarded as a distinct group of embryophytes that maintain the evolutionary bridge between ancient, algal stock from water bodies to the modern diversified groups of angiosperms.

Poor paleobotanical representation witnessed in bryophytes due to soft, herbaceous structural formations, is devoid of vascular supplies. In absence of ample paleobotanical evidence, origin, and evolution of bryophytes has been formed by comparing two opposing phylogenetic hypotheses, as follows:

i. Origin of Bryophytes from Algal stock or the Progressive Evolution Hypothesis:

This theory suggests that bryophytes are descendants of algae. Origin and evolution of an advance group of embryophytes, i.e., amphibian land plants as bryophytes have taken place as a result of progressive diversification of simple forms of algae. Based on the earnest effort of transmigration from water to land, simple structural forms and presence of the mentioned common features between algal communities and bryophytes, lead various scientists to promulgate different algae as the ancestor of bryophytes in a different time.

Common features:

a. Amphibian nature of plants.
b. thalloid gametophyte, cellulosic cell wall.
c. presence of Chlorophyll a, b; carotene α, β; starch
d. filamentous protonema and flagellated sperms
e. fertilization takes place in presence of water.

Lignier (1903) postulated the presence of "Prohepatics" as a hypothetical, common basal stock of algal origin as giving rise to bryophytes and pteridophytes in the due course of diversification. Bower (1908) has referred to the "Chlophycean stock" of bryophytes and pteridophytes. Fritsch (1916, 1945) postulated that the heteroicous thallus of Chaetophorales (e.g., *Trentepohlia*) group of algae gave rise to the erect part of land plants by stunting the ramifications of prostrate forms and is referred to as the heteroicous branching pattern of bryophytes (e.g., filamentous protonema of mosses, conforming prostrate system; umbrella-like gametophore of moss forms of erect system). In the Thallasiophyta and the "subaerial transmigration," Church (1919) postulated that planktonic marine algae (e.g., members of Chlorophyceae) were transformed into well diversified, benthic algae (e.g., members of Phaeophyceae) as a result of the gradually receding seawater levels, ultimately forming an extinct hypothetical mother-stock, "Thallasiophyta," which gave rise to the contemporary bryophytes and pteridophytes.

In support of his hypothesis, Church (1919) has referred to features like-

a. enhanced metabolic power to produce more starch in Chlorophyceae members,
b. a step closer to the structural trinity (like leaves, stem, and root) of the members of Phaeophyceae (like leaf-like blades, stems like stipe and root-like holdfasts),
c. an alternation of sexual (haploid) and asexual (diploid) generations; in-situ fertilization etc. should be taken into consideration to validate this hypothesis.

Smith (1955) considered that Bryophytes originated from the Ulotrichales group of Chlorophyceae. Whereas Mehra (1969) considered that both bryophytes and pteridophytes originated from the hypothetical ancestral stock: "Proto-archegoniate," originated from the algae.

ii. Origin of Bryophytes from the Pteridophytean stock or the Regressive Evolution Hypothesis:

This theory suggests that bryophytes are the descendants of pteridophytes; the origin and evolution of bryophytes, the diversified group of embryophytes, and nonvascular plants took place as a result of progressive and gradual degeneration of pteridophytes, the early form of vascular plants.

It has also been recognized as the "Regressive Evolution Hypothesis" as it supports the view of gradual degeneration of a complex form of plant life (Pteridophytes) to yield simple forms of plant life (Bryophytes) (Wettstein, 1903-1908; Scott, 1911; Kidston and Lang, 1917; Kashyap, 1919; Takhtajan, 1953; Christensen, 1954; Cronquist et al., 1966).

The hypothesis has been postulated on the basis of identity resemblances of the sporophytes of fossiliferous, extinct pteridophytes, the primitive vascular plants of Rhyniopsida of the Silurian-Devonian period and the diversified sporophytes of the extant, diversified group of bryophytes, comprised of some mosses (e.g., *Funaria, Sphagnum* etc.) and hornworts (e.g., *Anthoceros, Dendroceros* etc.).

The common identity features:

a. The members of Rhynopsida and the representatives of bryophytes are found devoid of roots, devoid of leaves, and aerial shoots have borne terminal sporangia.
b. The presence of photosynthetic tissues and stomata has been noticed in the capsular regions of pteridophytes, and also in the capsular regions of bryophyte members (e.g., *Funaria, Anthoceros* etc.)
c. The presence of specialized sterile tissues, known as Columella were found inside the sporangium of the extant bryophyte members, also noticed in the sporangium of the extinct members of Rhyniposida (e.g., *Horneophyton*).
d. The cell layer of sporangium jacket of *Dendroceros* sp. of the Anthocerophyta group of bryophytes has been found to be spirally thick and is identical to the tapetal layer of the jacket of the sporangium of *Horneophyton* of Rhyniopsida, the extinct member pteridophytes.

The striking resemblance of the structural formations, as well as relict features of sporangia of extant members of Anthocerophyta (*Dendroceros crispus*), bryophytes and extinct members of Rhyniopsida (e.g., *Horneophyton*), and pteridophytes, have caused the supporters of the regressive evolution hypothesis to be stuck with the pteridophyte origin of bryophytes. The further extension of this hypothesis could explain the formation of simple forms of liverworts of bryophytes from complex form of pteridophytes, by progressive degeneration of complex structural forms.

According to the "*Coleochaete* model" of molecular biologists (Mishler et al. 1994; Kenrick and Crane 1997b; Qiu et al. 1998; Graham et al. 2012), primitive bryophytes originated from the "*Coleochaete*" type of soil algae, having close proximity to the early tracheophytes. According to the stereotypic taxonomic classifications (classical taxonomic formats) of the nineties and late twenties, Bryophyta has been assigned in between the positions of algae and pteridophytes.

The contemporary cladistics analysis (mainly on the basis of the ultrastructure of cells and molecular biology) of bryophytes has been studied under three categories:

a. Liverworts (Hepatophya),
b. hornworts (Anthocerophyta) and
c. mosses (Bryophyta).

While it's still in a gray area whether bryophyte is the progenitor of vascular plants or not, the recent analysis on molecular systematics of Bryophytes (Qiu et al., 1998) has lead the plant phylogenetics to consider this group of plants as early land plants, transmigrated from water. According to Kenrick and Crane (1997b), these three groups of Bryophytes formed an important evolutionary bridge to narrate the origin and evolution of ancient plants on Earth.

With the phylogenetic analysis based on molecular review, Mishler et al. (1994) ascertained that all three embryophyte sub-groups are related to each other paraphyletically and further ensured that the liverworts and charophytes (algal group) maintain a close evolutionary relationship (sister group). Tracheophytes (vascular plants) and mosses maintain a close evolutionary relationship and hornworts maintain the evolutionary bridge between mosses and liverworts. A comparative review of the algal origin of bryophytes as well as the progressive evolution of bryophytes and the pteridophyte origin of bryophytes or regressive evolution of bryophytes, algal origin, and the evolution of bryophytes has been found tangible.

Due to the absence of lignified, vascular tissues the scanty, fossil remnants of bryophytes have mostly been recognized by the vegetative fragments of gametophytes of bryophytes (e.g., *Pallavicinites devonicus* belonging to Jungermanniales) from the Devonian deposit of New York, United States, and bears structural resemblance to the extant members of *Fossombronia, Metzgeria, Pallavicinia* etc. The fossiliferous spores of extinct and ancient bryophytes, not clearly distinguishable from vascular plants, have been recognized as "Sporae dispersae," most likely originated in the Ordovician and Silurian periods and dominated the Earth during the Carboniferous-Pleistocene period. The paleobotanical evidences of bryophytes have been reviewed herewith:

a. Fossil Liverworts: The fossil representatives of Marchantiales: *Marchantiolites*, the predecessor of the contemporary *Marchantia*s, are from the Cretaceous deposit (Montana) of United States; whereas one of its fossil representative: *M. porosus* has been reported from the Jurassic deposit of Sweden. The extant descendants, *Marchantia* of the Old world, is considered as the successor of *Marchatites*, the fossil representative of Marchantiales reported from a Triassic deposit in South America.

The fossil liverwort *Diettertia* has been reported to be from the Cretaceous deposit and shows its affinity towards the contemporary extant descendants of Jungermanniales.

Ricciopsis, the thalloid hepatic and fossiliferous predecessor of extant genus *Riccia* has disjunctively been found from the Deccan Intertrappean flora of India and the Jurassic deposits of Sweden (Sheikh and Kapgate, 1982).

The phylogenetically unique, fossil bryophyte: *Naiadita lanceolata* has been reported from the Triassic deposit of England, the United Kingdom, showing its common evolutionary relationship with Marchantiales, Sphaerocarpales, and Calobryales, when vegetative and reproductive features were taken into account (Harris, 1938; Schuster, 1966).

a. Fossil Hornworts: Almost no fossil remnants of thallus or gametophytes of Anthocerophyta has been reported so far except some fossil spores (round shaped, trilete and distinct cingulum) are from the Cretaceous deposit of (Maastrichtian) North America, having an identical spore architecture in the contemporary hornwort *Phaeoceros* (Krassilov and Schuster, 1984). The sporangial compression (characterized by a number of sporangial stalk bearing capsules, more than one layered sporangium, and presence of columellar stalk) of bryophyte *Sporogonites* fossils reported from the Devonian deposit, supposedly been considered as the predecessors of modern hornworts, but some of the paleobotanists considered it as part of the sporophytes of extant moss (Taylor et al., 2009).
b. Fossil Mosses: The diversified moss flora has been discovered in the Permian deposits of Siberia, Russia; the fossil representatives: *Bajdaieira, Buchtia, Intia, Polyssaiuria,* and *Uskatia* have been considered under the order Bryales; whereas the extinct members like *Jungajia, Protosphagnum* and *Vorentannularia* have been grouped under the Prosphagnales order and phylogenetically the *Sphagnum* is considered the evolutionary descendant of the extinct genus *Protosphagnum* (Neuberg, 1960).

The impression of vegetative parts of extinct bryophyte *Muscites* is thought to be reported from the Carboniferous deposits (France) in Europe and the Triassic deposits of

Africa (Townrow, 1959; Renault and Zeiller, 1885) and the extinct *Muscites* have a structural resemblance to the extant members of the modern Moss.

Phylogenetically, unique, diverse moss flora from Upper Permian deposits in Russia have shown an affinity towards the extant modern representatives of bryophytes: Dicranales, Hypnales, Funariales, Pottiales, and etc. (Ignatov, 1990).

The well-preserved gametophyte of the extinct moss *Mercera augusta* was found in the Permian deposits of Antarctica and careful observations of its structural features have helped establish its evolutionary relationship with the modern members of Bryophyta: Bryidae (Smoot and Taylor, 1986).

The fossiliferous predecessor, *Aulacomnium heterostichoides,* was found in a well-preserved submerged condition from the Eocene clay deposits (freshwater lake in British Columbia) in Pacific Canada (North America), retaining a phylogenetic relationship with its modern representative, *Aulacomnium heterostichum*, disjunctively distributed over Eastern Asia and North America.

Longfengshania, the pre-Cambrian plant fossil has been found to be from China, which has been noticed as structurally symmetrical to the fossil bryophyte *Sporogonites* (Zhang, 1988) but the time of its appearance on Earth has created evolutionary ambiguity to recognize this plant fossil as the primitive representative of bryophytes on Earth.

Chapter 8

EVOLUTION OF TRACHEOPHYTES: "PTERID-O-PHYTES"- AN ERA OF POST-MIGRATORY DIVERSIFICATION, IN THE QUEST FOR "SPERM-A-TO-PHYTES"

The bodies of pteridophytes are well built with the ideal structural trinity for green plants; roots, stem, and leaves with a well-connected vascular structure of xylem and phloem. They also have a well-defined alternation of generations in their life cycles and are characterized with independent sporophytes (not dependent upon gametophytic generations like bryophytes). This makes them distinguished from the primitive group of embryophytes, popularly known as "bryophytes."

The unique features, such as the liberation of female gametophytes from sporangiums, make pteridophytes distinct from spermatophytes. The preliminary features of any photosynthetic transducer of aquatic origin for survival and long-term widespread dispersal on land, need to be acclimatized to facing challenges like an abrupt shift in habitat and changing environmental factors such as changes of temperature, low levels of humidity, variable pH levels, shifts from physiological desiccation to physical desiccation, and etc.

The physical and physiological expression of acclimatization of pteridophytes was primarily noticed in the protective structural features of the reproductive units. The bi-layered formation of peripheral spores (inner intine, desiccation sensitive and degradable layer of spores in acidic pH and the outer layer or exine of spores, desiccation-resistant to acidic pH) is required to survive unfavorable conditions and disperse. The presence of cuticles in pteridophytes protect it from being exposed to desiccation due to water loss, and chemical or mechanical injuries in the land environment which is more complex than the simple aquatic environment.

The absence of stomata in the gametophytes of bryophytes and the presence of stomata in pteridophytes could be interpreted as the evolution of stomata having a direct relationship with the formation of the vent of the cuticle layer, necessary for better gaseous exchange of land plants. However, paleobotanical evidences of the pre-Devonian period suggest that the primitive vascular plants without stomata cannot establish a direct link between the appearance of stomata and the appearance of the terrestrial vascular plant. Rather, a coevolution of the cuticle layer and the stomata should be considered as an important feature for the transmigration of amphibian transducers from water to land. The aquatic plants could excrete the by-products of photosynthesis in a simple way and did not have any storage tissues. On the contrary, land plants store the by-products of photosynthesis in the form of non-toxic, secondary metabolites stored in tracheids, to form the earliest form of conducting, mechanical tissues for ancient terrestrial vascular plants. They were remarkably different from aquatic plants (having a carinal canal, vallecular canal, and rudimentary vascular systems) dependent upon the permeability of cell walls for passing fluids.

The "origin of pteridophytes" is a debatable "hot potato" to the two schools of thought, as one group is in favor of algal origins whereas the other claims that they originated from a bryophyte mother-stock. Referring to the structurally well-developed, fossilized, marine algae of Phaeophyceae (e.g., *Crocelophyton, Nematothallus* etc.), the phylogenetics in the nineties like Lignier (1903), Arnold (1947), Arber (1950), Andrew (1961), Stewart and Rothwell (1993), Kenrick and Crane (1997b) promulgated that pteridophytes originated at the time of "transmigration" of algal communities from ocean to land during the Silurian Period.

Postulating a progenitor-like hypothetical green algae with bottom root-like anchoring structures for gathering nourishment from the ground. It has aerial, photosynthesizing, vertical, dichotomous, parenchymatous tissues and isomorphic alternation of generations (e.g., *Rhynia* sp.). In the hypothesis, "Thallasiophyta and the sub-aerial transmigration," evolutionary phycologists, Church (1919) claimed that the well-diversified marine algae (seaweeds), "Thallasiophyta," the polyphyletic progenitor of land plants, has given rise to both group of tracheophytes: bryophytes and pteridophytes. Another group of scientists of the nineties; Lotsy (1909) and Fritsch (1916, 1945) claimed that pteridophytes originated from filamentous green algae like *Chaetophora*; though Mehra (1969) postulated that members of Chaeotophoraceae polyphyletically gave rise to the hypothetical mother-stock, "Proto-archegoniate group," giving rise to two different groups: psilophytes and lycopodiaceous pteridophytes. The available paleobotanical evidence of evolution and diversification of pteridophytes (Figure 10), and non-availability of earlier fossil evidence of bryophytes, make it difficult for phylogeneticists to establish the bryophytic origin of pteridophytes, but that could be due to a technical issue like the lack of preservation of early bryophytes (Mukherji and Ganguly, 1964). Highlighting the "Anthocerotean ancestry," Bower (1890, 1908), in his earlier postulation, "Strobilar Hypothesis," conjectured that the evolution of well diversified, independent sporophytes is the result of progressive sterilization of sporogenous tissues; it has further been claimed that gradual changes of the Anthocerotean predecessor gave rise to *Lycopodium selago* having numerous sporophylls on its body surface without any formation of compact strobilus-like structures.

Later, Bower (1935) and Zimmermann (1930, 1938) propounded that bryophytes (tracheophytes-1) and pteridophytes (tracheophytes-2) evolved from the ancient hypothetical land plants, possessing archegonia. Eminent phylogeneticist, Cambell (1891, 1905), firmly proclaimed that pteridophytes originated from "Anthocerotean" bryophytes, which was initially criticized due to the existence of wide gaps of morpho-anatomical features between the two groups but discovery of the dichotomously branched sporophyte of Psilotales without roots and leaves refuted these criticisms. Moreover, embryological resemblances between these two groups validated the earlier claims of Cambell (1891, 1905). Smith (1955) supported the Anthocerotean origin of pteridophytes with a few modifications like the shifting of meristems from the base to the apex, and spore formation at the apex of the branches that restricted the growth of the aerial shoot.

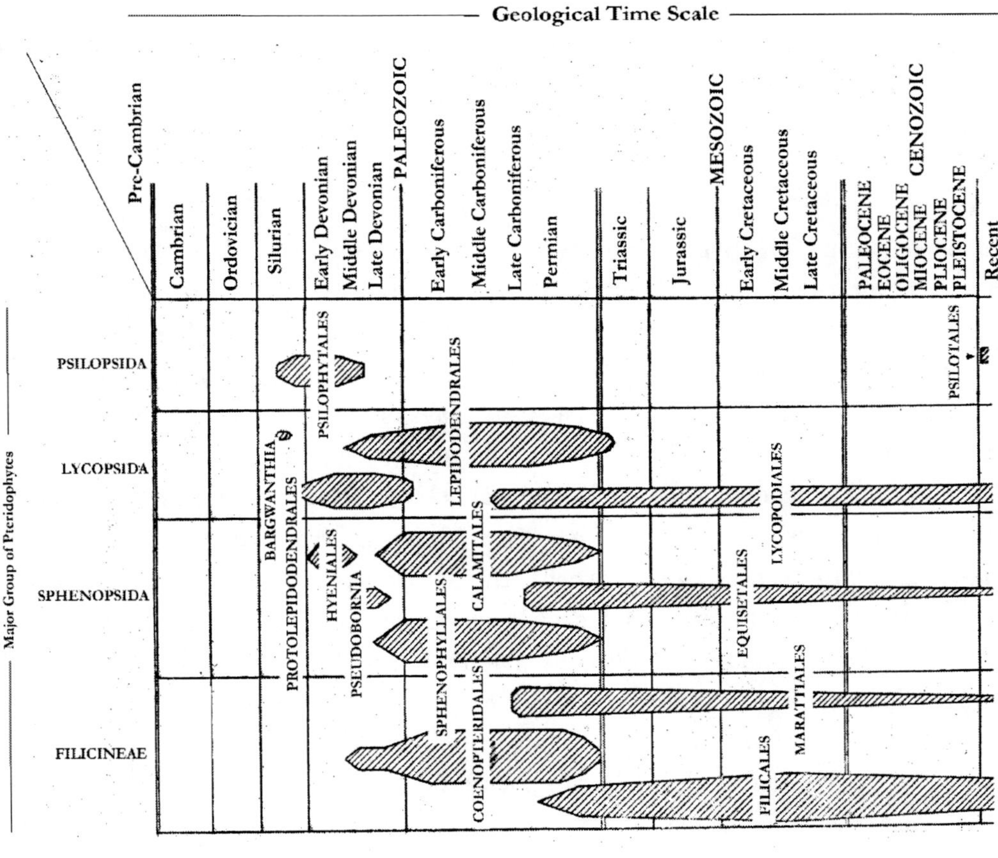

Figure 10. Evolution and diversification of Pteridophytes in the geological periods of time on the Earth (Mukherji and Ganguly, 1964).

The fossil pteridophyte: *Horneophyton*, having parenchymatous corms like body and sporangial structures suspended on the stock of columella tissues (considered as a closed form of vasculature), was considered as the connecting link between the early forms of vascular tracheophytes; *Anthoceros* and *Rhynia*. According to "Protocorm Hypothesis," the scientist Treub (1890), postulated that there is a transitional evolutionary phase between gametophytes- a dependent, undifferentiated sporophyte of bryophyte to a well-differentiated, functionally independent sporophyte- pteridophytes, recognized as protocorm, having root-like rhizoids and leaf-like prophylls. Though scientists Bower (1908), Holloway (1910) and Browne (1913) have considered it's perennating structure to be capable of surviving unfavorable conditions and the experimental observations of Wardlaw (1955) ascertained that high carbon/nitrogen ratios triggered the formation of protocorms and delayed the formation of differentiated sporophytes. Irrespective of any explanations, the carbon dominated early atmospheric conditions created an overall unfavorable scenario for transmigration of plants from water to land, and they needed a shock absorbing transitional state to survive the huge shift to be differentiated into the sporophyte dominated plant world further down on the evolutionary trail, and "protocorm" is that phase on the evolutionary trail. In the "Phyton Hypothesis," Celakovsky (1874), states that its phyllids (leaves), rather than caulids (stem-like structure) play a key role in the formation of sporophyte and the caulid (stem-like structure) is the result of the fusion of basal regions of leaves.

The "*Coleochaete* model" was propounded by molecular biologists Mishler et al. (1994); Kenrich and Crane (1997b); and Qiu et al. (1998), where they studied the ultrastructure with biochemical assays, molecular configurations (5S/16S RNA sequence, gene sequence data from plastids, mitochondrial and nuclear genes). It was noticed that the primitive bryophytes and pteridophytes originated from the "*Coleochaete*" type of soil algae (Fritsch, 1916), thus having a close resemblance to the early tracheophytes. Hence, *Coleochaete* (belonging to the order Coleochaetales), a unique land alga, has finally been recognized as the progenitor of land plants and plays the key role in the transmigration of plants from water to land.

The paleobotanical evidences from the coal deposits of Euramerican provinces of the northern hemisphere ascertain that tall (30-50m), arborescent lycopods (e.g., *Lepidodendron, Stigmaria, Lepidocarpon*) with aerial axes (the stem-like structure) with functional stomata-like aerating tissues, called parichnos, with tall (10m long) arborescent Sphenopsids (e.g., *Calamities*), evolved mainly in the Early Devonian Period. They anchored themselves in the shallow water-logged swamps with specialized root-like subterranean axes, called rhizomorphs also popularly known as Stigmarian roots of Lycopods and subterranean rhizomes of Sphenopsids. Most paleobotanists are in favor of the hypothesis that palaeoclimatic changes in the Carboniferous period tend to subside the swamp to peat mass (supposed swamp basins that dried out due to high rate of vaporization or sudden drought) and lead to natural compaction of the soil. The initial challenges of abrupt changes of habitat and climatic conditions pushed the arborescent lycopods and sphenopsids to the edge of the swamp basin. However, the gradual subsidence of the aquatic swamp basin to a compacted, solid land mass, could not support the rhizomorph in performing the physiological and mechanical functions in the vastly changed scenario to survive so those massive tree-like lycopods and sphenopsids disappeared in the course of transmigration from water to land and proved that transmigration happened as a "spatiotemporal coarse filter," helping evolution and diversification of terrestrial plants.

The members of rootless, leafless (or with rudimentary leaves without veins), most primitive pteridophytes with terminal sporangia at the tip of the branches were found from Silurian deposits, belonging to the "Psilophytales." They have supposedly given rise to advanced groups of pteridophyte members, belonging to "Psilotales." Based on the top-down evolutionary format of primitive to advance forms of sequential arrangements, the pteridophytes were broadly classified (Smith, 1955, Zimmerman, 1959, Sporne, 1962; Stewart & Rothwell, 1993; Smith et al., 2006) into six classes.

According to Stewart & Rothwell's (1993) composite system of classification of the plant kingdom (which is a further extension of the work of Sporne, 1962)[15], plantae, pteridophytes, and gymnosperms were categorized under the same division; Tracheophyta or vascular plants, and flowering plants under the sub-division: Angiospermatophyta. According to Sporne (1962) the six groups of pteridophytic plants are as follows:

A. Psilopsida
B. Psilotopsida
C. Lycopsida
D. Sphenopsida
E. Pteropsida
F. Progymnospermopsida

A. Class- Psilopsida: The sporophytes are rootless but possess subterranean rhizomatous stems with rhizoids and are leafless (or with rudimentary leaves without veins) plants bearing dichotomously branched aerial axes. Anatomically the stem is protostelic; it has a conspicuous protostelic stem having xylem (with annular tracheids), surrounded by phloem, the fossiliferous primitive pteridophytes with terminal sporangia were found on the tip of the branches. It was categorized under three orders:

1. Order: Rhyniales: The paleobotanical evidence from Europe, North America, Australia and South Asia report, that members of this group are ancient, terrestrial, vascular plants, evolved in Early Silurian period and existing till the Upper Devonian or Early Carboniferous Period. *Cooksonia* was considered the earliest tracheophyte (vascular plant) on Earth (Banks, 1975). Kidston and Lang (1917, 1921) discovered the ancient vascular

[15] Who considered cryptogams: Pteridophyta as a division, like cryptogamic Bryophytes as a division and division: Spermatophyta was divided under Gymnospermae and Angiospermae.

fossil plants of the Silurian-Devonian period: *Rhynia major, R. vaughanii, Horneophyton lignieri* and *Asteroxylon mackiei* from the Rhynie Chert beds of Scotland. The critical observations of Edwards (1986) determined that *Rhynia major* is a nonvascular plant but its lifecycle was considered similar to those of pteridophytes. This creates a link between ancient pteridophytes or early tracheophytes and non-vascular cryptogams like bryophytes. This group has further been classified into five families:

i. Rhyniaceae: Fossil remnants: e.g., *Rhynia* (aerial ax and capsule), *Horneophyton* (Sporophyte).
ii. Cooksoniaceae: Fossil remnants: e.g., *Cooksonia* (aerial axe and capsule).

2. Order: Trimerophytales: The members of this group were considered as the connecting link between microphyllous and megaphyllous members and it was reported from the Devonian deposits of Gaspe Bay, Canada. They have a single family, Trimerophytaceae, having the fossil plants: *Psilophyton, Pertica, Trimerophyton*, and etc. (fragmentary plant body).

3. Order: Zosterophyllales: This ancient fossiliferous group of plants of the Silurian-Devonian Period was considered as the progenitor of microphyllous leaves as well as the connecting link between phyllid (leaf-like structures of bryophytes) and microphyllous leaves of pteridophytes. It was further categorized into two families:

i. Zosterophyllaceae: Fossil remnants e.g., *Rebuchia*, *Zosterophyllum* (branching axes with kidney-shaped sporangia).
ii. Gossilingiaceae: Fossil remnants e.g., *Crenaticaulis, Gossilingia, Kaulangiophyton* etc. (fragmentary plant body).

B. Class- Psilotopsida: The living members of Psilotopsida are structurally similar to the fossil members of Psilopsida of the Silurian-Devonian period. It has only one order, Psilotales, under this class.

1. Order: Psilotales: It has one family, Psilotaceae (including the family Tmesipteridaceae), having two genera: *Psilotum* (*P. nudum* and *P. complanatum*) and *Tmesipteris* sp. (having 10 to 13 species). The living descendants of this group of pteridophytes, also known as living fossils, were restricted to South-East Asia (Indonesia), Australia, and the Polynesian region (New Zealand and the Pacific islands). While the classical concept is to consider these family members as descendants of the Silurian group of pteridophytes, Psilophytales, recent molecular studies suggest that pteridologists hypothesize inclusive evolution (structural reduction) which has given rise to this group from the members of *Filicales* (fern groups).

This group of pteridophytes were characterized by a rootless, subterranean rhizomatous plant body with aerial chlorophyllous, dichotomously branched, with chlorenchymatous epidermis and frequently occurring stomata. It has an *actinostelic* structure with a core of phloem with exarch xylem (comprising of inner metaxylem and peripheral protoxylem). The sporangia (also known as synangia) were arranged in triads in the axils of scales of the aerial appendages. The gametophyte (germinated from spores), is called prothallus in its early stages and in the mature stage, it forms archegonia (having egg cells in it) and antheridia (bear multi flagellate sperms).

C. Class- Lycopsida: The sporophyte of this group is well differentiated into leaves, stem and root system. The microphyllous leaves are present leaving a leaf trace but devoid of leaf gaps in its vascular supply; sporangia are found on the adaxial surface of sporophylls that sometimes form strobilus like formations; this class was categorized under five orders:

1. Order: Prolepidodendrales: The paleobotanical evidence from Pragian to Emsian deposits of Euramerica and from Victoria, Australia, ensured that the members belonging to this order evolved during the late Silurian to early Devonian Period and flourished in the early Carboniferous Period, then disappeared in the late Permian Period of the Paleozoic Era.

Prolepidodendrales, the extinct, primitive pteridophyte predecessors of Lepidodendrales and Sellaginellales orders, were characterized by the presence of arborescent habits with wood and bark formations (quite unusual in arborescent lycopods). The horizontal rhizomatous part carrying erect branched axes, covered by "enations" (e.g., *Asteroxylon*), the primitive form of leaves, without vascular supply in it or rudimentary leaves forked at the tip, with vascular supply (e.g., *Drepanophycus*), formed as a result of plantation of a branching system. The "enations" of *Asteroxylon*, were considered as the progenitors of microphyllous leaves as well as an evolutionary transition from the penultimate form of linking between tracheophytes and non-tracheophytes. Normally, the features of heterospory and presence of ligulate leaves are correlated with pteridophytes like homospory and iligulate leaves, but *Leclercquia* are the extinct members of Prolepidendrales acting as an evolutionary transition and therefore possess homosporous features with ligulate leaves. Its members have further been categorized under three families:

i. Drepanophycaceae: Fossil remnants: e.g., *Baragwanathia, Drepanophycus*.
ii. Asteroxylaceae: Fossil remnants: e.g., *Asteroxylon* (plant body, stem).
iii. Protolepidodendraceae: Fossil remnants: e.g., *Colpodexylon, Leclercquia,* and *Prolepidodendropsis*.

2. Order: Lycopodiales: The living members of Lycopodiales can be distinguished from the rest of the orders belonging to Lycopodineae by homosporous habits and

possess iligulate leaves. The sporangium of the members of this group developed from 5-10 cellular masses, located in the axil of the sporophyll and function as sporogenous tissues to give rise to the sporangium. This is also known as the Eusporangiate[16,17] way of sporangium development, also noticed among vascular plants. The members of this group were characterized by homothallic or monoecious (male and female sex organs growing on the same thallus) conditions and they have biflagellate spermatozoids, which take part in sexual reproduction for both algae and bryophytes. While the arrangement of the vascular bundle comprises of xylem and phloem associated with parenchyma and the stele is Protostele (devoid of pith), the formation of protostele varies from species to species and the phyllotaxy, stomatal arrangements, etc. also vary. It was further categorized into the single family:

i. Lycopodiaceae: There are around 400 species of *Lycopodium* distributed over tropical and subtropical regions and some members have also been reported to be from arctic and alpine regions of both the hemispheres. Pritzel (1902), has further categorized them into two groups-
   a. Urostachya (the terrestrial members are erect whereas the epiphytic members are pendant- the dichotomously branched aerial part) and
   b. Rhopalostachya (typically terrestrial with erect aerial parts branched dichotomously in the juvenile stage followed by monopodial branching in the matured stage). e.g., extant plants: *Lycopodium, Phylloglossum,* and extinct remnants: *Lycopodites*.

This group of plants has adapted a number of structural formations to meet the evolutionary challenges:

a. Stelar architecture:

1a. Plectostele: The segmented phloem tissues are surrounded by patches of xylem tissues. e.g., *Lycopodium campanulatum, L. clavatum.*
2a. Actinostele: Star-shaped xylem and phloem tissues are at the branch-tip of the star, embedded with xylem. e.g., *L. selago, Selaginella serratum.*
3a. Mixed protostele: Xylem and phloem tissues are scattered arbitrarily. e.g., *L.cernuum, L.laterale.*
4a. Stellate protostele: Stellate form of xylem, single cylinder fragmented to form multi-strands. e.g., *L.annoniticum, L. lucidum.*

---

[16] Jura: In Latin "Juria" means forest.
[17] On the contrary, Leptosporangiate way of sporangium formation was initited from a single cell, developed in the axil of sporophyll.

b. Phyllotaxy: The arrangement of leaves on stem vary-

1a. Helical: e.g., *Lycopodium clavatum.*
2a. Opposite decussate: e.g., *L. complanatum.*
3a. Whorled: e.g., *L.cernuum.*

c. Morphological diversity of leaves: Though isophylly (similar shapes and sizes) is predominant, heterophylly (foliar dimorphism) has also been noticed-

1a. Isophylly: e.g., *L. clavatum, L. lucidum.*
2a. Heterophylly: e.g., *L.complanatum, L.volubile.*

d. Stomatal arrangements: Though stomata are formed in both surfaces (amphistomatic) in most of the members of *Lycopodium*, in some species they are restricted to the lower surface (hypostomatic) of the leaves.

1a. Amphistomatic: e.g., *L. phlegmaria, L. serratum.*
2a. Hypostomatic: e.g., *L. companulatum, L.volubile.*

3. Order: Lepidodendrales: The extinct members of pteridophytes (except the extant members belonging to Pleuromiaceae), belonging to Lepidodendrales were found in the Europe, North America and East Asian deposits of the Upper Devonian and Triassic Periods of the Palaeozoic Era. They flourished during the Carboniferous Period and began declining during the Upper Permian Period. The fossiliferous seed-like structures are reported to be from the Carboniferous deposits of United Kingdom. The 10cm-40m tall tree-like pteridophytes are characterized by a straight trunk (of 1-2 m. diameter) having vascular supplies of protostele with exarch protoxylem and cambium; subterranean, dichotomously branched root-like structures (also known as Stigmarian root system); linear or acicular, microphyllous leaves with ligulate leaf-base; heterosporous habit. Furthermore, it has an arrangement of megasporophylls which form strobilus containing megasporangium (having four megaspores and containing female gametophyte permanently inside the sporangium), surrounded by a protective layer, integuments raised from megasporophyll found in *Lepidocarpon* (also called false seed), a well-diversified megasporangium as well as seed-like structures of *Miadesmia*, and biflagellate antherozoids are also present. It was further classified into four families:

I. Lepidodendraceae: Fossil remnants e.g., *Lepidodendron* (fossil wood), *Lepidostrobus* (fossil fruit) *etc.*
II. Sigillariaceae: Fossil remnants e.g., *Sigillariophyton* (foliar structure), *Siglariostrobus* (female sporangium) and *Mazocarpon* (fruit-like structure) etc.

III. Bothrodendraceae: Fossil remnants e.g., *Bothrodendron* (mature sporophytic stem).
IV. Pleuromiaceae: Fossil remnant e.g., *Pleuromeia* (herbaceous stem with leaf scar).

4. Order: Isoetales: The sporophytic herbaceous members of Isoetales supposedly originated in the Middle Triassic Period and are structurally characterized by ligulate, microphyllous, tufted, awl-shaped leaves that grow on the tuberous axis along with tufted rhinophores that were found at the basal-end of its stem-like axis. Anatomically the vascular bundle in the tuberculate stem form the protostele with a secondary growth of cambium; the terminal end of the leaf acts as the photosynthesizing location whereas the basal region acts as the sporangium bearing region or sporophyll; sporophylls are lax or compacted into strobila (so the whole plant body itself is like a functional strobilus); the eusporangiate way of spore formation in heterosporous sporangium (where sporogenous tissues are intercalated with trabecula[18]); predominantly heterothallic (archegonia and antheridia form on different prothallus); spermatozoids are multi-flagellated; fertilized ovums form oospore (structurally and functionally alike to seeds of Spermatophyta), a diploid, sporophytic body and embryo is endoscopic without any suspensor attached to it like the other Lycopsids. The majority of the members of the Isoetales are hydrophytes, acclimatized to grow in waterlogged swamps, while a number of members are acclimatized to grow in the dry, terrestrial habitat. This kind of ecological elasticity points prompts us to review the evolutionary hypothesis of "transmigration of amphibians in the Carboniferous Period." Isoetales has only one family Isoetaceae with three genera: Fossil remnants: e.g., *Nathorstiana, Stylites* and extant member: *Isoetes*.

5. Order: Selaginellales: The members of this order belong to one family Selaginllaceae having extinct species: *Selaginellites*. The extant genus: *Selaginella,* also known as resurrection plants as they can survive, like some xerophytic plants, in extreme conditions, as they become dormant and "come back to life" with the onset of favorable climatic conditions, e.g., *Selaginella bryopteris*. They comprise of 700 species, distributed over tropical and subtropical regions and some temperate regions of both hemisphere.

On the basis of structurally distinguishing features, Hieronymous (1902) categorized the *Selaginella* into two groups:

I. Homeophyllum (the plant body is erect, radially constructed), e.g., *Selaginella selaginelloides, S. rupestris* etc. and
II. Heterophyllum (the plant body is prostrate, dorsiventrally constructed) e.g., *S. helvetica, S. martensii*. This group of plants is characterized by microphyllous, ligulate leaves, a leafless rhizophore structure which gave rise to root like

[18] A part of sporogenous tissues of sporangium transformed into sterile cells (the fertile ones form mega or microspores) to provide nutritive support (functionally alike endosperm, which provide nutrition to seed to give extra life-span, required for extra time related to cover seed dispersal and germination related issues) to the spores to make it viable for long time.

structures, anatomically protostele, with the aggregation of sporophylls to form heterosporous strobila, heterothallic (dioecious) gametophytic bodies (two different forms of sexual bodies, e.g., antheridia and archegonia, on two different thalli), and sunken stomata. All these features tend to direct them to be a diversified group of pteridophytes on the evolutionary trail to acclimatize to the terrestrial habitat and survive in the absence of enough moisture, one step closer to the evolution of seed-bearing plants or spermatophytes.

D. Class- Sphenopsida: The paleobotanical evidence from the Euramerican coal deposits and deposits of Angaraland and Gondwanaland, have proven the evolution of the group members of Sphenopsida, which evolved during the Devonian Period, flourished during the MiddleCarboniferous Period, and declined during the Permian Period of the Paleozoic Era. The members of this group are represented by around 18 extinct forms (e.g., *Calamities, Equisetileties* etc.) and one extant genus e.g., *Equisetum*, (also known as a living fossil). The members of this group were characterized by branched stem with nodes and internodes, whorls of reduced leaves appearing from nodes, thickly layered sporangiums grown on plate-like appendages called sporangiophores, exosporic-gametophytes, and multi-flagellate antherozoids with embryos, exsoscopic and devoid of suspensors. It has further been divided into five orders:

1. Order: Hyeniales: It has a single family Hyeniaceae with one genus *Hyenia*.

2. Order: Sphenophyllales: The fossiliferous members of Sphenophyllales evolved during the Devonian Period, flourished during the Carboniferous Period and disappeared during the Early Cretaceous Period. In the Carboniferous forest, these undershrub members of the group were characterized by slender stems with nodes and internodes, ridges and furrows; anatomically the stem is a protostele, with whorls of different shapes of leaves (also known as verticels) formed at the nodes of the aerial stem; either loosely arranged strobila or terminal cone-shaped sporangium and having two segments, and the apical sporangium sac (containing homospores) suspended on top of the lower sterile podium. It was categorized into two families:

i. Sphenophyllaceae: Fossil remnants: e.g., *Sphenophyllum, Eviostachya* (Leaves and stem).
ii. Cheriostrobaceae: fossil remnants: e.g., *Cheirostrobus* (strobilus)

3. Order: Calamitales: The Fossiliferous members of Calamitales were reported from Euramerican deposits and the fossil remnants of Angaraland and Gondwanaland. They evolved during the Devonian Period, flourished during the Carboniferous Period and disappeared during the Tertiary Period. The tall (around 10m. long) arborescent sphenopsids were characterized by subterranean horizontal rhizomes, with nodes and

internodes, and adventitious root-like structures attached to the nodes of rhizomes, the vertical aerial shoots attached to the horizontal rhizomes, and the whorl of aerial shoots attached to the nodes. The aerial shoots and stele in them were found to be constricted to the point of attachment on the nodes of the rhizome. Anatomically, it has endarch siphonostele and its secondary growth was related to cambial growth, microphyllous leaves forming in a whorl at the nodes of aerial shoots; the cones of this group had a central axis with a whorl of sporangiophores bearing a number of sporangia, protected by sterile appendages called bracts. It has further been divided into two families:

i. Astrocalamitaceae: Fossil Remnants; e.g., *Asterocalamities* (stem).
ii. Calamitaceae: Fossil remnants: e.g *Annularia* (Leaf), *Calamities* (Stem), *Palaeostachya* (Strobilus).

4. Order: Equisetales: The palaeobotanical evidence ascertained that this living fossil with members of Equisetales (popularly known as horsetails), appeared during the Devonian Period, and are still living in the Quaternary Period.

This order has only one family; Equisetaceae; which includes the extinct form of Equisetites and extant genus of *Equisetum*, with 30 species distributed worldwide except in Australia and New Zealand. On the basis of distinguishing structural formations, the family members were categorized (Duckett, 1979; Hanke, 1989) under two subgenera: a. *Equisetum* (annual branched aerial stem, blunt leaves, plate-like lamellae on gametophytes, etc.) e.g., *Equisetum arvense, E. sylvaticum,* and b. Hippochaete (perennial, unbranched stem, pointed leaves, column-like lamellae on gametophytes, etc.) e.g., *E. laevigatum, E. scirpoides*. The small, herbaceous, sporophytic plants (18-50 cm high) of *Equisetum* are characterized by horizontally creeping rhizomes, attached to erect, aerial shoots with nodes and internodes (with ridges and furrows), branched monopodial forms; the stem surface is protected with cuticular coatings and siliceous materials; anatomically the stem is siphonostele; the aerial shoot shows hydrophytic (e.g., presence of carinal canals, vallecular canals, pith cavity, reduction of vascular supply etc.) and xerophytic adaptations (e.g., ridged and furrowed internodes, sunken stomata, scaly leaves, photosynthetic tissues on the stem, etc.). The cone or homosporous strobilus contain spores with hygroscopic attachments called elaters, responsible for optimizing spore dispersal, are borne on the tip of sporangiophores attached to the aerial shoots.

The presence of hydrophytic and xerophytic adaptations and unique spore dispersal mechanisms have opened up an evolutionary preferential edge in the transmigration process into the living fossil *Equisetum,* to go one step closer to the seed formation of spermatophytes.

E. Class- Pteropsida: The members of Pteropsida (popularly known as ferns) were distinguished from other members of pteridophytes by the presence of megaphylls. Paleobotanical evidence has proven that this group of well-diversified pteridophytes with over 9000-13,233 species of worldwide in epiphytic, terrestrial and epiphytic forms, evolved during the Devonian Period of the Palaeozoic Era. They flourished during the Late Permian and survived till the present Quaternary Period (Hassler, 2017). The mature sporophytes of Pteropsida bear the distinguished subterranean stout rhizome with roots, stems and dimorphic leaves: sterile or vegetative brown-colored scaly leaves protecting the rhizomes and with spirally arranged fertile, green megaphylls (which perform photosynthesis and act as sporophylls), called circinate ptyxis[19]. Anatomically, the structural formation of the stele in the stem of the different members of ferns, is well diversified and ranges from protostele, siphonostele, solenostele, dictyostele to polycyclic stele, and etc. The acclimatizing elasticity of the Pteropsida members towards habit and habitat diversity, leaf formation, sporangial formations, style diversification, etc., provided an evolutionary edge for a better opportunity of diversification and higher opportunity of survival to cope with spatiotemporal changes on the Earth's surface and direct the evolutionary trail to evolve flowering plants. It was categorized under four sub-classes:

E-I. Subclass. Primofilicales: The paleobotanical evidences partly from Europe (Belgium, Germany, Scotland) and North America implied that the members of this group of plants appeared in the MiddleDevonian Period and survived till the Lower Carboniferous period. Its name itself points out that this is the progenitor of the modern diversified group of pteridophytes: Flilcales (Reimers, 1954; Sharma, 2006; Sporne, 1962). It was broadly categorized under two orders:

1. Order: Cladoxylales: This group is an all-time topic of phylogenetic controversy, and the prime reason is the extinct taxon, *Pseudosporochnus*, which was considered under the Psilophytales in the early Nineties, till Arnold (1947) placed it under the Coenopteridales, Smith (1955) placed it under the Protopteridales, but Scheckler (1973) then considered it under the Cladoxylales based on its stem anatomy. The unique feature of this taxon is that it has no clear distinction between its stem and its leaves. The members of this group were characterized by an erect trunk with a crown of aerial, dichotomously spread branches, bearing lateral appendages or fronds with a bilateral symmetry, (having four rows of pinnate leaves) and a pair of sporangia on the branch tip; anatomically the stem is polystelic. It has a single family: Cladoxylaceae: Fossil remnants, e.g., *Cladoxylon*, *Pseudosporochnus*.

2. Order: Coenopteridales: the extinct members of this group evolved during the Upper Devonian and Lower Permian Periods. The members of this group were characterized by small tree ferns with stout trunks, protestelic bearing alternate pairs of branches and scale-like protecting structures known as "aphlebiae," found at the joints of the branches on their trunk; sporangiums (either homosporous or heterosporous) are found on the tip of the

[19] The pinnately compound leaves carrying primarily homosporous (some aquatic members are heterosporous) sporangia at the margin of pinna or at the abaxial surface.

branches; the petiole trace of the modern ferns are adaxially curved, whereas the petiole trace (vasculature) of this group is found abaxially curved.

i. Zygopteridaceae: Fossil Remnants: *Ankyropteris*.
ii. Stauropteridaceae: Fossil Remnants: e.g., *Stauropteris*.
iii. Botryopteridaceae: Fossil Remnants: e.g., *Botryopteris* (sporangia).

E-II. Subclass. Eusporangiatae. It was further classified under two orders:

1. Order: Marattiales: The two hundred extant species of Maratriales, are disjunctly distributed over Indo-Malayan regions, including the West Pacific regions, Polynesia and Madagascar, and the American regions (West Indies, United States, Mexico). The fossil remnants of *Angiopteris* were retrieved from the Jurassic deposits of Yorkshire (Van Konijnenburg- Van Cittert, 1966). The members showed a eusporangiate type of sporangium development where a group of sporangia, called sori, is found in free form e.g., *Angiopteris* or fused forms to form synangia e.g., *Marattia* that grow on the abaxial surface of a leaf; the sporangium wall acts as tapetum. The matured sporophyte of this group of pteridophyte ranges from 30 cm to 5 m high. They are characterised by rhizomatous root-stock, stem is protostelic in early phases but siphonostele (vascular tissues passing through the node diagonally, called commissural strand) in the matured condition bears the presence of tannin filled cells, mucilage canals (Smith et al., 2006) found in the stem cortex, leaves and roots; symbiosis found between roots and endophytic mycorrhiza, which provide nourishment to septate roots on the rhizome; leaves (a pair of stipules at leaf base) are either simple (e.g., *Danaea*) or compound, e.g., *Angiopteris*, (pinnately or palmately) with circinate ptyxis; gametophyte is homothallic; endophytic fungi provides nourishment to the gametophytes. It was further categorized into five families:

i. Asterothecaceae: Extant: e.g., *Asterotheca, Scoleopteris* etc.
ii. Angiopteridaceae: Extant: e.g., *Angiopteris*.
iii. Marattiaceae: Extant: e.g., *Marattia*.
iv. Danaeaceae: Extant: e.g., *Danaea*.
v. Christenseniaceae: Extant: e.g., *Christensenia*.

2. Order: Ophioglossales: The members (around 50 herbaceous species) of this group of pteridophytes are widely distributed over the tropical and temperate regions across the two hemispheres. It has one family, Ophioglossaceae (Sporne, 1962; Smith et al., 2006), with four genera: *Botrychium, Helminthostachyus, Ophioglossum,* and *Mankyua*. The members of *Ophioglossum* are characterized by subterranean rhizomes with an adventitious root system and are covered with sterile scaly leaves; anatomically, it has a

protostele (devoid of pith) at the bottom of the young part or siphonostele (with pith) in the matured part (upper part), often solenostele (due to one leaf gap in the siphonostele). Some dictyostele (due to there being more than one leaf gap), sometimes meristele (due to multiple fragmentations of dictyostele); annual ovate shaped leaves grown on this perennial plant, characterized by a thin petiole and a wide lamina with reticulate venations, mid-vein, and the petiole (with one leaf gap with two leaf traces) is attached to the base by a smaller stipule; leaves are amphistomatic (present in both surfaces); the (homosporous) sporangiferous fertile spike (sporangia are formed in two rows) formed at the juncture of petiole and lamina; the outer layer of sporangium possesses stomata and the innermost layer of sporangium acts as tapetum, homothallic; symbiosis of non-photosynthesizing gametophytes with endophytic fungi (provide nutrition to it); having multi-flagellate antherozoids; and an exoscopic embryo without a suspensor. The factors like the diversification of formation of the stele, annual leaf formation on the perennial structure, multi-flagellate antherozoids, a symbiotic relationship between endophytic fungi and gametophytes, and an exoscopic embryo, makes this group of plants phylogenetically closer to flowering plants. On the basis of repetitive, dichotomous branching patterns of vasculature in the frond of Osmundales, Chrysler (1945); Holttum and Sen (1961), established its affinity to the members of Coenopteridales of ancient groups of pteridophytes: Primofilicales; hence Nishida (1957) considered Osmundales as an evolutionary transition between the primitive, radial form of frond bearing Primofilicales and the modern plain frond bearing members of Filicales.

The petiole (connected to the fertile spike), was considered as a modified form of stem or phyllomorph (Nozu, 1950; Nishida, 1957), considered as an interface between the stem and the leaf, and it was further conjectured that its dichotomous branching gave rise to the reproductive spike on one end and the vegetative leaf on the other (Kato, 1990).

E-III. Subclass: Osmundidae: Though contemporary works of Smith et al. (2006) put Osmundales in the class of Polypodiopsida (erstwhile Filicopsida) and other workers like Bower (1935); Holttum (1949); Rashid (1999) considered the family Osmundaceae under the Order Flicales, it was classified under the sub-class: Osmundidae, (Reimars, 1954; Pichi-Sermoli1958; Sporne, 1962; and Sharma, 2006), subsequently under the single order Osmundales due to the presence of common features of Eusporangiate and Leptosporangiate pteridophytes. Paleobotanical evidence (e.g., *Thamnopteris*) proved that it evolved during the Upper Permian Period, flourished during the Mesozoic Period (a number of fossil remnants were discovered from Mesozoic deposits e.g., *Osmundaties*) and the extant members (also known as living fossils), e.g., *Osmunda, Leptopteris,* and *Todea*, are still alive in the Quaternary Period. It has a single family: Osmundaceae.

1. Order: Osmundales: The members of this group are characterized by terrestrial, small, tree-like plants with erect, hard stems, covered with uniseriate hairs, with a dense crown of leaves; anatomically, stem siphonostelic with isolated xylem strands and

sclerenchyma tissues (providing mechanical support); presence of stipules at the base of the petiole; stalked sporangia forming at the margin of fertile leaves or abaxial surfaces of pinnules; a rudimentary annulus and archesporial tapetum is present on the single layered sporangial wall; archegonia and antheridia are grown on the same thallus; and antherozoids are present. The presence of eusporangiate characteristics in this family member is the endodermis poorly developed in the petiole; the presence of stipules at the petiolar base; the presence and development of numerous sporangia (in eusporangiate form) do not form sori, and the antheridia and archegonia are projected on thalloid formations. The presence of leptosporangiate features in Osmundaceous members is the presence of archesporial tapetum and a rudimentary annulus; the presence of a single-layered sporangial wall; absence of endophytic fungus in prothallus; vertical division of zygotes, etc. This group of terrestrial pteridophytes (popularly known as Flowering ferns) is well adapted to (physiologically and mechanically) the shady, moist forest floors in the bogs with compacted soil to maintain a transitional status between the eusporangiate and leptosporangiate pteridophytes, acquiring the mechanical strength of small trees from supportive tissues as well as defined sexual reproductive structures, required for further evolutionary progress and diversification.

E-IV. Subclass: Leptosporangiatae: The members of leptosporangiate ferns are mostly mesophytes, hydrophytes, and few members show xerophytic adaptations, and were considered as the most advanced group of pteridophytes, found distributed worldwide in moist, shady tropical and sub-tropical regions of both the hemispheres. Paleobotanical evidence shows that the members of these group appeared in the Permian Period, diversified in the Tertiary Period and existed till the Quaternary. It was further classified under three orders:

1. Order: Filicales: The members of Filicales (also known as true ferns) are perennial plants, supposedly evolved during the Early Permian, flourished during the Lower Cretaceous Period and survived till present the Quaternary Period. There are around (Copeland, 1947; Simth et al., 2006; PPG-I, 2016[20]; Hassler, 2017) 260 genera and 8,321

---

[20] PPG-I: Pteridophyte Phylogeny group (2016), Comprised of: Eric Schuettpelz, Harald Schneider, Alan R. Smith, Peter Hovenkamp, Jefferson Prado, Germinal Rouhan, Alexandre Salino, Michael Sundue, Thais Elias Almeida, Barbara Parris, Emily B. Sessa, Ashley R. Field, Andre Luis de Gasper, Carl J. Rothfels, Michael, D. Windham, Marcus Lehnert, Benjamin Dauphin, Atsushi Ebihara, Samuli Lehtonen, Pedro Bond Schwartsburd, Jordan Metzgar, Li-Bing Zhang, Li-Yaung Kuo, Patrick J. Brownsey, Masahiro Kato· Marcelo Daniel Arana;Francine C. Assis, Michael S. Barker, David S. Barrington, Ho-Ming Chang, Yi-Han Chang, Yi-Shan Chao, Cheng-Wei Chen, De-Kui Chen, Wen-Liang Chiou, Vinicius Antonio de Oliveira Dittrich, Yi-Fan Duan, Jean-Yves Dubuisson, Donald R. Farrar, Susan Fawcett, Jose Maria Gabriel y Galan, Luiz Armando de Araujo Goes-Neto, Jason R. Grant, Amanda L. Grusz, Christopher Haufler, Warren Hauk, Hai He, Sabine Hennequin, Regina Yoshie Hirai, Layne Huiet, Michael Kessler, Petra Korall, Paulo H. Labiak, Anders Larsson, Blanca Leon, Chun-Xiang Li, Fay-Wei Li, Melanie Link-Perez, Hong-Mei Liu, Ngan Thi Lu, Esteban I. Meza-Torres, Xin-Yuan Miao, Robbin Moran, Claudine Massi Mynssen, Nathalie Nagalingum, Benjamin Øllgaard, Alison M. Paul, Jovani B. de S. Pereira, Leon R. Perrie, M_onica Ponce, Tom A. Ranker, Christian Schulz, Wataru Shinohara, Alexander Shmakov, Erin M. Sigel, Filipe Soares de Souza, Lana da Silva Sylvestre, Weston Testo, Luz Amparo Triana-Moreno, Chie Tsutsumi, Hanna Tuomisto, Iv_an A. Valdespino, Alejandra Vasco, Raquel Stauffer

species in this order that are characterized by perennial homosporous pteridophytes, well distinguished with roots, a subterranean rhizomatous stem and diversified hypostomatic fronds[21] with circinate ptyxis attached to the rhizome by a rachis, and sori at the ventral margin of pinnae. They are well connected at the distal end of the leaf veins; false indusium are present; sporangiums are well protected by structural formations of a thick layer of an outer annulus, inner thin stomium, and parenchymatous tissues; archesporium have a protective tapetal layer, covering inner sporogenous tissues; they have a scale leaf (also known as ramenta) covered rhizomatous stem, well developed sclerenchymatous tissues with an outer and inner cortex of parenchymatous tissues and diverse types of stele monoecious gametophyte; multi-flagellate antherozoids are present; the embryo remains well protected with a layer called the calyptra until germination is initiated. The diversified reproductive structures, better vascular supplies, and mechanical support helped this group of plants get better evolutionary support to meet the external challenges and equipped them with a better reproductive mechanism to follow the footprints of evolutionary progress, which was witnessed among flowering plants in the supreme level. The Filicales have further been categorized in ten families:

i. Schizaeaceae: Fossil Remnants: e.g., *Klukia*, Extant: e.g., *Lygodium, Schizaea* etc.
ii. Gleicheniaceae: Fossil Remnants: e.g., *Gleichenities*, Extant: e.g., *Gleichenia*.
iii. Hymenophyllaceae: Extant: e.g., *Hymenophyllum, Serpyllopsis* etc.
iv. Matoniaceae: Fossil Remnants: e.g., *Metonidium*, Extant: e.g., *Matonia*.
v. Dipteridaceae: Fossil Remnants: e.g., *Clathropteris*, Extant: e.g., *Dipteris*.
vi. Cyatheaceae: Fossil Remnants: e.g., *Coniopteris*, Extant: e.g., *Cyathea, Dicksonia* etc.
vii. Dennstaedtiaceae: Extant: e.g., *Asplenium, Dryopteris, Pteridium, Nephrolepis* etc.
viii. Thelypteridaceae: Extant: e.g., *Thelypteris*.
ix. Adiantaceae: Extant; e.g., *Adiantum, Cheilanthus* etc.
x. Polypodiaceae: Extant: e.g., *Platycerium, Polypodium* etc.

2. Order: Marsileales: The members (popularly known as water ferns) of this group of pteridophytes, growing on shady, shallow, water-logged conditions, primarily on the edge of water bodies, have three extant genera: *Marsilea* (of 60 species), *Pilularia* (of 5 species) and *Regnellidium* (of 1 species); around 66 species are distributed in tropical Africa and

Viveros, Alan Weakley, Ran Wei, Stina Weststrand, Paul G. Wolf, George Yatskievych, Xiao-Gang Xu, Yue-Hong Yan, Liang Zhang, Xian-Chun Zhang, and Xin-Mao Zhou.

[21] The hypostomatic frond structurally ranges from decompound or unipinnately compound to multipinnately compound.

Australia except *Regnellidium* which is endemic to Brazil. The matured sporophytes of this group of plants is characterized by tufted roots attached to the rhizomatous stems, long-petioled leaves and heterosporous sporangium carried by specialized structures, known as sporocarp. It has a true indusium; the rhizomatous stem containing the outer amphiphloic siphonostele, outer layer of cortex with parenchyma tissues[22], innermost cortex of parenchyma tissues, intercalated by the middle cortex of sclerenchyma tissues and the central pith is parenchymatous for the members with hydrophytic adaptations, and sclerenchymatous for the members with xerophytic adaptations. The cortex and stele support the better conductivity of sap. The formation of male gametophytes of Marsileales is identical to the primitive pteridophytes like Lycopodiales and Selaginellales, and the formation of oospore by fertilizing the archegonium by multi-flagellated antherozoids has resemblance to the members of Isoetales.

According to the "Leaf segment or Laminar Hypothesis" (Bower, 1889; Eames, 1936; Smith, 1955), the sporocarp of Marsiliales is the lateral, bean-shaped segment of a leaf rather than the whole leaf; whereas the "Whole leaf or Petiolar Hypothesis" (Johnson, 1933) defined sporocarp as a modification of the lamina and the petiole and further stated that the marginal cells of lamina were developed into sporangia. The third hypothesis, popularly known as the "phylogenetically synthesized new organ" (Bierhorst, 1971), states that sporocarp originates from the specialized cells of the vegetative pinna (not found in other pteridophytes) and it was clarified as a differential expression of genome where sporangial ontogeny is expressed, followed by the ontogeny of pinna. Features like the appearance of mechanical support and vascular supply, grated sporangium, the presence of predominant hydrophytic adaptations and less xerophytic adaptations, sporocarp formations, the presence of true indusium etc., indicate that an evolutionary experiment of structural and functional diversification was an integral part of this group of pteridophytes to keep the evolutionary progress on-going. This group has further been divided into two families:

i. Pilulariaceae: Extant: e.g., *Pilularia.*
ii. Marsileaceae: Extant: e.g., *Marsilea, Regnellidium.*

3. Order: Salviniales. The extant members of this group of free-floating, heterosporous pteridophytes with hydrophytic adaptations, were found in tropical and sub-tropical freshwater bodies of Africa and Asia. On the basis of this evidence, it was ascertained that this group of pteridophytes (e.g., fossil remnants of *Azolla*) evolved around the Early Cainozoic Era. The mature sporophyte of this group is characterized by a root-less plant body of which the rhizomatous stem is covered with multicellular hairs and densely clustered sessile leaves (function as root systems) and submerged, dissected leaves; the

[22] The parenchyma tissue with air vacuoles were found in some of the members of pteridophytes with hydrophytic adaptations, called aerenchyma.

lower surface of the dorsal lobes of the floating leaves of *Azolla*. There were some large cyanobacterial cavities, *Anabaena azollae,* fixing nitrogen as well as maintaining symbiotic relationships; the rhizomatous stem has siphonostele, noticeable hydrophytic adaptations like reduced vascular tissues, fragmented xylems; the sporocarp contains either the male or the female sporangium and the submerged leaves bear sporocarps and indusium forms the inner-wall of the sporocarp; the antherozoids are multi-flagellated. The functional developments, structural developments of reproductive bodies, embryonic developments of these aquatic ferns, symbiotic relationships, and hydrophytic adaptations, give this group of pteridophytes a better opportunity to survive, diversify, and progress through the trail of plant evolutions. This group of pteridophytes has further been divided into two families:

i. Salviniaceae: having extant genus: *Salvinia* with 12 species.
ii. Azollaceae: having extant genus: *Azolla* with 7 species.

F. Class- Progymnospermopsida: On the basis of morphological features of sporophytes and reproductive structures (like spore formations), the extinct plant fossils from the late Devonian deposits of New York, United States, were initially considered as fossil pteridophytes, but further review of the anatomical structure of the stem revealed its resemblance to the wood anatomy of the extinct fossils of the gymnosperm *Callixylon,* due to the presence of pycnoxylic patterns of secondary wood formations. The presence of reproductive (like the characteristic of sporangia bearing sporophylls) and vegetative (like arborescent habit and wood anatomy of gymnosperms) characteristics of pteridophytes were considered to be of immense evolutionary importance in establishing an evolutionary link as well as for being the penultimate class of early seed-bearing spermatophytes (i.e., gymnosperms), which were known as "Progymnospermopsida" (Beck, 1960). The branching patterns of the fructification of the extinct members of Progymnospermopsida was characterized by trees with well- developed vascular bundles, presence of xylem and phloem tissues and the presence of vascular cambium; the axes were clustered loosely and branched monopodially, the reproductive structures were borne by sporangiophore like formations, dichotomously branched in the initial phase followed by pinnate type branching, the tips of which bear sporangiums (either homosporous or heterosporous). The paleobotanical evidence indicates that this group of plants disappeared during the Carboniferous Period. A group of phylogeneticists has considered this Progymnospermopsida group of plants as the progenitor of seed-bearing vascular plants or primitive spermatophytes, supposedly evolved from the Devonian group of extinct pteridophytes: Trimerophytes or Trymerophytopsida. Beck (1960) has further categorized this group of plants under three orders:

1. Order: Aneurophytales: The paleobotanical evidences (Krausel and Weyland, 1923, 1926, 1929; Goldring, 1924; Lecllerca and Bonamo, 1971) from the Upper Devonian deposits of North America (New York) and Europe (Belgium, Germany etc.) have established that the predominant members (e.g., *Aneurophyton, Rellimia, Tetraxylopteris* etc.) of this group of plants evolved; the earliest fossiliferous member of this group, *Rellimia* (the erstwhile *Protopterydium),* had loosely arranged axes and fronds. Sporangiums are spirally arranged on the aerial axes (branches) attached to the trunk, forming crowns. Contrastingly, the fronds are spirally arranged on the aerial axes or branches in *Aneurophyton* and *Tetraxylopteris* and the branches are arranged on the trunk in an opposite decussate way. On the basis of these comparative structural features, phylogeneticists infered that the members belonging to Aneurophytales in the Late Devonian Period possibly gave rise to the ancient members of the seed fern group members belonging to Lyginopteridales. The fossiliferous trunk of *Aneurophyton* (about 13m long and 1-1.5m in diameter) and *Rellimia* were characterized by their triangular or trilobed protosteles with the central protoxylem, surrounded by secondary woods (comprised of medullary rays, tracheids and bordered pits) of the pycnoxylic type. In *Tetraxylopteris*, tetra-lobed protostele were found with the rest of the similar anatomical features in *Aneurophyton* and *Rellimia*. The relict member *Aneurophyton* has a crown head comprising of loosely arranged fronds (upto 3m long), dichotomously branched on the same plane (identical to the way the fronds of the members of Filicales are arranged) on the top of the trunk, and the tips of the branches were subtended by an oval-shaped sporangia with a pointed tip. Most paleobotanists support the dichotomously branched structure with sporangiums as fertile leaves (as it was found in one plane), but Arnold (1947) postulates the rachis of the frond-like structures like the stem and the 'Y' shaped pinnule of the dichotomously branched pinnae as leaves. Due to the 3-dimensional arrangement of the formation of fronds (opposite decussate), the fertile and vegetative frond-like structure was considered as a stem-like structure rather than a leaf-like structure.

All types of transitional forms like the helical arrangement of frond formations to the three dimensional form (decussate form), pro-leaf-like frond formation to pro-stem-like structures, variations of protostellar formations, presence of composite characters of the members of gymnosperms and pteridophytes, made these group of plants and the timing of their evolution (through age-dating process) during the Devonian Period, helped draw the missing foot-prints, to complete the evolutionary trail. These group members potentially constitute the penultimate stage of early seed-bearing plants (evolved during the Lower Permian Period). The members of the order of Aneurophytales belong to the family: Aneurophytaceae: Fossil Remnants: e.g., *Aneurophyton, Rellimia, Tetraxylopteris, Propterydium* etc.

2. Order: Protopityales: Paleobotanical evidence revealed that the sole genus of Protopityales, *Propitys* appeared in the MiddleCarboniferous Period and disappeared in

the Permian Period. The stem genus, *Protopitys* (*P.buchiana*), is characterized by the trunk having a diameter of 43 cm, having pith, encircled by metaxylem in the center, subsequently surrounded by wood fibres and tracheids forming the pycnoxylic type of secondary wood; anatomically, it has medullated protostele and leaf traces originating f rom the opposite end of the elliptical form of pith, which has encouraged the phylogeneticists to establish its resemblance to the members of Cycadopsida. The stellar structure of *Protopitys* has phylogenetically brought it closer to the pteridophytes, whereas the presence of pycnoxylic wood has shown its affinity towards the members of Coniferopsida of gymnosperms. The discovery of the heterosporous member *P. scotia* (Walton, 1957), inspired pteridologists to establish their close proximity to the advanced group of pteridophytes. This intermediate form of the extinct genus, *Protopitys*, in between the advanced tracheophytes and the early spermatophytes, was classified under the family Protopityaceae and supposedly show a parallel evolution with the family members of Aneurophytaceae (e.g., *Rellimia, Tetraxylopteris*). Aneurophytales, in terms of the identical pattern of formation of anatomical and reproductive structures, seems to be phylogenetically related to early spermatophytes.

3. Order: Archaeopteridales: According to the presence of unique combinations of vegetative characteristics of gymnosperms and reproductive characteristics of pteridophytes, the 15 fossiliferous members of *Archaeopteris*, found in the Late Devonian deposits of North America, the Russian federation, and other parts of Europe, were classified in the Pityales order (Beck, 1960) under Progymnopsida. The members of *Archaeopteris* are characterized by the presence of large dimorphic fronds (upto 1m) in the same plane, bipinnately compounds, devoid of midrib but the lateral supply of dichotomously branched vasculature ends in each pinnule. Mostly, the sporangia replaced the pinnules of lower pinnae of fertile fronds, sometimes occurring in the adaxial surface of some species.

On the basis of paleobotanical evidences, till the Mid-Nineties, the members of *Archaeopteris* were considered as early forms of seed plants, close to Gymnosperms on the basis of their homosporous habit. However, the discovery of the heterosporous members of *Archaeopteris* (e.g., *A. latifolia, A. macilenta* etc.); Phillips et al., (1972), Bonamo (1975); made them a close ally of an advanced group of pteridophytes, which has an evolutionary link with seed plants. The fossiliferous members: *Actinoxylon, Eddya, Sidirella* etc. were considered as different genera under the Archaeopteridaceae, but the critical observations of Scheckler (1976, 1999), tend to reconcile the fossil remnants as different stages of development of the sole genus, *Archaeopteris*. This prompted him to consider it under the family: Archaeopteridaceae, under the order of Archaeopteridales.

Chapter 9

EVOLUTION OF ANCIENT SPERMATOPHYTES, "GYM-NO-SPERMS": PEEPING THROUGH THE TEMPORAL MIRROR

Opposing the idea of angiospermy, the scientist Robert Brown (1827) proposed the concept of gymnospermy, and that was the beginning of the conflict of perceptions. The conflict was between the supporters of the monophyletic origin versus the supporters of the polyphyletic origin of gymnosperms. The perception of the polyphyletic origin prevailed due to the availability of ample fossil evidence.

Among the spermatophytes, the gymnosperms are segregated from the angiosperms into separate artificial groups, mainly based on their seed characteristics. Irrespective of the perceptional bickering of the phylogenetic origin of gymnosperms and the composite segregatory approaches of classification, ultimately epigraphic evidence ascertains that the so-called "naked seeded plants" (gymnosperms) dominated the Earth in the Devonian and Silurian Periods. The age-dating of "closed seeded plants" (angiosperms) point to its origin in the boundary of the Jurassic/Triassic Period.

Hence, paleobotanically, gymnosperms are more primitive than angiosperms. The reproductive body of gymnosperms, popularly known as cones or strobilus, it is unisexual in nature (either a male or female cone), arranged with a whorl of sporophylls. Critical observation of the female cone revealed that carpels of the female cone, as well as the female flower, never fused to form ovaries (which helps to form "fruit" in higher spermatophytes). The megasporangium as well as the ovule, are borne by megasporophylls in an exposed condition, which has a structural resemblance with pteridophytes. Thus, gymnosperms such a spermatophytes, don't bear any "*true fruit*" (devoid of the closed-chamber of the ovary), but have an exposed seed. As the ovule inside the fruit is found biologically advantageous, in terms of the evolutionary progress, gymnosperms were logically placed as a transitional form, between pteridophytes and angiosperms.

Until the end of the 19th Century, spore-bearing organs, known as sporangia, were borne by leaf-like structures of vascular plants, called sporophylls. According to the old notions of plant anatomists in the 70s and 80s, if ovules are functionally alike to sporangia, ovules should be borne by some sporophyll-like leaves. Even this perception was further strategically ascertained by Von Goethe (1790), in his "Metamorphosis of Plants," where he has considered that all lateral appendages of the vascular spermatophytic plants (e.g., leaves, sepals, petals, bracts, androecium, gynoecium etc.) are leaf-like structures modified from leaves, gonophylls, etc.

Von Goethe's (1790) postulation narrowed down the key components of any vascular spermatophytes into three groups: roots, stem, and leaves. Later, it was found that a distinct group of spermatophytes (members belonging to Coniferales and Taxales) of the gymnosperms do not corroborate to the old concepts such as Goethe's postulation. Critical studies of some fossilized pteridophytes devoid of leaves and with the presence of sporangium on its stem-like appendages, make the old perception absolutely unacceptable. Rather, the observation of the presence of reproductive organs on leaf-like appendages on a group of plants and the presence of reproductive organs on stem-like organs of a group plants could support the pteridophytic (sporangial formation) and a gymnospermous (seed formation) group of plants.

Challenging the classical spermatophyte-centric approach, with the three part morphological identity (root, stem, and leaf) of vascular spermatophytes prevailing, a culminating hypothesis was promulgated on the basis of the collective ideas and interpretations of scientists (Thomas, 1932; Lam, 1948; Zimmerman, 1959; Riedel, 1978; Rieppel, 1988), popularly recognized as the supporters of the "New Morphology," of the late 90s and early 2000s, also known as the "Telome Hypothesis." According to this hypothesis, the approaches of evolutionary analysis of morphological formations were stressed through a bottom-up analysis for the "phylogenetic trees" approach followed by the classical top-down model of "morphogenetic analysis" of the early 80s. The Telome Hypothesis tries to explain the origin, development, and diversification of different forms of complex organs of plants from the simple forms of multicellular building blocks, known as "Telome." These were formed in the early plants of the Devonian Period. It is an earnest approach to explaining the morphogenesis of a higher group of plants. Telome is nothing but a single, vascular, cylindrical axis (protostele), growing from the apical meristem tissues of the plants. Tellomes are categorically divided into two groups:

a. vegetative tellomes or phylloids and
b. reproductive tellomes or sporangia.

Zimmerman (1959), the flag bearer of the "New Morphology" movement, proposed an extended part of telome at the junction of the vegetative and reproductive telome, known as the "mesome." Perceptually telomes and mesomes are developmental differentiations of the same formations. The Telome theory gained popularity among the plant morphologists, paleobotanists, and molecular biologists and have a huge impact in understanding the origin of the similar types of organ systems in Pteridophytes (e.g., Lycopodiales, Filicales etc.) and Gymnosperms. While the Telome theory is found very effective in explaining the morphogenesis of vascular cryptogams and early vascular phanerogams or primitive spermatophytes (gymnosperms), it has not been found suitable in the morphogenesis of angiosperms (the advanced spermatophytes).

As per the Telome theory, the concept of homology is considered as an organ system that explain the morphogenesis of all vascular plants but it has failed to do so in its angiosperms; the critics of Telome Hypothesis, (Niklas, 1997; Stein, 1998) argued that as a theory it could describe anything but did not have enough predictive power to explain any complex forms of morphogenesis (like complex organs such as the underground stems, phylloclades, pneumatophores, spines, and etc.), in advanced vascular plants.

The pattern of straight-forward gymnospermous pollination: like the transfer of pollen grains in the micropylar region of the ovule, is to some extent similar to the pteridophytes, where vectors play a key role in the pollination of angiosperm flowers. However, if we look at the hypothesis on the co-evolution of insects and angiosperms, it would be convincible, that the group of insects evolved much later than the geological ages of the origin and dominance of gymnosperms. Unique fertilization patterns were witnessed in gymnosperms, like motile spermatozoids, that play a key role in the primitive groups of gymnosperms (like the Cycadeoid group) similar to the lower groups of plants like algae, bryophytes, pteridophytes, and etc. Whereas, the non-motile gametes play key roles in an advanced group of gymnosperms (like Gnetalian groups) and has a noticeable resemblance to more complex forms of the transducer: angiosperms. The fertilization patterns of gymnosperms have also left a trail of origin and diversification of the progress of evolution in the plant kingdom, supposedly starting from unicellular simple forms to multicellular complex forms in the due course of time.

Two different morphological forms of leaves were noticed in pteridophytes, known as "microphylls," characterized by the leaves with unbranched central veins and the leaf trace has not been associated with the leaf gap. This type of microphyllous leaves were noticed in the species of *Isoetes*, *Lycopodium,* and *Selaginella*. While "megaphylls" were noticed among the members of *Marattia, Marsilea* and *Pteris* containing a network of veins and the leaf trace associated with leaf gap, likewise, Gymnosperms were noticed with two different morphological leaves: microphylls (found in Coniferales), needle-like narrow leaves with one or few venations and megaphylls (found in Gnetales), broad leaves with multiple veins and veinlets from branched venations.

Scientist Dawson (1859) discovered fossil remnants of *Psilophyton robusticus* from the Upper Silurian deposit of the Goose Bay, Canada, which was further re-discovered taxonomically by Hopping (1956) as *Trimerophyton robustica* (Trimerophytopsida), a fossilized, homosporous pteridophyte having trimerous branching and lateral sterile appendages. 'Telome tussles' (webbed formations which might have given rise to megaphylls) that resemble megaphyllous pteridophytes (e.g., the megaphyllous members of Sphenopsida and the members of Cladoxylales and Coenopteridales of Filicopsida) and progymnosperms, appearing in the Early Devonian Period. Hence, Trimerophytes are considered as a connecting link in evolving megaphyll from microphyll. Though the broader leaves with dichotomous venations (like Ginkgoales) do not reconcile morphologically to any of the forms, contemporary paleobotanical evidence ascertained its microphyllous origin. While, pteridophytes and gymnosperms have both these morphological forms of leaves, the lack of evidence (either structural or paleobotanical) makes it difficult to conclusively determine the microphyllous descent of gymnosperms, originating from the microphyllous stock of pteridophytes (like scale leaves of Lycopodiales that gave rise to middle leaves of Coniferales) or megaphyllous gymnosperms originating from megaphyllous pteridophytes (like broad leaves of Gnetales originated from the branched fronds of Dryopteridaceae).

Plant anatomists noticed a unique form of vasculature: "Solid Protostele," which is comprised of a central core of xylem and phloem and endodermis in its periphery. A segment of xylem was lignified prior to the completion of the elongation phase, considered as "Protoxylem"; whereas the diversified form of lignifications (like- reticulate, scalariform, pitted etc.) took place in the post elongation phase of some xylem, designated as "Mesarch." Based on the relative position of protoxylem with respect to metaxylem, it was categorized under three forms of protostele:

a. Exarch (where protoxylem encircled metaxylem and the development of metaxylem is centripetal),
b. Endarch (where metaxylem encircled protoxylem and the development of metaxylem is centrifugal) and
c. Mesarch (where the development of the outer metaxylem is centrifugal and the development of protoxylem in the center is centripetal).

The "solid protostele" represents a connecting link between fossiliferous, primitive members and others belonging to Filicales (primitive Pteridophytes), simple forms of gymnospermous trunk possessing a solid protostele (Eames, 1936; Sporne, 1991). Paleobotanists outlined a hypothetical gymnosperm mother plant having protostele of the mesarch type of xylem in the center, encircled by the secondary wood at the periphery.

Later, it was noticed that *Heterangium grievii* had an identical resemblance to the hypothetical gymnosperm mother plant, except the presence of parenchyma cells in metaxylem patches. The existence of "Protostele" was evidenced among the primitive fossils of gymnosperms, existing in the Upper Devonian to Upper Carboniferous Period.

Based on the anatomical studies of secondary woods of gymnosperms, two different structural forms were noticed:

a. Pycnoxylic (wood fibers are dense with scanty wood rays) wood bearing gymnosperms of commercial timber values and
b. Manoxylic (sparsely dispersed softwood fibers with parenchyma cells) wood bearing gymnosperms having no commercial values of its timber.

The fossil evidence of early gymnosperms (which have some structural resemblance with pteridophytes) also ensure the presence of the two different wood structures among gymnosperms or progenitors of gymnosperms, as pycnoxylic formations were noticed in the *Lyginopteris* and *Calamopitys*; whereas, *Eristophyton* and *Mesoxylon* were noticed with pycnoxylic wood formations and structural resemblances to the fossil gymnosperm *Mesoxylon,* and an advanced group of living members of Coniferales. This helps to draw a line of continuity of evolution among the early spermatophytes (Frolin, 1951, 1963).

However, the above findings are not so useful in a direct phylogenetic analysis of this wood anatomy, it is an added feature to the symmetrical forms of seeds and morphological variations of leaf forms to attribute in the taxonomic analysis to distinguish a form from the other artificially.

It is amazing to notice that features like the pycnoxylic wood anatomy are integrated within the bilateral symmetry of seed forms and microphyllous leaves in a group of gymnosperms, whereas the other groups have characteristic features like manoxylic wood anatomy, radial symmetry of seed forms and macrophyllous leaves. An assortment of characteristic features among gymnosperms inspired the plant phyllogenists to move on from a monophyletic form of evolution to a di-phyletic evolutionary trail, that ultimately ends in the polyphyletic evolutionary trail of spermatophytes.

In the review of phylogenetical studies on gymnosperms, a group of experts in plant evolution postulated that this group of plants progressed halfway through the evolutionary trail rather than a complete journey. Hence this group of plants should be regarded as a transitional form of evolution in the plant kingdom, to be recognized as "gymnosperms' rather than a distinct taxonomic group of "Gymnospermae" (Arnold, 1947, 1948).

Chapter 10

ORIGIN OF SEED HABIT ON THE PRIMITIVE EARTH: A PERCEPTIONAL OVERVIEW

In the Quaternary Period (current age of time) the visible dominance of vascular seed plants (comprising of gymnosperms and angiosperms) or phanerogams collectively over vascular cryptogams (pteridophytes), nonvascular cryptogams (bryophytes), diversified, unicellular or multicellular transducing algae were established in terms of structural and functional complexities rather than numeric diversities. The contemporary dominance of non-vascular forms of bryophytes over unicellular and multicellular algae were chiefly due to the structural and functional complexities and reproductive patterns (like the appearance of alternation of generations in the life cycle and dominance of gametophytic generations over the sporophytic phase).

Likewise, the dominance of vascular cryptogams (pteridophytes) over non-vascular cryptogams (bryophytes) is mainly due to the structural complexities and reproductive patterns (like the dominance of the sporophytic generation phase over the phase of gametophytic generations in the overall cycle of alternation of generations). In the penultimate phase of evolution of plant life, the dominance of was noticed over vascular cryptogams, partly due to the development of vascular tissues and mainly due to the reproductive efficiency (like viability, dormancy etc.) of naked seeds that were prioritized over spores.

Finally with the dominance of highly diversified groups of transducers, known as angiosperms, closed seeded plants were established on this blue planet presumably due to the inner structural advantage (exposition and changes in external environmental factors like temperature, moisture etc. are more likely to impact naked seeded plants as opposed to closed seeded plants) and the reproductive efficiency of seeds (like the presence of stored food in endosperms to supply energy, ensure the longevity of the seeds and their viability in the face of changes in temperature, moisture etc. and prolonged dormancy) which provided an edge to this plant group. A comparative review between vascular phanerogams and vascular cryptogams reveals that the reproductive efficiency of seeds over spores was favored in the course of evolution in the plant kingdom.

Cryptogamic reproduction was performed by spores, coming out of parental sporangium, followed by germination to form slender gametophytes and the survival of gametophytes and sexual reproduction was typically dependent upon the levels of moisture and temperature. While the female as well as the male spores remain inside the sporangium and form microgametophytes after settling on the stigma receptor or the micropyle region, sexual reproduction and fertilization, take place inside the integuments; hence small shifts in external environmental factors like moisture, temperature etc. would directly influence the varieties of sexual reproduction. The gymnosperms represent sporophytic generations characterized by a heterosporous condition, which indicates its transitional state between heterosporous cryptogams and seed-bearing angiosperms. The initiation of the evolution of seeds supposedly started with a homosporous form of cryptogams switching to the heterosporous state of phanerogams or seed plants. The presumptive perception about the origin of pollen from angiosperms took place from the microspore while the embryo-sac was produced from the megaspore. The gametophyte-producing smaller spores, which are identical in appearance to the megasporangia of pteridophytes, are supposedly in the homosporous state while the sporangia having a few number of large spores indicate a heterosporous formation of the seed habit.

Most botanists refer to the formation of the archegonium as originating from megaspores which remain inside the megasporangium in two species of *Selaginella*: *Selaginella apus* and *S. rupestris;* the archegonium forms the embryo with cotyledons and roots. So, the morphological, anatomical, palynological and embryological evidence to determine the origin and development of the seed habit of angiosperms or phanerogams from pteridophytes were substantiated by the connecting plant species of pteridophytes: *Selaginella* sp. It was found that two different types of gametophytes (distinguished as two opposite sexes or strains) were produced by the heterosporous pteridophytes, which features as a vestigial phase in the life-cycle of angiosperms.

In support of this hypothesis, the pteridophyte spore went through the phases of development and diversification to develop the phanerogamous seed habit, scientist Benson (1904, 1914) has explained that the seed-like structure of extinct gymnospermous groups of the Carboniferous Period: Cycadofilicales (which has the composite characteristics of Cycadeoidean gymnosperms and ferns), should be treated as a special synangium since all but one sporangium was sterile. It was stated by Benson (1904), that the fertile synangium was encircled by sterile sporangiums and furthermore, the progress towards forming the seed habit of phanerogams lead to the fusion of the encircled sterile sporangiums to form integuments of the early seed structure. Later, critical studies of the foliferous fruit-like structures of Cycadofilicales lead scientist Walton (1953) to reaffirm that the earliest version of sterile sporangiums had sterile branch tips.

Famous palynologist Thomson (1927, 1934) performed comparative studies on pollen spores and seed spores to come to the conclusion that seed plants are heterosporous and their heterosporous state evolved from their homosporous predecessor. Thomson (1927, 1934) further stated that it was in the heterothallic state, rather than heterospory to which the key characteristic features of the seed plants can be attributed.

When we try to visualize the reproductive bodies of pteridophytes and early spermatophytes, a flash of diverse pictures of sporangiums comes to our mind-

i. is it homosporous or heterosporous;
ii. do the micro-sporangium sit on microsporophyll or does the megasporangium sit on the megasporophylls;
iii. Are sporangiums located on the stem (stachyspermous) or on the leaves (phyllospermous).

All these features are apparently isolated issues but the common root of these 3 features is related to the transitional phase of the evolutionary changes of the reproductive units from spore formation to seed formation. The perception of homospory vs. heterospory can be well defined by reviewing the reproductive biology of pteridophytes.

Likewise, a brief study of the lifecycle of gymnosperms can narrate the distinguishing features of the megasporophyll vs. the microsporophyll reproductive body formations. The comparative studies of the habits of *Psilotum* (a primitive member of pteridophytes), revealed that the *Psilotum nudum* is an erect plant with triquetrouses (a modified form of the cylindrical stem), photosynthetic aerial shoots (which is the primary photosynthesizing organ) having rudimentary sporophylls and sporangia (synangium), formed in the axil of the rudimentary sporophyll. In another species, *P. complanatum*, the photosynthesizing aerial shoot is pendulous, and bilaterally symmetrical[23]; rudimentary sporophylls possess sporangium (synangium) in its axes and fossil predecessors of these groups of plants were discovered around 416-443 million years ago (in the Silurian Period) e.g *Cooksonia.* It has also been treated as the progenitor of the early spermatophytes or gymnosperms (Lang, 1937).

The stereotypic differentiation of phyllosporous (leaf-bearing) vs. stachysporous (stem-bearing) sporangium is somewhat an artificial approach. The development of reproductive organs, particularly the "progressive protecting mechanism" of naked seeds (megasporangium) in early spermatophytes (gymnosperm) were related to a set of evolutionary changes like the formation of integuments, cupules, arils, and perianths, and it was further stated that evolution of female cone-like structures, functionally like fruits, was the result of the sterilization of the fertile zones of ovules and their inversion at the same time (Anderson et al., 2007).

[23] The green aerial shoot of *P.complanatum* is bilaterally symmetrical and which is phylogenetically a transitional form between circular stem and bilateral leaves.

According to the fossil evidence of the progenitors of angiosperms, and backtracking extrapolation of the gymnosperm plants, it was concluded that ancient gymnosperms supposedly possessed solid primary wood (where secondary wood might or might not be present) and there hardly were any functional distinctions between the stem and leaves; the reproductive organs like pollens, spores, and naked seeds were borne on the tips of photosynthesizing aerial branches (Anderson et al., 2007). However evolutionary progress was not necessarily witnessed in the gradual development of reproductive body formations, even if it was witnessed in the transitional developments of the vegetative parts of the plant bodies. The vascular supplies and meshwork of veins were absent or found in rudimentary forms in thallophytes and early amphibian tracheophytes like gymnosperms. While the presence of microphyllous leaves of early tracheophytes (bryophytes) are rarely found, the simple form of venations is noticeable in advanced group of tracheophytes (pteridophytes), thus enabling better observations of supplying of ingredients for maximizing the photosynthesizing and respiratory potential of plants.

The gradual evolutionary changes of phylloid or leaf-like structures were witnessed in the thalloid chlorophyllous (algae) form of plants with no leaves and the primitive group of tracheophytes (bryophytes) that had microphylls with poorly formed or no venations, and vascular supplies to the advanced group of tracheophytes. Pteridophytes transforming from the state of microphyll type of leaves with preliminary venations and vascular supplies to the megaphyll form of leaves with conspicuous venations and vascular supplies bearing the early spermatophytes (gymnosperms) and further diversification of the megaphyll form of leaves with well-developed mesh-work of venations and conspicuous vascular supplies were finally observed in the advance group of spermatophytes (angiosperms). These structural change of events from the microphyllous form to the megaphyllous form concomitantly with vascular supplies and the gradual change of spore bearing structures to seed bearing structures, points towards an evolution of energy efficient biochemical simulative mechanisms to maximize the photosynthesizing and respiratory efficiency, 2-way energy trafficking (like the supply of energy procuring ingredients and the supply of energy to different parts of body) and better energy metabolism, energy buffering etc. for better reproductive performances (as flowering and fruiting of transducers requires huge quantities of energy), maintaining the bio-geo-chemical cycles in the biosphere (e.g., carbon sequestration), and maintaining the equilibrium state of gaseous components in the Earth's atmosphere (e.g., 21% of oxygen level) etc.

In support of the "evolution of seed habit of phanerogams from cryptogamous spore Hypothesis", Scientist Halle (1935) has construed that the terminal sporangium of the pteridophyte group, "Psilophytales," is the progenitor of the cupulate, fossiliferous seed of primitive phanerogams, evolved during the early Carboniferous Period. In the "The structure and evolution of primitive Seed-plants," scientist Sporne (1991) observed that the gymnosperms could be distinguished from pteridophytes as they possess an advanced form of reproductive units, “seeds,” of two different forms, radially symmetrical and bilaterally symmetrical, in the Cycadopsida and Coniferopsida groups respectively. In keeping with the “Diphylatic evolutionary trail of gymnosperms,” Sporne (1991) postulated the “evolution of seed habit” did not appear on the same evolutionary trail, but rather appeared in these two different groups separately.

Chapter 11

PREDOMINANCE AND EVANESCENCE OF "GYM-NO-SPERMS" AND THE DAWN OF "AN-GIO-SPERMS" IN THE JURASSIC GARDEN OF DINOSAURS

While it is a matter of controversy whether gym-no-sperms should be treated as a distinct taxonomic group, like the "Gymnospermae," or just an artificial group of plants (literally, a state of the polyphyletic group) that lost its evolutionary trail and gave up its journey through the evolution destined to reach Spermatophyta.

According to the classical framework of systematic classification of plants (Engler, 1926; Chamberlain, 1935; Pilger and Melchior, 1954; Mukherji and Gangulay, 1964), Gymnospermae were broadly categorized as Cycadophyta (Cycadopsida) and Coniferophyta (Coniferopsida). This is based on wood-structure, the structural form of leaves, and the symmetry of seeds; these three factors should be treated as the "characteristic trinity." The group Cycadopsida was further categorized into four orders:

a. Pteridospermales,
b. Bennettitales,
c. Pentoxylales and
d. Cycadales.

whereas the second group, Coniferopsida, was further divided into four orders:

a. Cordaitales,
b. Coniferales,
c. Taxales and
d. Ginkgoales.

Apart from these two groups (Cycadopsida and Coniferopsida), scientists found a third group: Gnetopsida, having the solitary order: Gnetales, with apparently no fossil evidence. Among three groups, the plants, belonging to Cycadophyta were comparatively smaller, unbranched trees, having pinnately compound leaves, stems with scanty type of Manoxylic wood, wide pith, thicker cortex region, possessing the simple type of cones containing seeds with radial symmetry. Conversely, the plants belonging to the Coniferophyta group were characterized by large trees, profusely branched, simple leaves (either needle-like or fan-like), stems with the dense type of Pycnoxylic wood, narrow pith, thin cortex region, simple male cones but compound female cones having seeds of bilateral symmetry. There is always an exception, but the 'characteristic trinity' was used to ascertain the di-phyletic origin of gymnosperms (the origin of two groups of gymnosperms from the pteridophyte stock possessing the characteristics of a diverse group of primitive pteridophytes). Determining the geological age of different fossil gymnosperms was performed from time-to-time to determine whether they evolved in the same time period, so that they could help hypothesize the monophyletic origin of gymnosperms. The age dating process on the fossil members of Pteridospermales (the leading representatives of Cycadopsida) and Cordaitales (the leading representatives of Coniferopsida) ascertained that Cycadopsida and Coniferopsida evolved at the same time (between the Lower Carboniferous and Upper Devonian Periods), which points towards the di-philatic origin of gymnosperms since the age-dating results do not help establish a successor–predecessor relationships between the two groups. The distribution of gymnosperms in the geological time-scale is presented in Figure 11.

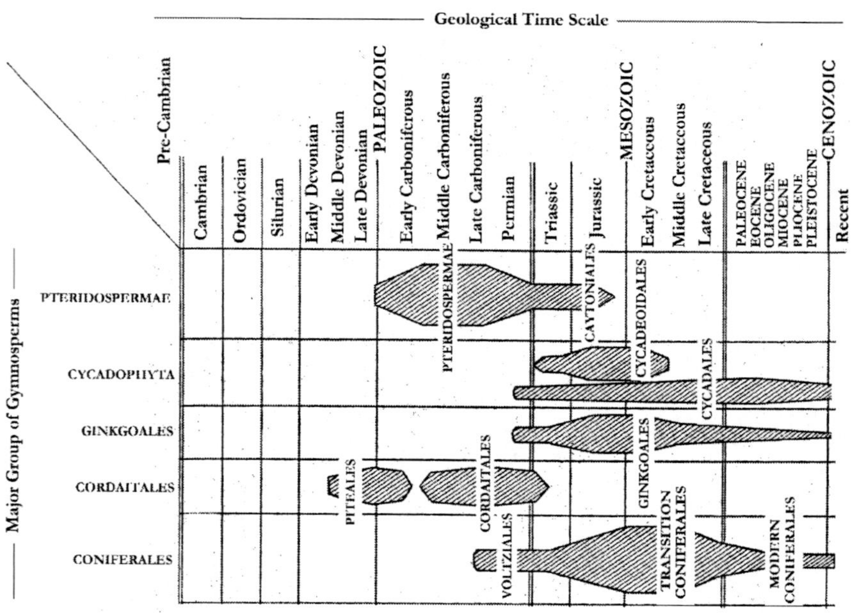

Figure 11. Evolution and diversification of Gymnosperms in the geological periods of time on the Earth (Mukherji and Ganguly, 1964).

The four orders belonging to Cycadopsida are as follows-

a. Cycadofilicales (Pterdospemales): The paleobotanical evidence from Antarctica, Australia, Central Europe, North and South America, Southern parts of Africa and peninsular India, ascertained that this group of gymnosperms appeared in Upper Devonian Period, dominated during the Carboniferous Period, started declining in the Permian Epoch, but existed until the Jurassic Period. Scientist Grand'Eury (1877) discovered a number of fossil members, assigned to the Myeloxylon genus, belonging to the Cycadofilicales order.

Renowned paleobotanist, Williamson (1887) recommended this group as a connecting link between cycads and ferns. Later Oliver and Scott (1903) considered this group as "Seed ferns" or Pteridospermae" and this was recognized as the "Pteridospermales." The seed ferns are characterized by the presence of megaphyllous leaves, well-oriented protostele exarch or mesarch formations with leaf traces, occasionally polystelic, manoxylic wood, fern-like leaves, occasionally pinnate, absence of a vegetative frond, no conspicuous distinguishing characters between male and female sporophylls, no cone formation, and the fronds (megasporophylls) bore primitive seeds (matured ovules) with spacious pollen chambers and a well developed vasculature. This "seed fern" groups have further been categorized into seven families:

1) Lyginopteridaceae: (fossil remnants) e.g., Stem: *Lyginopteris*, *Heterangium*; fronds: Seed bearing organ: *Eurystoma;* Pollen bearing organ: *Crossotheca.*
2) Medullosaceae: (fossil remnants) e.g., Stem: *Medullosa*; Fronds: *Neuropteris*; Seed bearing organ: *Polypterospermum*, Pollen bearing organ: *Aulacottheca.*
3) Calamopityaceae: (fossil remnants) e.g., Stem: *Calamopitys.*
4) Glossopteridaceae: (fossil remnants) e.g., Stem: *Vertebraria*; Fronds: *Glossopteris*, *Gangampoteris*; Reproductive organs: *Lanceolatum*, *Pluma.*
5) Peltaspermaceae: (fossil remnants) e.g., Frond: *Lepidopteris*; Seed bearing organ: *Peltaspermum*; Pollen bearing organ: *Antevsia.*
6) Corystospermaceae: (fossil remnants) e.g., Frond: *Xylopteris*; seed-bearing organ: *Spermatocodon*; Pollen bearing organ: *Pterorachis*. and
7) Caytoniaceae: (fossil remnants) e.g., Leaves: *Sagenopteris*; Seed bearing organ; *Caytonia*; Pollen bearing organ: *Caytonanthus.*

Out of these 7 families, the first 3 families existed in the Paleozoic Era whereas the remaining 4 families appeared in the late Paleozoic Era and existed till the Mesozoic Era.

b. Bennettitales (Cycadeoidales): The paleobotanical evidence from United States, Mexico, Europe, and India, ensured that the members of the group Bennettitales

(Cycadeoidales) arrived on Earth during the Triassic Period, became dominant during the Jurassic Period, and declined during the Cretaceous age.

Krausel (1949) discovered *Westersheimia* from a Triassic deposit of Austria, possessing ovules and inter seminal scales possessed by the pinnate segments of megasporophylls, which establish its link with primitive pteridosperms. Presumably, it originated from a pteridospermous stock and shows a parallel evolution with the Cycadales stock.

Due to the absence of visually distinguishing characteristics, the fossil remnants of the leaves of Bennetttitales in the Mesozoic Era have mistakenly been identified as leaves of the primitive Cycads; an abundance of the members of the Cycadales group in the Mesozoic Era is the reason why it is also called the "Age of Cycads." Until the beginning of the 90s, there was no clear differentiation between the primitive Cycads and the fossil members of Bennettitales, however, Thomas and Bancroft (1913) distinguished these groups by studying their leaf anatomy. They noticed the presence of "Syndetocheilic stomata" (where the formation of guard cells and subsidiary cells, originated from a single mother cell) in the leaves of Bennetitales, while the leaves of Cycadales possesses "Haplochelic stomata" (where formation of two guard cells originated from single mother cells and the origin of two subsidiary cells is the result of the modification of nearby epidermal cells). Hence, in the late Nineties, the scientists Thomas and Bancroft (1913), Florin (1933, 1936, 1963) and Sporne (1991), came to the conclusion that it is less likely that the member of Cycadales, as opposed to the Bennettiales, dominated the Mesozoic Era.

In 1870 the first the fossiliferous member, *Williamsonia* (a fossil form of leaf and reproductive organ of the Jurassic age), was reported from Yorkshire (Williamson, 1870). In 1932, eminent paleobotanist, B. Sahni, reconstructed the complete fossil tree of Bennettitales, known as *Williamsonia sewardiana* from Upper Gondwana. The Benettiatales were characterized by the presence of cycad-like xerophyllous leaves, endarch protoxylem, mesarch leaf traces, hermaphrodite or unisexual flowers, stalked ovules mixed with inter seminal scales, pollens in microsporangia inside the capsule and dicotyledonous seeds, etc. This order was further categorized under three families:

1) Williamsoniaceae: (fossil remnants) e.g., Whole plant: *Willamsonia*; Fronds: *Ptilophyllum*, Seed bearing organ: *Benneticarpus*; and the Pollen bearing organ: *Bennettistemon*.
2) Wielandiellaceae: (fossil remnants) e.g., Fronds: *Wielandiella*, *Williamsoniella*; and
3) Cycadeoidaceae: (fossil remnants) e.g., Male and Female strobilus: *Cycadeoidea*.

c. Pentoxylales: It is an enigmatic group of gymnosperms, discovered by Sahni (1948). The paleobotanical evidence, collected from a North-eastern part of India and New Zealand, suggests that it appeared in the Jurassic age (Sahni, 1948; Harris, 1962). The fossil plants are most likely small tree like bushes, characterized by polystelic stems, pycnoxylic wood, reproductive structures of stachyosporous (arranged on the stems), and lanceolate leaves arranged spirally on the stem.

Male reproductive organs comprised of a whorl of sprangiophores centripetally arranged on a disc, female organs had a morphological resemblance to the stalked mulberry bearing sessile seeds covered by a layer of integument. The presence of polystelic stems in Pentoxylales has shown its structural resemblance to the Paleozoic group of gymnosperms; Cycadofilicales (in the family of Medullosaceae). The arrangements of sporangiophores and ovules were found to be morphologically similar to Bennttitales but anatomical findings and the absence of inter seminal scales, distinguish them from Bennetitales.

Likewise, Pentoxylales have a partial anatomical resemblance to Cycadales where the leaf trace vasculature, and syndetochelic stomatal formation is concerned. The paleobotanists Lam (1952), Vishnu-Mittre (1953) and Sporne (1991), observed that the members of this group had a partial resemblance to the members of Cycadofilicales, Bennettiales, and Cycadales and had some unique characteristics, to consider this group phylogenetically related to those groups. Pentoxylaceae is the solitary family belonging to the Pentoxylales. e.g., Stems: *Pentoxylon*, *Nipanoxylon*; Seed bearing organs: *Carnoconites*; Pollen bearing organ: *Sahnia*.

d. Cycadales: The maiden appearance of this group of gymnosperms took place in the Upper Triassic Period, gradually dominated throughout the world in the Mesozoic Era, and started with only a Cycadaceae family members surviving.

Presently, the living successors (9 genera, comprising of 5 Oriental and 4 western genera and 75 species belonging to Cycadaceae) of this primitive group were restricted to the tropical and subtropical regions. Out of the 5 Oriental genera, *Bowenia* and *Macrozamia* are endemic to Australia; *Encephalator*s and *Stangeria* are restricted to South Africa, whereas *Cycas* is pan-endemic in its distribution from India, China, Japan, Australia and Polynesia and disjunctively distributed in Madagascar.

On the other hand, *Ceratozamia* and *Dioon* are confined to Mexico; *Microcycas* is endemic in Cuba; *Zammia*, one of the 4 Western genera, is restricted to Cuba and Mexico. According to the scientist Sporne (1991), the members of Cycadales, the dominant plant group in the Carboniferous Period, and their consequent group of plants, occupying around 0.4 million square miles, which is approximately 8% of the Earth's surface. This group of plants is characterized by unbranched (sometimes branched adventitiously) woody dioecious plants with manoxylic wood having co-axial vascular cylinders and its true vessels being absent; large pinnate leaves with a 'diploxylic' leaf-trace; terminal cones; scaly microsporophylls bearing sporangia in groups at the adaxial surface; mutiflagellated spermatozoids; megasporophylls bearing orthotropous ovules attached to the suspensor; and bearing dicotyledonous seeds. Cycadales were categorized into two families:

1) Nilssoniaceae: (fossil remnants) e.g., Leaves: *Nilssonia*; Seed bearing organs: *Beania*; Pollen bearing organ: *Androstrobus*.
2) Cycadaceae: e.g., Leaves: *Palaeocycas* (fossil remnants); *Cycas, Microcycas, Encephalartos, Stangeria, Bowenia, Microzamia, Zamia* (living) etc.

The presence of two vascular bundles, a hypothetical fusion of two integuments, surrounding the ovule of the members of Medullosaceae (Pteridospermales), help conjecture the formation of integuments, surrounding the Cycadoid ovule. Based on this analytical study on the primitive morpho-anatomical characteristic features, scientists Stopes (1905), Arnold (1953), Mamay (1976) and Sporne (1991), consider this group as descendants of the ancient Pteridospermae, evolved in the Palaeozoic Era and Cycadales being distantly related to the four orders belonging to Conferopsida (Cordaitales, Coniferales, Taxales, and Ginkgoales), characterized by pycnoxylic wood, seed forms with bilateral symmetry and microphyllous features.

e. Cordaitales: Paleobotanists revealed that the members of Cordaitales evolved in the Devonian-Silurian Period, diversified in the Upper Carboniferous Period, started declining in the Permian period and disappeared in the Lower Triassic Period. The fossil members, belonging to this group were reported from Australia, parts of Africa, South America, North America, parts of Europe (UK, France, Russia) and South Asia (China and India). During the Carboniferous Period, these groups of plants formed forests with around 30 m tall trees with a crown like branching pattern upwards. They have also been characterized by the spirally arranged grass-shaped or paddle-shaped leaves (100cm x 15 cm) on the trunk (Florin, 1936).

Based on the presence of ovules on the stem instead of on the leaf, Sahni (1948) classified this as the "Stachyospermae" group. Drawing a direct evolutionary relationship between Cordaitales and Pteridospermales is most unlikely, but the analogical resemblance of the seeds of Cordaitales with Medullosaceae (Seed Ferns) points towards an evolutionary trail of bilateral symmetry of seed habits from the radial symmetry of seeds. While the presence of pycnoxylic wood, bilateral seed symmetry, and microphyllous habit classify them under the Coniferopsida group, the presence of a pycnoxylic woody stem has pushed the phylogeneticists to consider this group as a connecting link between the

Cycadopsida and Conferopsida groups. Popularly, these groups of plants were recognized as "primitive conifers." This group has further been categorized into the following families:

1) Eristophytaceae: (fossil remnants) e.g., Stem: *Endoxylon, Eristophyton.*
2) Cordaitaceae: (fossil remnants) e.g., Stem: *Mesoxylon*, Leaves: *Cordaites*, Cone: *Cordaianthus*; Seeds: *Cardiocarpus.*
3) Poroxylaceae: (fossil remnants) e.g., Stem: *Poroxylon*, Seeds: *Rhadospermum.*

f. Coniferales: Palaeobotanical evidence ascertains that ancient Conifers belonging to the Coniferales group, appeared during late Carboniferous and early Permian Period, which points to its parallel evolution with Cordaitales; as the ancient predecessors of this group appeared when the members of Cordaitales already existed on the primitive the Earth (Florin, 1951). The primitive members of these group, comprised of 39 genera and 490 species (Chamberlain, 1935), flourished during the Cretaceous age and were primarily reported to be from the Northern hemisphere (Europe and North America) and the temperate regions of East and South-east Asia. The members of this primitive group were characterized by stout and branched woody plants with two different shoots (long and short), pycnoxylic woods, resin canals in needle-like leaves, a well-formed female cone, ovuliferous scales and unattached bract scales, winged pollen, no clear distinction of corpus and tunica in the mother cell of the shoot apex, and a long lasting free-nuclear form during embryogenesis. The primitive members of Coniferales were categorized into three families:

1) Lebachiaceae: (fossil remnants) e.g., *Ernestiodendron, Lebachia* etc.
2) Voltziaceae: (fossil remnants) e.g., *Pseudovoltzia, Voltziopsis* etc.
3) Palissyaceae: *Palissya, Stachyotaxus* etc.

According to the fossil evidences, the members of Lebachiaceae appeared in the Late Carboniferous Period, stayed for a short span of time and disappeared during the Upper Permian Period, succeeded by the members of Voltziaceae, which existed till the Triassic Period. The members of Palissyaceae existed for a very short span of time during the Mesozoic (Triassic and Jurrasic Periods) Era. A comparison of the female cones of Lebachiaceae and Voltziaceae members reveal that the fertile dwarf shoots of the latter group of the family show a conspicuous dorsiventral arrangement. The structural analysis of the modern group of Conifers could not establish a connecting link between the ancient group of Voltziaceae and the living modern conifer successors, however, the ancient Lebachiaceae group of the late Carboniferous Period bears a close resemblance to the living group of Cephalotaxaceae.

On the basis of the arrangement of ovules on the fertile dwarf shoot, the ancient Palissyaceae group (e.g., *Stachyotaxus*) is found to be a relative of the modern group of Conifers belonging to the Cephalotaxaceae.

The living descendants (around 550-570 species, under ca. 52 genera) of Coniferales (Florin, 1963; Sporne, 1991) were categorized under the following families:

1) Pinaceae: (living form) e.g., *Abies, Tsuga, Pinus, Larix* etc.
2) Taxodiaceae: (living form) e.g., *Cryptomeria, Sequia, Metasequoia* etc.
3) Cupressaceae: (living form) e.g., *Cupressus, Thuja, Juniperus* etc.
4) Podocarpaceae: (living form) e.g., *Phyllocladus, Podocarpus* etc.
5) Cephalotaxaceae: (living form) e.g., *Cephalotaxus*.
6) Araucariaceae: (living form) e.g., *Agathis* and *Araucaria*.

On the basis of comparative analysis of primitive morphological and anatomical characteristics of the living successors of Coniferales, the members belonging to the family of Podocarpaceae are considered as primitive and the members belonging to the families of Cupressaceae and Taxodiaceae are considered as advanced, whereas the members belonging to the families of Araucariaceae, Cephalotaxaceae, and Pinaceae were found to be somewhere in between.

g. Taxales: Till the first half of the Nineties, the members belonging to the Taxaceae family were classified under the Coniferales order. Based on the arrangement of the solitary, terminal ovule on the fertile branch of fertile fossil representatives of Taxaceae, *Taxus jurassica* of the Jurassic deposit and *Palaeotaxus* in Triassic rock deposit, Sahni (1920) classified the genera *Taxus, Torreya* and *Cephalotaxus* under the Taxales group.

The fossil remnants of *Torreya* were found in the deposits of the Middle Jurassic Period. Florin (1951, 1958, 1963) has considered that the terminal solitary ovule of some members of Podocarpaceae is formed from the reduction of a number of ovules has categorized *Cephalotaxus* as a true conifer belonging to the Coniferales order. The Taxales group comprises of one fossil genus (*Palaeotaxus*) and five living genera (*Amentotaxus, Austrotaxus, Pseudotaxus, Taxus,* and *Torreya*) consisting of around 20 species distributed over Europe, North America and parts of South and South East Asia, forming the Taxaceae family. This group of gymnosperms was characterized by branched trees with spirally arranged linear leaves, pycnoxylic wood, the absence of resin canals in wood and leaves, a solitary terminal ovule, a terminal fertile dwarf shoot, the presence of peltate microsporangiophores, and dicotyledonous seeds. The presence of peltate microsporangiophore is a common link between the pteridophytes (*Equisetum*) and the Taxads (Florin, 1963; Anderson et al., 2007). According to Nakai (1938), the sporangiophores of Taxads end in a synangium of 3-4 fused sporangiums, which is a reminiscent reproductive organ of pteridophytes (*Psilotum nudum*). Based on the non-distinctive characteristics between the fertile and sterile appendages of the cones of the members of Taxales and *Cordaianthus penjonii*, Florin (1948) opined that Taxales evolved parallelly to the Cordaitales group of gymnosperms.

h. Ginkgoales: The members of Ginkgoales appeared on the ancient Earth in the Permian Period, flourished during the Lower Jurassic Period, and started declining in the Upper Jurassic Period. Once, this group of gymnosperms was distributed across the globe of the ancient Earth, but today there exists only a solitary living representative, *Ginkgo biloba,* which was found in the inaccessible parts of Western China and appeared around 200 million years ago during the Triassic age (Crane, 2013). The earliest fossil representatives of the *Cheirocladus longicherius* group, *Trichopitys heteromorpha,* was discovered from the deposits of the Lower Permian Period (Florin, 1949; Mukherji and Ganguly, 1964; Naugolmykh, 2006). The members of the Ginkgoales were subdivided into four families:

1) Trichopityaceae: (Fossil remnants) e.g., *Trchopitys*.
2) Ginkgoaceae: (Fossil remnants) e.g., *Arctobaiera, Baiera, Eretmophyllum, Ginkgoites, Spenobaiera, Windwardia*; (living successor) *Ginkgo*.
3) Psymophyllaceae: (Fossil remnants) e.g., *Psygmophyllum, Psygmophyllodendron.*
4) Cheirocladaceae: (Fossil remnants) e.g., *Cheirocladus*.

The features like the distinctive anatomical characteristics and foliar dimorphisms have helped distinguish the fossil genus *Eretmophyllum* (having multi-veined fan-like leaves) from *Arctobaiera* and *Windwardia* (having strap-like leaves with lesser number of veins). The *Trichopitys* fossil member and the living fossil *Ginkgo biloba* can be distinguished on the basis of their characteristics like their short and long shoots. The presence of morphological features like the petiolar structure of leaves, link the fossil members of *Ginkgoales; Baiera, Ginkgodium; Ginkgoites;* and *Psygmophyllum* to the living successor *Ginkgo biloba*.

The members of this group of plants were characterized by branched trees with dimorphic shoots: long shoots with bi-lobed, fan-shaped, multi-veined leaves with an apical notch and dwarf shoots (also known as spurs) with undivided leaves and few veins; pycnoxylic wood; 2-10 ovules growing on the axil or terminal points of the branches; large sized seeds with two layers: outer fleshy layer and the inner stony layer; catkin like male reproductive organs possessing micro-sporangiophores grown in the axillary position of branches, and each microsporangiospore possessing 2-10 microsporangia; and motile, flagellar antherozoids. *Gingo* and *Pinus* have a strong morpho-anatomical resemblance where wood anatomy, developmental stages of male prothallus and structures of pollen grains are concerned. On the other hand, the developmental phase of pollen grains, motility of antherozoids, and the structure of female gametophytes of *Ginkgo* were found identical to *Cycas*. On the basis of comparative studies of primitive characteristics, the conventional outlook has points out the affinity of Ginkgoales to the Cycadales group. Florin's (1933, 1936, 1949, 1951) conclusive observations on the basis of phylogenetic analysis, have ascertained that the members of Cordaitalis, Taxales and Ginkgoales belong to the parental group "Coniferopsida," and all these groups seemed to have evolved parallelly in the period of the Upper Devonian to the Lower Carboniferous age.

i. Gnetales: Lack of paleobotanical evidence before the Tertiary Period, classifies them as an advanced group of Gymnosperms. The fossils from the pollen of Ephedra were found in the Eocene deposits, whereas some pollen like structures of *Ephedra* and *Welwitschia* were reported to be from the Permian age (Delevoryas, 1962). This order comprises of three families (Sporne, 1991; Pearson, 1995):

    1) Gnetaceae: comprised of the solitary genus *Gnetum* having 34 species;
    2) Welwitschiaceae: contains the single genus *Welwitschia* comprised of a solitary species and
    3) Ephedraceae: the only genus *Ephedra* having 32 species.

With the exception of the solitary species, *Welwitschia mirabilis* (which was found in the Namibian desert), distributed over South–West Africa, most of the Gnetales species were distributed widely across the eastern and western hemispheres. As this unique group of gymnosperms have a very close morphological and anatomical proximity to the group of angiosperms, Gnetales were classically considered as progenitors of angiosperms in the nineties. Even in the age of molecular phylogeny oriented cladistic analysis, *Ephedra* was considered to be the phylogenetic as well as the gymnosperm predecessor of the angiosperm representatives of *Casuarina*. The characteristic features of the members belonging to this group of gymnosperms are as follows: The vessels present in the secondary, pycnoxylic wood; the arrangement of flowers in its strobilus; the presence of perianth in male flowers and some envelopes around its ovules; the presence of lengthy micropylar tube etc.

The presence of vessels in wood was considered as the connecting link between Gnetales and Angiosperms but critical observations ascertained that the Gnetalian vessels are of the "pitted type" whereas the Angiosperm vessels are of the "perforated type" and fossil evidence suggests that some of the ancient angiosperms are "homoxylous" (without vessels in the wood). Hence, the vessels aren't a connecting link. The presence of two vascular bundles in the leaves of *Ephedra,* and the members of Pteridospermales and Ginkgoales helped establish a distal relationship among these groups. To establish the relationship between the ancient groups of gymnosperms, 'Benettitales', and the representative of Gnetales, *Welwitschia*, scientists Takeda (1913) and Rodin (1958) have considered "Syndetochelic stomata" to be the link. Later it was found as a modified form of the haplocheilic stomata, but the cupular disc which gave rise to a whorl of 6 sporangiophores in Welwitschiaceae, show its affinity to the primitive group of gymnosperms, Benettitales (Sporne, 1991).

Presence of successive zonations of xylem and phloem tissues in *Cycas* and *Gnetum* helped discern the affinity of this group to the early group of gymnosperms, Cycadales (Florin, 1933). It was noticed that features like the presence of syndetochelic stomata, and the secondary thickening of some climbing species of angiosperms, happened due to the formation of a number of co-axial cylinders of xylem and phloem from successive cambial mother cells, which is a connecting link between the members of Gnetales and the angiosperm liana (Maheshwari and Vasil, 1961). The female gametophytes of Welwitschiaceae are tetraspores (gave rise to the linear spore tetrads) and these unique features are absent in other gymnosperms, reminiscent of angiosperms.

While ancient spermatophytes or gymnosperms arrived in the Devonian age of the Palaeozoic Era and some of the living successors of this predominant, extinct group of plants survived in their primitive form (as living fossils). Some of the diversified descendants till now (i.e., in the Quaternary Period), have noticeable features which indicate the dominance of primitive gymnosperms in the primitive forests (with 20-30 feet long, slow-growing evergreen trees) on the ancient Earth during the Jurassic age of the Mesozoic Era (Figure α1.), popularly known as the "Mesozoic World" or "Jurassic Gardens", concomitantly known to us as the "Age of Dinosaurs," due to the large dominating group of animals identified as reptiles. The energy flow in the forest ecosystems of the "Mesozoic World" or the "Jurassic Park," cannot be conceptualized without a basic understanding of the prey-predator relationships of dinosaurs (Consumers) and their food plants (Producers). Paleontologists Barrett and Willis (2001) postulated the "Co-evolution Hypothesis" to explain the evolution and diversification of early spermatophytes (gymnosperms) and dinosaurs. Going further Bakker (1978) recommended that "dinosaurs invent flowers." According to Bakker (1978), the dominant group of small and medium-sized, herbivorous dinosaurs (e.g., Sauropods, Stegosaurs, etc.) consumed bulk quantities of gymnosperm trees (e.g., Pinus, Conifer, etc.) and the overfeeding habit of these herbivorous dinosaurs affected the populations of the slow-growing, evergreen trees, and their regeneration, paving the way for fast-growing, deciduous, angiosperms. Reciprocally, the increased supply of a new type of food (primitive angiosperms), helped in the evolution of super-sized, vigorous herbivorous dinosaurs (e.g., Horned dinosaurs and Hadrosaurs) in the Cretacious age (Hummel et al., 2008). In the late nineties and early twenties, the improvised sampling methods of dinosaurs and the examination of plant fossils of the Mesozoic Era failed to validate the "Co-evolution Hypothesis" of Bakker (1978).

Recently the eminent ecologist Leslie (2011) established a specific correlation between the development of protecting tissues, and the size and shape of the cones of the family members of Araucariaceae, to the dominance of the herbivorous members of sauropod dinosaurs, around 300 million years ago during the Jurassic age, chiefly due to the availability of enough foods and favoritism of sauropods towards the new food. If the divergence of angiosperms was considered as the progress of evolution, then it was initiated with the disappearance of a diversified group of gymnosperms in the Mesozoic age and the large, predominant groups of reptiles, dinosaurs, played a cardinal role in this evolutionary transition before disappearing from the Earth due to the Permian mass extinction. The extinction of the dicynodont and therapsid groups of dinosaurs is evidenced by the paleontologic, paleoclimatic and paleopedological evidence, studied in relation to the Permian/Triassic boundary event of mass extinction (Retallack *et al.*, 2003). It has further revealed that the physicomechanical activities related to the dynamics of feeding and body movements of herbivorous dinosaurs played a key role in the evolution of seed plants, comprising of diversified gymnosperms and early angiosperms (Barrett and Willis, 2001). The contemporary geological, paleoclimatic, paleopedological, paleontological, and paleobotanical studies affirm that the giant sauropod, *Alamosaurus,* shared the same habitat as the arborescent dilleniid group of angiosperms, and anatomical studies of the fossiliferous trunk (presence of excessive ray parenchyma) of this tree confirmed that the Dillenid group of plants were better equipped to sustain being trampled by large reptiles (Wheelar and Lehman, 2000).

The trampling and overfeeding of the leafy foliages of *Ginkgo biloba*, tender shoots of cordaites, conifers of the gymnospermous group of plants by duckbill, hadrosaurians, ceratopsians and archosaurian group of herbivorous dinosaurs that inhabited coniferous forests, supposedly opened up the ground for the ecological succession of angiosperms (Fricke and Pearson, 2008; Miller, 2008). The studies on the coprolites of red-sarcotestae indicate that the fruits of the early seed plants were adapted to be viable in the digestive tract of these herbivorous dinosaurs to germinate from the excreta dispersed over a wide area. The investigations of coprolites combined with paleoclimatic and paleopedological investigations have revealed that duckbill dinosaurs fed on the shoots of conifers, the leaves of woody dicotyledonous plants (Miller, 2008) and the leaf fragments of *Pennistemon* (belonging to Monocotyledonae) were also found in the Neocomian (Early Cretaceous) coprolite (Friis et al., 2006). Hummel *et al.*, (2008) and Leslie (2011), have tried to focus on the evolutionary pressure on the structural evolution of seeds of spermatophytes, supposedly induced by the changing food habit of herbivorous dinosaurs.

In his extrapolation, Leslie (2011), suggested that the evolution and diversification of a group of animals like dinosaurs (reptiles), birds, insects, or small mammals (rather than dinosaurs themselves), apply an evolutionary pressure (in the form of new plant food sources) on the whole plants (vegetative parts) and seeds (reproductive parts), ultimately resulting in the invention of defense mechanisms (like the formation of thick barks, thorns, bitter tasting phytochemicals, metal crystals, toxic chemicals, formation of a protective layer around the ovules, etc.), tantamount to an evolutionary response against these pressures. Hence, it appears that the divergence of Mesozoic angiosperms is an anti-herbivorous trait (Zavada and Mentis, 1992; Barrett and Willis, 2001; Tiffney, 2004).

On the other hand, the critics of the "Cretaceous evolution of angiosperms Hypothesis", contemporary paleobotanists, molecular botanists and evolutionary biologists referred to the picture of Kirkland et al. (2002), published in the "Survey notes" of the Utah Geological Survey Publication, United States, apparently a still picture of the impression of a dinosaur's (belonging to the category theropod) tracks. Careful observations of this "impression of dinosaur tracks" would help find a megaphyllous leaf impression, which has an identical resemblance to the leaf (26 cm wide) impression of the living member of the *Nuphar* sp. genus belonging to the Nymphaeaceae, with complete impressions of the petioles (of 18 cm long) and leaf veins. Furthermore, the dinosaur track-line indicates that a particular member of theropods trampled the leaf under its feet on its way from the marshy Lake Dixie to the mud bank and dragged it a few steps to leave a stretch mark of its veins on the fossil site, squashing the soft tissues of the lamina but leaving the shape of the petiole unaffected. A few number of dragged lamina impressions of that "lily" helped the paleobotanists contemplate the morphological resemblance, venation patterns, mechanical rigidity and thickness of the fossil object to the leaves of "Nuphar" (Kirkland et al., 2002).

The cladistics-molecular-phylogenetic tree of the basal angiosperms helped predict that Nymphaeaceae was supposed to evolve during the early Jurassic or the late Triassic period. Kirkland et al. (2002) reported water lily-like fossil forms of leaves of basal angiosperms in the Jurassic deposits of Whitmore Point, Utah, United States, during the fossil excavation process of dinosaurs and other reptiles.

The eminent paleontologist, Andrew Milner retraced the trackways of theropods, sauropods, and biofacies (fossil bones of dinosaurs) across the eastern Riverside drive, adjacent to Lake Dixie (Milner et al., 2004) and noticed that the subterranean rhizomes (supposedly of the arborescent form of pteridophytes) from the same fossil site were embedded in the sedimentary, siltstone lake-bed and stratigraphically found around 1.7 meters above the fossilized "water-lily" impression. This indicates the existence of the fossil predecessor of "water-lilies" (Nymphaeaceae members) during or before the Jurassic Period, and the existence and evolution of primitive angiosperms (members of Nymphaeaceae) and tree forms of pteridophytes, were found overlap in some respects. In this regard, it should be mentioned that the identity of fossil representatives of "*Sanmiguella*" (on the basis of leaf impressions) were erroneously diagnosed by the "Cretaceous school" due to inadequate identifying features, but critical observations on a similar fossilized leaf, petrified from the "Chinle formation" in the Upper Carnian as well as Upper Triassic Period (Cornet, 2002). The paleobotanical evidence was found to be mostly missing, supposedly due to the combined effects of the Triassic-Jurassic mass extinctions, fall-out and poor preservation of angiosperm fossils on the rock surface (Miller, 2008; Milner et al., 2004).

Chapter 12

PALEOBOTANICAL EVIDENCE, PHYLOGENETIC INTERPRETATIONS: COEVOLUTION OF ANGIOSPERMS AND THE ROLE OF ANIMALS IN THE POST-CRETACEOUS REGIME

In order to understand the origin and evolution of the "Keystone transducers", which are the angiosperms in the plant kingdom of the post-Cretaceous regime on Earth, the grey areas related to angiosperm phylogeny were identified here:

i. Protected ovule in closed carpel and dimorphic floral architecture,
ii. Radially symmetrical flowers,
iii. Bilaterally symmetrical flowers and
iv. Fleshy fruits, nuts, and seeds.

The emergence of the advanced reproductive biological features mentioned above acts as a selective, evolutionary pressure on the genetic profile of "angiosperms" to opt for "outcrossing" in nature (Dilcher, 2000). If outcrossing were to be chosen as a process to reap the harvest of "short-term outcomes" of the "pressurized impact of evolution in nature" in the post-Cretaceous regime, then the stereotypic mode of wide dispersal of seeds and gametes in a short time span might not were possible by utilizing water or wind-borne vectors. The evolution, dispersal, vast distribution and colonization of angiosperms in the post-Cretaceous regime on Earth has also pointed towards a "Coevolution of animals" (particularly insects) as an added advantage to achieve the objective of out-crossing of angiosperms in nature to maximize the effects of evolution in a short period of time.

Stebbin's (1950) earlier effort to review the vegetative characteristics (e.g., leaves, stems) of fossils to match their extant genera to draw a phylogenetic synthesis (also recognized as the "Paradigm in angiosperm paleobotany" in 1940), was later considered to be an artificial approach. As artificial identification, as well as the physical matching process was followed by phytogeographical and paleoclimatic investigations of such fossil remnants, which considers those fossils as early predecessors of an extant member rather than an unbiased consideration of some of those species as extinct members. Stebbin's (1974) effort to focus on "The Nature and Origin of Primitive Angiosperms" was primarily based on comparative structural (morphological and anatomical) analysis of living angiosperms rather than a paleobotanical review of the fossil remnants (which are either megafossils or microfossils). Instead of the 1940's coarse line of analysis, "Paradigm in angiosperm paleobotany in 1970" engaged in the consideration of features like-

a. Ultrascopic morphological and anatomical features,
b. Cuticular structures and vein architecture of leaves,
c. Fruit morphology,
d. Seed architecture, etc. of fossil remnants (mesofossils were prioritized over mega or micro-fossils) and extant angiosperms.

And these helped consider most of the primitive angiosperms of the Cretaceous and Tertiary Periods, as extinct, with a few phylogenetic extant predecessors (Dilcher, 1974; Hickey, 1973; Roth and Dilcher, 1979). This new shift of paradigm (70's in comparison to 40's) of paleobotanical review of the origin of angiosperms, dealt with an intensive review of character-based data, backed up by cladistic analyses rather than artificial floristic approaches (stressed on phytogeographical and paleoclimatic reviews).

"Paradigm in angiosperm paleobotany in 1970" is a unique start, that paved the way to make a worldwide database (e.g., a wide collection of leaf-vein architecture slides in different paleobotanical labs). It provided an opportunity to analyze a diversified database (having millions of data on particular characteristics of extant and extinct members) either to establish the evolutionary bridge between them or discard the established and biased evolutionary relationships, based on the results of cladistics analyses.

The key feature of plants like encased ovules, as well as the closed carpel of angiosperms, were regarded as the main feature to segregate angiosperms from the rest of the plant groups in the plant kingdom. Apparently, the function of the closed carpel is to ensure the protection of ovules to optimize seed formation from a fertilized egg and minimize the chance of herbivorous predation, pest infestation etc., i.e., to set up the mechanical barrier. Factors like a high rate of divergence of insect groups and increasing diversity rate of herbivorous predators have not made this theory convincible as the chance of breaking this barrier is getting higher in the course of aggression of advanced groups of insects and explosion of a huge number of herbivorous predators. From an evolutionary perspective, the formation of closed carpels should primarily be considered as contrivances of cross-pollination in opposition to self-pollination.

In order to oppose self-pollination in flowering plants, which was considered as the ultimate choice of nature in favor of evolution by Darwin (1853), a few specialized physical, cytogenetical, phytochemical, developmental organogenetic, and biomechanical impediments were developed in flowering plants:

i. Dicliny: For unisexual flowers, dicliny is the ultimate way of fertilization.
ii. Dichogamy: In these bisexual flowering plants the maturation of stamens and carpels occured at different times so self-pollination was not possible. It has two types-
   a. Protogyny: The maturity of carpels is found to be earlier than stamens' (e.g., members belonging to Annonaceae, Magnoliaceae etc.).
   b. Protandry: The maturity of stamens is found to be earlier than carpels' (e.g., members belonging to Asteraceae, Malvaceae etc.)
iii. Heteromorphism: In some bisexual, flowering members the height of the style was noticed to be incompatible with respect to the stamen (called heterostyly), while the anthers of different heights with respect to carpels, were called heteroanthy (e.g., the members of Primulaceae).

The dichogamy and heteromorphism in the anther-carpel relationship in a flower should be considered as a developmental organogenetic impediment in favor of cross-pollination; supposedly controlled by structural and functional homeotic genes at a molecular level (ThieBen and Becker, 2004; Glover, 2007).

iv. Self Sterility: In a number of flowering plants, the carpel of a flower cannot be pollinated by the pollen of the same flower or the same strain due to self-sterility (e.g., Solanaceae members). This unique genetical impediment has prioritized cross-pollination through cytogenetic incompatibility.
v. Herkogamy: In some bisexual flowers, the physical barrier between carpel and anther was found to be so wide that self-pollination is almost impossible without the help of insect pollinators (e.g., Caryophyllaceae and Brassicaceae members). The chance of cross-pollination is much higher than self-pollination in insect driven pollinations.

The above mentioned five categories of impediments against self-pollination as well as contrivances for cross-pollination should be considered as secondary attributes for the evolution of flowering plants as cross-pollination is favoured. Except for all these situations, exudation of some phytochemicals has also been noticed from the opening of the carpel to prevent self-pollination by stopping the germination of the pollen tube inside the ovary (Endress, 1995). The closure of the carpel of a bisexual flower should be considered as a primary attribute to the evolution of angiosperms (Dilcher, 1995).

According to Dilcher (2000), the closed carpels act like a mechanical barrier for most flowering plants in preventing self-pollination. It ensures cross-pollination without any interference of petals, sepals, and stamen. As insect pollinators play an important role in cross-pollination, it is conceivable that flowers as a whole, carpels, and stamens have also evolved to be capable of attracting insect vectors by luring them with vibrant colors, fragrances, nectar, pollen, etc. If the evolution of the closure of a carpel is not considered as an isolated evolutionary incident (like the defense shield for a seed), then it should be considered as a part of the evolution of floral features (Dilcher, 2000).

Coevolution of plants and animals: The successful occurrence of "Coevolution" of plants (particularly angiosperms) and animals (particularly groups of insects, mammals, and avifauna) ascertains the maximization of evolution and diversification of certain groups of plants (e.g., angiosperms) conceivably evolved during the Cretaceous Period, as it enhances the dispersal of seeds to a vast area. Apparently, the incidents of coevolution of plants and animals in the post-Cretaceous Period was enhanced by the evolution and diversification of angiosperms, witnessed in the paleobotanical histories of the last 135 million years, though it was noticed that some insect pollinators helped pollinate a group of naked seeded plants (mainly seed ferns like *Cycadeoidea, Medullosa, Caytonia, Williamsoniella* etc.) during the Paleozoic Era (Dilcher, 2000).

While conventional hypothesis has always portrayed coevolution as a defense response to insect predation pressure on plants, a group of evolutionary biologists and phylogeneticists consider this mechanism as an added reproductive advantage for the angiosperms in favor of evolution, which involves structural changes in flower architecture (e.g., lowers with radial symmetry or flowers with bilateral symmetry). It was observed that in the early Cretaceous Period, the impact of coevolution on flowering plants with radial symmetry was less (as animal-induced gamete exchange possibilities in this group of plants was less likely) than the plants bearing flowers with bilateral symmetry (which supposedly evolved after 60 million years of the evolution of angiosperms). It was the later type of floral arrangements that changed the behavioral patterns of insect groups (Dilcher, 2000). Paleobotanical evidences exhibit a wide array of facts like-

a. Cellular upgradations, structural modifications (in morphological and anatomical complexities),
b. Diversification of habits,
c. Transmigrations and ecological adaptations,
d. Alterations of the dynamic environment,
e. Better reproductive mechanisms (dominance of sexual outcrossing over asexual reproduction),
f. Evidence and interpretations of coevolution etc. to narrate the footprints of the evolution of plants.

The two conflictive evolutionary hypotheses, related to floral evolution are- i. Progressive elaboration by "synorganization" and fixation at a structural level in favor of evolution; ii. Reduction or Simplification of loss of fixation at a structural level, also known as reversals (Endress, 2001, 2010). According to Endress (2011), the evolutionary changes of morphogenetic floral features (from simple to complex) leading to the diversification of angiosperms, are asymmetrical (found a higher rate of progress in one clade than in another clade), where ecological and physiological constraints act as "stabilizing forces" and this trend of "pervasive convergence" attributes a conservative status to the history of floral evolution. Thus, causing a delay in the period of divergence of "radiation of angiosperms" to some extent. According to the phylogenetic analysis on the evolutionary development of flowering plants, different angiosperm phylogenetics (evolutionary biologists), like Donoghue et al., 1998; Endress, 2001; Rudall and Bateman, 2004; Sargent, 2004; Marazzi and Endress, 2008; Preston and Hileman, 2009) observed that "floral symmetry" is an important indication to the visualization of evolutionary progress, where "Monosymmetry" (with a single symmetrical plane) is supposed to be more advantageous over "Asymmetry" (without any symmetrical plane) as pollinators can optimize the pollination as well as maximize the chance of diversification, naturally. The chance of progress of molecular evolution of the pollinators in terms of pollinator-plant specificity is supposed to be expressed at the gene level.

Phylogenetically, the complex level of synorganization was observed in "highly nested clade" like "core eudicots" (that went through a series of adaptive radiations). Whereas, the simplified levels of synorganization or reductive structural evolution was observed among "basal angiosperms." At the molecular level, the development of floral evolution went through combinations of both or predominance of one over the other and that was found in different orders and clades in their modes of diversification. The synorganization of perianth and androecium was witnessed in the highest extent in the monosymmetric flowers in Fabales, monocots (e.g., Orchidaceae), core eudicots (Lamiales) (Endress, 1994; Prenner, 2004; Bello et al., 2010). But a loss of petals (e.g., in the family of Cyperaceae, Rubiaceae, Apiaceae) or entire perianths (e.g., in the Hydatellaceae of ANITA grade) have also been noticed (Rudall et al., 2007; Borg et al., 2008; Nuraliev et al., 2010).

The simple form of synorganization of petals and stamen lead to the formation of smaller petal appendages, supposedly responsible for nectar protection and canalization (Brown and Terry, 1992; Endress, 2010). The simple fixation at the structural level or fusion of stamen with petal tubes to form "Stapets" were found in Asteroids (in Asterales, Gentianales, Solanales, Ericales etc.); whereas, the complex fused form of "Corona tube" was noticed in between the perianth and androecium in some monocots (e.g., Amaryllidaceae. Velloziaceae) and some core eudicots (in the family of Apocynaceae) (Kunze, 2005; Endress, 2010; Sajo et al., 2010). The highest level of synorganization between androecium and gynoecium in floral evolution is the evolution of the "Pollinaria" and it has diverged at least a couple of times, once in monocots (e.g., Orchidaceae) and once in the eudicots, Gentianales (Apocynaceae), throughout the evolutionary history of angiosperms (Endress, 2011). The Pollinaria is comprised of Pollinia (the compact pollen grains) and a translator, glue or clip-like structure, (holding 2-4 pollinia to the body surface of the pollinator) (Fishbien, 2001; Pacini, 2008) to optimize cross-pollination. If the "evolution of Monosymmetry" is the preliminary effort towards floral evolution, "evolution of Pollinaria" is the ultimate stage of floral evolution both of which are directly or indirectly expediting the angiosperm radiation (Rudall and Bateman, 2002; Gorniak et al., 2010). All these forms of structural developments in floral evolution are related to the diversification of the coevolution of plant-pollinator mutualism and the apparent radiation of angiosperms in a short span of time.

As dependence on the diversified, conflictive hypothetical interpretations on origin and evolutions, dispersal and diversification of plants are still noticed. It forces the world of phylogeneticists and evolutionary biologists to hang around in their labyrinths of conjectured truths of the evolution of angiosperms.

Chapter 13

PALEO-ICEHOUSE, PALEO-GREENHOUSE AND HYPOXIA: EVOLUTION OF ANGIOSPERMS THROUGH THE BOTTLENECK OF FIVE MASS-EXTINCTIONS

"To begin to connect the ecology with early radiation of angiosperms, we must understand the ecological starting point for angiosperm evolution."-Field and Arens (2005)

The paleoecological cycle of this blue planet has gone through a diverse form of catastrophic incidents, like glaciation, flood, volcanic eruptions, earthquakes, global warming, and etc. The evolution of this galactic world; with so many catastrophes, but the impact of five mega incidents were found to be so severe that they almost pushed the living organisms at that time to the brink of extinction and forced them to reascend the journey of evolution of life, again and again.

Though there are unequivocal observations on the occurrence of Five mass-extinction events, but differences of opinion are there to adjudge the time of the occurrence of those "Five mega extinctions" according to the interpretations of different geoscientists. The initial mass-extinction event began at the Ordovician-Silurian boundary (DeSLR) of the Paleozoic Era (Kolbert, 2014) between 420-460 million years ago, immediately after the amphibian migration of plants from water to land as well as the evolution of land plants. This mass extinction event wiped out most of the aquatic plants and primitive amphibian land settlers. The reason of this mass extinction event has mostly remained unknown, as the relevant quantitative data, related to the mass extinction event, like the occurrence of a global carbon cycle anomaly, or a global oxygen level depletion, or whether a hypoxic atmosphere was created or not, are missing. Hence Paleo-Icehouse effect has consequently been considered as the first one. Among these major global catastrophes, the first one was recognized as the Paleo-Icehouse during the Upper Devonian and Lower Carboniferous

Periods (DeCARB) and rest of the five were recognized as Paleo-Greenhouse followed by the Five mass-extinctions that are chronologically:

a. Carbon cycle anomaly in the Upper Permian Period (GuCCE);
b. End Permian extinction also recognized as the "Great Dying" (EPE);
c. Triassic-Jurassic boundary carbon cycle event (TrCCE);
d. Early Cretaceous-Late Barremian biogeochemical perturbation (BaCCE) and
e. Kretaceous (Cretaceous)-Tertiary asteroid impact (K-T Event) (McGhee, 1996; Berner and Kothavala, 2001; Huey and Ward, 2005; Knoll et al., 2007; Miller, 2008; Retallack and Jahren, 2008).

Kolbert (2014) mentioned that the DeSLR was the first mega extinction while others preferred to consider the DeCARB as the maiden one. Furthermore, Kolbert (2014) did not recognize GuCCe as a recognized mega-extinction event, rather she considered EPE as the third mega-extinction event followed by the BaCCE or the End–Cretaceous Mega extinction as the last one in her great treatise on the disappearance of the species "Sixth Extinction." However, most of the biological conservationists and geoscientists consider the K-T or Cretaceous-Tertiary mega extinction as the latest one. Regarding the number of mass-extinction occurences, it is a matter of debate whether this blue planet went through "Five Mega extinctions" or "Six Mega extinctions." However, the current rate of disappearance of species in the regime of the Anthropocene is an inconvenient yet undeniable truth.

If we were to inquire about the common issues in all these mass-extinction incidents, it would be noticed that the carbon dioxide, oxygen level, and methane gas in the paleoclimatic atmospheres radically changed at the ground level during these events. This lead to abrupt changes in the global temperature, resulting in the change of the profile of terrestrial biomes, by altering the biodiversity on the subsequent biomes (Looy et al., 1999; Berner and Kothavala, 2001; Huey and Ward, 2005; Retallack et al., 2006). The oxygen content in the paleo-atmosphere during these mass-extinction cataclysms (Miller, 2008; Kolbert, 2014) were presented in Figure 12. Irrespective of the nature of the habitat (whether the species belong to aquatic or terrestrial biomes) or type of organisms (whether it belongs to the plant or animal kingdom), more or less most of the living organisms went through survival challenges of different mass extinctions. Though, there is no quantitative data on the effect of mass extinction on plant groups (e.g., number of species or order or families made extinct with the occurrence of the major mass extinctions), there are well-established facts on the quantitative losses of marine diversity (in the family level) that happened during mega extinctions (Kolbert, 2014). This reveals that the "Lower Cretaceous end-Barremian biogeochemical perturbation" had the highest impact on the loss of biodiversity on Earth. If we consider at least 10 species belonging to each family, it's probable that huge numbers of species were lost forever during this mass extinction event.

Except the Devonian Icehouse effect and the "thermal minima" (when hypoxia was found as the challenging issue for living organisms to survive in a scattered arrangement of terrestrial biomes), only a few of the "thermal maxima" in the Mesozoic and Cainozoic Eras, were found to be the main issues to store-up excessive content of carbon dioxide in the paleo-atmosphere. Meanwhile the volcanic activity driven torment on the sea-floor lead to the releasing of lethal volumes of methane in the ocean (Palfy et al., 2001). The experiments on the simulative model of the "global carbon cycle" appraisal revealed that around 9000 gigatons of carbon dioxide was released in the atmosphere and around 5000 gigatons of methane was released in the ocean during the Central Atlantic Magmatic Province (CAMP) (Beerling and Berner, 2002). This kind of carbon emission in the atmosphere and the ocean had a huge impact on the aquatic and terrestrial biomes. It caused the hypoxia-driven mass-extinction of animals and plant life in those times on the terrestrial biomes (Berner, 1999). The paleoecological, paleoclimatic and evolutionary consequences of each mass extinction was evidenced by the wipeout of the continuity of evolution due to a "maximum number of extinctions of plant and animal species" including seed plants, as well as insect pollinators of those seed plants. Yet, a few of those mutualists survived the adverse paleoclimatic conditions (Miller, 2008) to wait for the reversal of the favorable climatic conditions and to accelerate the mode of divergence with consecutive attempts to survive and appear in new forms. They eventually were successful at one point and the critical analysis and review of the whole thing helps comprehend some critical issues like the radiation of flowering plants (angiosperms) and their relation with coevolution of insect mutualists. The impact of these global cataclysms is mentioned here:

i. DeCARB: Around 365 m.y.a., the Earth witnessed the Paleo-Icehouse mass-extinction event in the Farsnian of the Devonian Period when the fragments of the continental plate moved towards the South pole, i.e., the Antarctic region. The Upper Devonian DeCARB event created the Paleo-Icehouse (super cold conditions lead to the condensation of atmospheric gases) drive to a hypoxic condition, lead to reducing the oxygen level in the paleoatmosphere to make it 13% (Berner, 1999; Berner and Kothavala, 2001; Ward, 2006).

   At the end of the Paleo-Icehouse effect and at the beginning of the Carboniferous Period, the paleoclimatic reversal warmed up the paleo-atmospheric conditions thus resulting in an increase of oxygen upto 31%. It helped the expansion of herbivorous arthropods and terrestrial plants after the DeCARB (Berner and Kothavala, 2001; Labandeira, 2006).

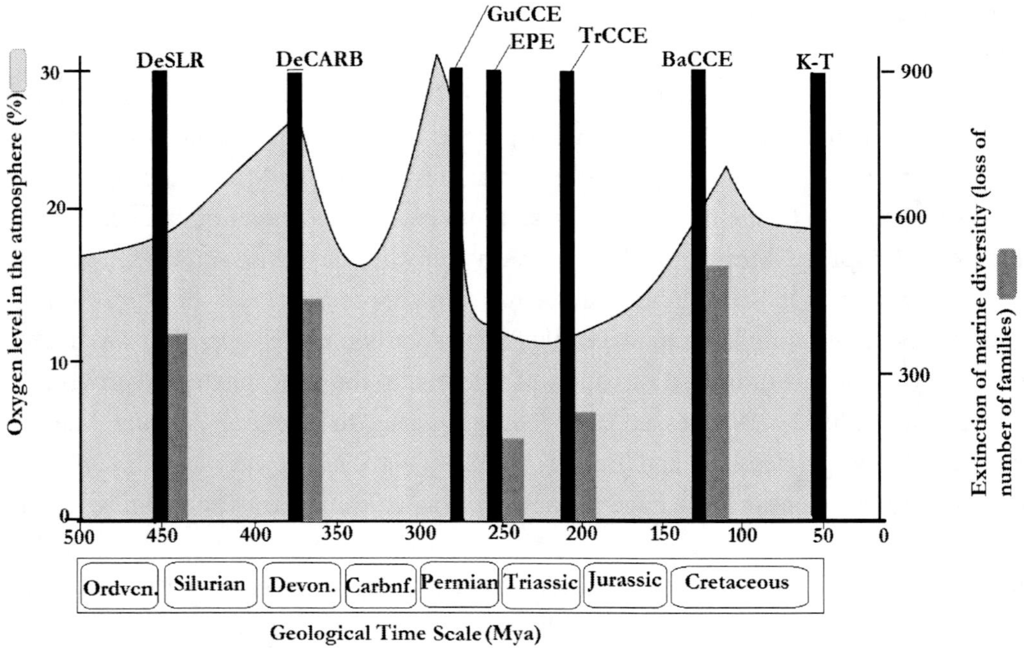

Figure 12. The global oxygen level in different mass extinction cataclysms (MOMIIMI after Elizabeth Kenna, 2019 and M. Miller, 2008).

ii. GuCCE: The end of the Guadalupian age concomitantly occurred with the global carbon isotope anomaly during the Upper Permian Period. The Earth witnessed a series of actions in this phase. The fossilized, terrestrial coal reserves were ignited by a lava flame, then erupted from volcanoes, as a result excessive volumes of carbon monooxide and carbon dioxide raised the atmospheric carbon volume; at the same time the methane clathrate storage under the sea-bed was released to add huge volumes of sulfur (in the form of sulfur dioxide) and carbon (as methane reacted with ozone in the stratosphere to form carbon dioxide), this caused an excessive carbon volume in the paleo-atmosphere or "hypercapnia" and the oxygen levels drastically dropped to create severe hypoxic conditions (Retallack et al., 2006; Knoll et al., 2007).

Noticeably, the oxygen levels of the regions adjacent to Panthalsa became 12%, going further down from the pre-GuCCE event at the 31% level; this lead to the creation of a huge physiological stress of survival for aquatic and terrestrial plants and animals (Knoll et al., 2007). The excessive methane, sulfur dioxide dust in space and scorching heat lead to the creation of a galactic gas cloud during the end of 600,000 years and it lead to the end of the Paleozoic Era, thereby reducing the opportunity of survival of most plants and animals, though a few of them survived in isolated refugia in the Earth biome. Moreover, it has slowed down the divergence of life on Earth due to severe Greenhouse effects in combination with hypoxic conditions (Knoll et al. 2007; Miller, 2008; Retallack, 2009).

iii. EPE: Around 251-252 million years ago, at the end of the Permian Period, there were reports of an apocalypse, triggered by the collision of the Earth with an incoming comet resulting in a tremor at the continental shield, followed by a resumption of severe volcanic eruptions after s 600,000-year interval of time after the GuCCE (Erwin, 2006; Knoll et al., 2007). The rupture of the sea-basin rock-shield created the way out for molten lava forming epicornical lava lakes (known as Siberian traps) and trapped carbon monoxide and carbon dioxide gases with sulfur dioxide and huge volumes of methane gas from the methane clathrate deposits under the sea-rock bed. Eventually, voluminous methane was released in the paleo-atmosphere (Retallack and Jahren, 2008). The released methane in the atmosphere was ignited by the fire created by the hot molten lava and the open burning of coal pits which added huge amounts of gaseous carbon in the atmosphere and mixed with the other gases and dust particles to form flammable, dense gaseous clouds. This lead to frequent solar flaring in the paleo-atmosphere of the Earth in the Upper Permian Period; as a result, the paleoatmospheric condition of the Earth in the Upper Permian period was extremely hot with prevailing hypoxic conditions (Wang, 2000; Rees, 2002; Benton and Twichett, 2003). The paleoclimatic, paleophysiological, paleoecological and biostratigraphic investigations have indicated that the consecutive effects of GuCCE followed by EPE lead to the mass-extinction known as "trigger and kill," of plants and animals in most of the terrestrial and aquatic ecosystems (e.g., Cathaysian floras took refuge in the desolate refugia of the arid landscapes of the Upper Permian period) on the Earth in the Upper Permian- Lower Triassic boundary (Gastaldo et al., 2005; Grauvogel-Stamm and Ash, 2005; Knoll et al., 2007). The biostratigraphic studies of fossil sites at Cathys, rocks of Greenland and paleosols of Antarctica in the late Permian and early Triassic boundary, showcased the existence of vascular plants like Gigantopterids, cycads, ginkgoes, peltasperms, ferns and some other conifers (Wang, 2000; Looy et al., 2001). They went through prolonged drought conditions (thick cuticular leaves indicate this) even post-EPE made the environment unsuitable for insects, tetrapods, and other invertebrates (Retallack, 2009).

iv. TrCCE: The "Triassic-Jurassic boundary carbon cycle event" was considered as the most severe mass extinction event that created the highest level of threat to the survival of living biota on Earth. This particular mass-extinction event was designated as the "Super-Greenhouse event" that almost pushed most of the animals and plants to the brink of complete extinction (Beerling and Berner, 2002; Berner, 1999) during the transitional phase from the late Triassic Period to the early Jurassic Period on the geological timescale. In fact, a brief history of plant life was found to be missing in the paleo-equatorial zone "because of the pause in the accessible deposits in that climactic belt" (Scotese, 2005). The segmented identity of paleofloristic elements of paleo-equatorial regions during the Triassic-

Jurassic climatic regime was paleobotanically discovered and identified (Reyre, 1973; Weber, 1985; Cornet, 1993) from the dispersed deposits around the paleo-equatorial regions of:

1) Santa Clara formation of Sonora, Mexico: The palaeoflora, dominated by tropical late Triassic tree-fern.
2) Richmond basin of Virginia, United States: palaeoflora primarily characterized by the tropical late Triassic tree-fern.
3) Saharan region of North Africa: palaeoflora was characterized as primarily subtropical.
4) Middle East (e.g., Saudi Arabia and Egypt): palaeoflora was characterized as primarily tropical, acclimatized to grow on dry laterite soils.

During the transitional climatic regime as well as the developmental phase, the whole paleo-equatorial region was geologically characterized by far and wide coverage with newly formed rift basins (predominantly Triassic rift basins), which were also recognized as the "red basin" with seismic lines. During this developmental phase of paleofloristic changes and as a result of continental drifting, geological evidence indicates that Pangaea gradually drifted in the north-ward direction; concomitantly, the equatorial climatic regime (which was primarily recognized as the subtropical paleofloristic element) gradually shifted southward. This shifting continued till the northern part of South America and Northern Africa moved from subtropical to arid climatic zones in the Middle Jurassic period (Cornet, 2002). The "climactic barrier" of the Middle Jurassic Period has hypothetically been considered as the main impediment for southward migrations of basal angiosperms. The major "Paleo-Greenhouse" condition (sudden flux of inorganic carbon dioxide gases in the paleo-atmospheric condition) evolved in the Cretaceous/Tertiary boundary, which has scientifically been proven by the presence of "large negative excursion in delta C^{13}" and that sudden, inorganic flux of CO_2 was considered to be the prime cause of global warming induced paleo-Greenhouse condition in the Cretaceous/Tertiary boundary period (Whiteside et al., 2004).

The mass extinction of the Triassic/Jurassic boundary in the paleoequatorial region, which was evidenced by the fossils of extinct plants and animals and Iridium anomaly, was found, to some extent, identical to the sudden surge of the temperature curve noticed in the Cretaceous/Tertiary border (Cornet, 1993; Fowell et al., 1994; Cornet and Waanders, 2006; Whiteside et al., 2004). The identical surge in the paleotemperature curve, as well as the palaeo-Greenhouse condition due to the high content of CO_2 in the paleoatmosphere during the Permian/Triassic border, was noticed to a lesser extent (Scotese, 2005).

According to Olsen and Kent, (2000), the sudden surge of CO_2 in the atmosphere lead to abrupt changes in the paleoclimatic conditions and put the living organisms in the paleo-equatorial area under tremendous stress. Most of the living organisms could not cope with the super-hot and severe flood conditions and disappeared in the bottle-neck of mass-extinctions. Barrett and Willis (2001) have considered this CO_2 induced global warming as a fine filter system to push the species options to adapt to the changing scenario or move on their way to extinction. The contemporary research in the field of atmospheric science is under the process to estimate the atmospheric salvage and decreasing of global temperatures in the early Jurassic Period; apparently, this global warming induced arid climatic barrier was considered as the main impediment of paleo-floristic migration between Gondwana and Laurasia. However, quantitative estimation of CO_2 in the paleoatmospheric level during the Jurassic period is yet to be decided, the composite studies like the PALEOMAP project of Christopher Scotese (2005); Whiteside et al. (2004) and contemporary research on the "Milankovitch lake cycle stratigraphy," indicated that "Volcanism" was a key factor for the earlier "Global warming" incidents.

Apparently, the Triassic/Jurassic mass extinction, as well as the Jurassic Green House incidents, don't have any direct relationship with the determination of the time of evolution of angiosperms on Earth. The sudden appearance of fossil records of angiosperm fossils in the Cretaceous Period in the absence of contemporary molecular data lead the paleobotanists and evolutionary biologists to consider the Cretaceous Period as the "Age of birth of angiosperms." However, factors like-

a. Amelioration of the climate in the late Jurassic Period,
b. Jurassic desertification of tropics,
c. Global warming induced mass extinction of early evolved angiosperms of the Permian period,
d. Chance of survival of epibiotic dispersed refugia (e.g., existence of *Amborella*),
e. Fossil evidences of Triassic angiophytes,
f. Molecular investigations,
g. Cladistics analysis and
h. DNA sequencing of ANITA, etc. have helped evolutionary biologists start accumulating evidence to counter the "Cretaceous Origin Hypothesis" and dig down the deep-rooted phylogenetic origin of angiosperms.

It is necessary to explore the Triassic Period to solve the deep-rooted mystery behind the birth of angiosperms on this blue planet. In this regard, the recent molecular-botanical observations based on cladistic analyses have ascertained *Amborella* as the extant basal angiosperm, which occurred in the mountain ridges of New Caledonia (Sampson, 1993).

Regarding the Cretaceous radiations of angiosperms, Axelrod (1970) postulated that primitive angiosperms originated on the mountain and gradually receded to the low-lands mainly due to climatic shifts and it has some resemblance to the occurrence of *Amborella* in New Caledonia with antithetic observations of "Cretaceous radiations of angiosperms." The critical review of the Triassic flora of Greenland has revealed that the change of tropical forest type of vegetation of this region started changing due to the Greenhouse effect of TrCCE and most of the plant species disappeared at the end of the TrCCE event (McElwain et al., 1999; 2007).

BaCCE: On the basis of biostratigraphic, biogeochemical and paleoclimatic evidences, it was ascertained that the "Lower Cretaceous end-Barremian biogeochemical perturbation" presumptively occurred in the Upper Barremian period of the Lower Cretaceous Period and it was also most likely predicted that the BaCCE event expedited the diversification of some monocots and magnoliids (Heimhofer, et al., 2005). According to Barrett and Willis (2001), the high content of carbon dioxide in the Cretaceous Period might be responsible for the radiation of angiosperms, whereas the primary effort for radiations of flowering plants could add-up more carbon dioxide in the paleoclimatic atmosphere and this higher content of gaseous carbon would accelerate the rate of species turnover in a short period of time on Earth. It has also been assumed that the continuous cycle of "thermal maxima" acted like a cascade of rotations to maintain the constant warm climatic conditions (without any icehouse effect), required for speciation of angiosperms along with vertebrates and invertebrates (Barrett and Willis, 2001; Friis et al., 2006, Miller, 2008).

K-T event: the Kretaceous-Tertiary asteroid impact event took place around 50 m.y.a. in the Upper Cretaceous-Lower Tertiary borderline when the Earth was supposedly impacted by asteroids. The huge collision resulted in the releasing of huge heat energies in the atmosphere (Miller, 2008). The effect of the K-T event on the living biota of the Earth is not fully known, but it might be a coincidence that in the Upper Cretaceous Period, diversification of angiosperms was noticed. The abundance of arborescent habits was succeeded by herbs and shrubs and a vast area of the biome was occupied by the savanna grassland ecosystems as the extrapolation has supposedly been supported by the increasing number of vertebrates and grazing animals.

The reason for this mass-extinction event and the palaeo-ecological impact of this event is not known and overlaps with the earlier Cretaceous extinction and has also not been recognized by a number of scientists and environmentalists. Even the eminent writer Ms. Elizabeth Kolbert has not mentioned this incident in her contemporary, investigative literature on mass extinction events, "The Sixth Extinction" (Kolbert, 2014).

Apparently, it seems that factors like the paleo-Icehouse effect, paleo-Greenhouse effect, hypoxia, elevated levels of gaseous carbon, etc. since the Upper Devonian to the Upper Cretaceous Periods, acted as a potential force or selection pressure to drive coevolution. Repeated efforts of radiation of angiosperms which could better be referred tp as the "Cascade Evolutions," to complete the radiation of angiosperms took place over the due course of time. The biosynthesis as well as the molecular evolution of key biochemical, enzymes, hormones, chemical indicators, phytoecdysones, etc. of insects and plants (Labandeira, 2006; Miller, 2008) supposedly promote an explosion of radiation of angiosperms and the mutual insect pollinators.

Chapter 14

COEVOLUTIONARY MUTUALISM OF PLANTS AND RESIDENT INSECTS: ROLE OF ANGIOSPERMS IN RADIATION

> "One possibility is that these atmospheric variables have direct physiological consequences on the selection and turnover of particular plant clades globally, which in turn elicit an associational response from selected clades of insect herbivores."— Labanderia, 2006

The co-existence of invertebrates and primitive vascular plants in the Mesozoic Era was proven by different scientists at different times (Thien et al., 1985; Friis et al., 2000; Zherikhin, 2002; Krassilov et al., 2003) and the coevolutionary interactions supposedly evolved through the independent line of evolution. The fossil evidence of arthropods and seed plants were reported from the Upper Triassic Period (Ash, 1997). Apparently, the shifting of paleoclimatic conditions from super cold glaciation–house to extremely hot Greenhouse had nothing to do with the origin and evolution of angiosperms. However, Miller (2008) observed that the interactions between plants and invertebrates since the Upper Devonian-Lower Carboniferous and Upper Permian Periods were mutualistic and coevolutionary as the hypoxic conditions of the paleo-atmosphere during the post-mass-extinction regime in the Permian Period, supposedly expedited the evolution of a small body of insect groups and seed plants with bisexual cones.

The coevolution of insects and seed plants from the Carboniferous to the Jurassic Period was more mutualistic as its survival-strategy was oriented by, rather than the direct influence of, paleoclimatic changes. The well-preserved plant organs, belonging to Monocotyledonae and invertebrate fossils from the same fossil site during the Miocene age of the Paleogene Epoch indicate the existence of a mutual insect-angiosperm relationship (Ramirez et al., 2007).

Bartett and Willis (2001) supported the mutualistic coevolutionary mode of divergence in the members of tetrapods and flowering plants in the Mesozoic Era but they consider the coevolutionary hypothesis between the divergence of flowering plants and herbivorous dinosaurs, as poorly evidenced. Though, it has also been claimed by the critics that "Tight coevolution" among plant-animal relationships had hardly been substantiated yet (Tiffney, 2004). Sometimes, the coevolutionary relationship between the seed plants and resident insects were criticized in order to support the mutual relationship between the monopodial shrubs (small tree) and resident insects. The contemporary mutual relationship of *Zamia furfuracea* (a Cycadeoid plant) and Weevils *Rhopalotria mollis* (Coleoptera) of insect groups in Holocene, were referred to herewith (Norstog and Fawcett, 1989).

According to Miller (2008), molecular coevolution of host plants and resident insects, developmental toolkits were initiated for the first time, immediately after the DeCARB as the cold temperature and hypoxic conditions induced a severely cold selection pressure. The survival spree of predators with the low oxygen levels (around 13%) might have forced them to take refuge from the plants for food, oxygen, and shelter.

It is construed that the paleo-Icehouse conditions in Upper Devonian-Lower Carboniferous (DeCARB) event which created a hypoxic atmosphere, put some physiological stress on the respiratory systems of the invertebrate group of plecopteran insects, the resident insects of some seed plants. Coincidentally, the seed plants evolved during this time period (Labandeira, 2006; Rothwell et al., 1996). Eventually, the DeCARB hypoxic paleo-atmospheric impact drove the molecular evolution of Hox genes and the haemocyanin group of respiratory enzymes (with hexamerin food storage and multimeric protein) among the primitive arthropods, stonefly and seven clades of hemimetabolous and holometabolous insects (with multimeric protein potential) that diverged most likely induced by the DeCARB (Hagner-Holler et al., 2007). Contrarily, at the end of the GuCCE and the TrCCE mass extinction events, the coevolution of angiosperms and resident insects became more likely (Miller, 2008; Bonneton et al., 2003) along with the molecular evolution of phyto-enzymes, various transcription factors and developmental genes of angiosperm plants (e.g., ClassIII HD-ZIP, APETALA, AP2, ARF, DELLA, MIKC-type, MADS-box, PIN, GRAS, KNOX, LEAFY, etc.). The high levels of gaseous carbon and sulfur in the paleo-atmosphere or hypercapnia during the GuCCE and the EPE is supposed to decelerate the respiratory potential of plants and resident invertebrates to cope with the low oxygen (hypoxic) levels and hyper-acidic air to survive (Knoll et al., 2007).

According to Beerling and Berner (2004), the hypercapnia in the Silurian Period might act as an environmental factor, precluding the formation of megaphylly, divergence of well-developed roots and shoot and the reversal of favorable paleoclimatic conditions lead to the formation of the structural trinity (root, shoot, and leaves) in the plants. This helped develop the physiological ability to go for a higher rate of divergence and achieve better structural modifications to attain arborescent forms from the initial shrubby habit.

In the "Florigen Hypothesis" of the eminent plant physiologist Mikhail Chailakhyan (1937), "Florigen" was postulated as a molecule, identical to the hormone, which enhance/trigger flowering in the angiosperms. According to the earliest perception of the "Florigen Hypothesis," exposition of leaves of flowering plants to sunlight creates a leaf-generated stimulus followed by its attainment the active meristem transcending through phloem tissues for inducing flowers. The biochemical nature of Florigen is under investigation but the existence of it was experimentally established. Physiologically, *Nicotiana tabacum* var. MM[24] is a short-day plant whereas *N. sylvestris* is a long-day plant. Naturally, growing of *N. sylvestris* in short day conditions do not yield any flowers. Surprisingly, it was observed that the leaf grafting of *N. tabacum* var. M.M. to *N. sylvestris* under controlled growth of short day conditions induced flowering (Zeevart, 1976).

This kind of transferring of the biochemical inducer (i.e., Florigen) would normally be transferred from any species of angiosperms to another species (neither species-specific), irrespective of the physiological identities of the plants (nor physiologically plant specific), irrespective of plants growing in a short day, long day or day neutral conditions, in any period of the time. In the Florigen Hypothesis, Chailakhyan (1985) observed two different categories of flowering hormones- Gibberellins (e.g., Ga5 and Ga6) and Anthesin, initially synthesized on leaf surfaces, they transcend through phloems to shoot up to the apex to induce flowering, and King et al. (2001) supported the role of Gibberellins in inducing flowering. Chaillakhyan's (1985) observation on transferring of floral induction from induced flowering plants to flower non-flowering plant (Kim and Sung, 2013), created the stepping stone for considering the role of genetic factors as well as the divergence of the molecular tool-box in flowering plants, evolved from the mother-stock of early spermatophytes in the course of evolution, which was likely completed in two stages-

a. Preliminary shifting from the predominating vegetative state to the floral state.
b. Firm affirmation of formation of flowers in angiosperms, which is integrally related to the wide diversification of angiosperms in the post-Cretaceous period on terrestrial biomes.

Ayre and Turgeon (2004), observed in their contemporary research related to the interdisciplinary field of plant physiology and molecular biology, the isolated a protein "CONSTANS," found in the leaf veins and helps carry the important biomolecules through the phloem. CONSTANS promotes the gene expression of "galactinol synthase" either by itself or by interacting with an intermediary protein, called the FT protein, by signaling the florigen to provoke flowering. Further investigations of the contemporary molecular biologists on the "laboratory plant," "*Arabidopsis thaliana*, *Brassica* sp. and other plant species have helped open the molecular toolbox to recognize the gene like- APETALA1,

[24] Maryland Mammoth.

FLC, flo, LEAFY, VIN3, WAP1 etc. to convert the Shoot Apical Meristems (SAM) to Floral Meristems (F.M.) and control the timing, duration, frequency and volume of flowering (Parui and Mondal, 2016). Hence, evolutionary divergence of angiosperms from the primitive spermatophytic group and early radiation of flowering plants in Early Cretaceous was related to this Florigen Hypothesis and the important phylogenetic issues like the origin and diversification of the molecular tool-box related to the molecular genetics of flowering.

According to the "Shrub lifeboat Hypothesis", (Miller, 2008) the Palaeozoic seed plants and trees ensured the survival of struggling insects and its larva within the cavities and air sacs of the different organs (e.g., lamina, fruit like structures, strobila etc.) of the tree with an optimum level of oxygen in comparison with the hypoxic condition outside. These strategies adapted to the end of every mass extinction event (e.g., DeCARB, GuCCE, EPE, TrCCE, and BaCCE) and those refugia might help the insects and tetrapods to sustain hyperthermia, hypoxia, hypercapnia and other adverse conditions on Earth, that the plants and insects went through. Furthermore, the identical patterns of divergence and diversification of insects, flowering plants and occurrence of different mass extinction events between the Devonian and Tertiary Periods has determined the plant-insect co-evolution theory (Labandeira and Eble, 2000; Zheng et al., 2014), as presented in Figure 13.

According to Labandeira (1998), the expansion of terrestrial herbivores were integrally linked to the global events of mass extinction and it occurred in four phases:

i. Phase 1: The transitional shift of the "Silurian hothouse world" to the "Devonian-Carboniferous ice house world" lead to the attribution of a stress selection force to the expanssion of herbivores (Labandeira, 1998).
ii. Phase 2: The meltdown of ice in the Carboniferous Period lead to the rise of oxygen level 31% in the atmosphere which lead to the expansion of terrestrial herbivores followed by the extinctions at the end of the Permian Period due to the "hothouse effect" and hypoxic conditions (presence of 12% oxygen in the atmosphere) (Labandeira, 2006).
iii. Phase 3: The third expansion of herbivores started in the Upper Triassic Period and continued till the middle of the Cretaceous Period and this phase went through huge mutualistic interactions leading to the radiation of angiosperms (Labandeira, 2006).
iv. Phase 4: The current, ongoing phase of the expansion of herbivores was initiated in the Upper Cretaceous Period and has continued till the Holocene of the Quaternary Period (Labandeira, 2006), where a change of habit of angiosperms from arborescent to herb and shrub types, and diversification of smaller-sized herbivores (e.g., small mammals, insects, birds etc.), played a key role in the diversification of angiosperms across the Earth.

The mass extinction events exerted a selective pressure for developing mutualism as well as the coevolution of plant-herbivore divergence and diversification. The prevailing hypercapnic and hypoxic conditions after the GuCCE event, induced the hemi and holometabolous types of phytophagous insects to be dependent upon the Permian seed plants (even on Vojnovskyalians and Gigantopterids) to sustain extremely hot, hypoxic conditions (Miller, 2008).

Immediately after the commencement of the EPE, the vascular, seed plants survived the unfavorable conditions of the EPE and the insects tried to simulate the GuCCE survival strategies, but these two groups were immediately faced with another mass extinction event, the TrCCE, which reduced their opportunity for survival. The surviving host plants and insects developed another molecular evolution, as they supposedly started secreting a number of biochemical substances (Truman and Riddford, 2002; Glover, 2007; Miller, 2008), whenever the insects trampled the surface of the shoot apical meristems or SAM of host plants and those chemicals were popularly known as signaling molecules.

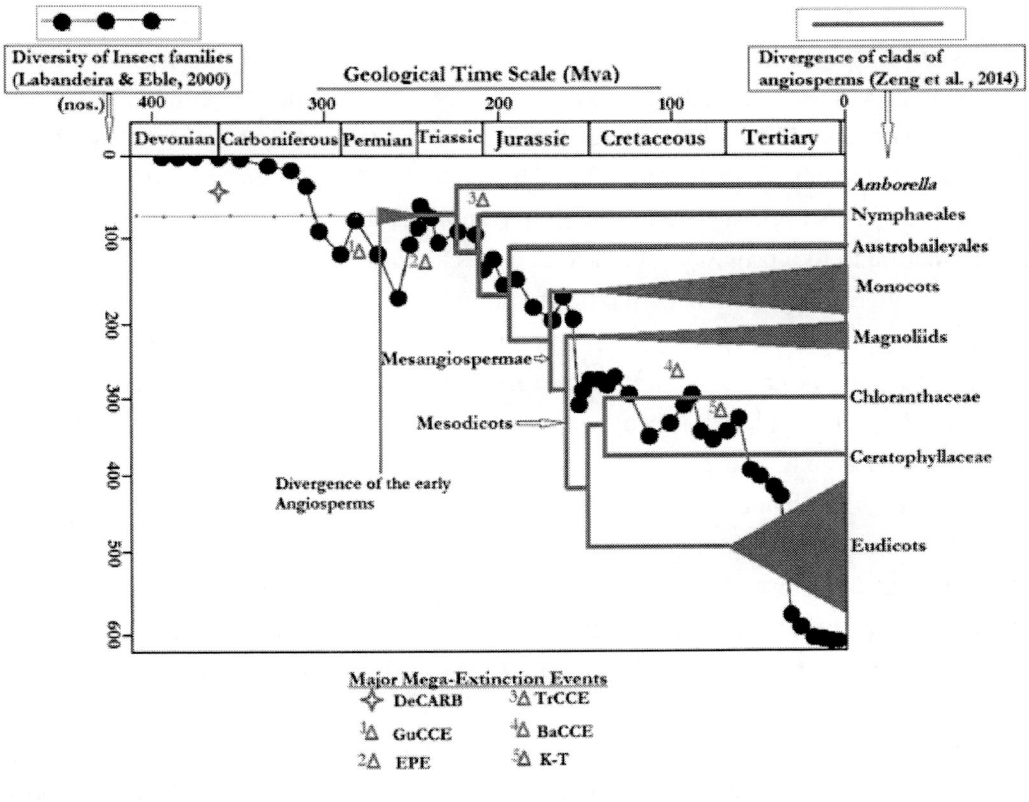

Figure 13. The divergence and diversification of insects, primitive flowering plants and mega extinction cataclysms on Earth.

Eventually, this kind of reciprocal secretion of the signaling molecules by host plants and resident phytophagous insects, was found as a result of mutualistic molecular evolution. This most likely took place inside the "shrub lifeboats," where signaling phytoecdysone hormones (produced as a result of the trampling actions of insects at the SAM region of those plants) in seed plants of the Permian period induced gene functions of insects (particularly the hexamerins) (Rogers et al., 1997; Gellon and McGinnis, 1998). It could be assumed that the hexamerin protein storage genes of insects was probably induced by hypoxia to diverge smaller sizes of insects from larger sizes of the ancestral stock.

The resident insects and arthropods on the host vascular seed plants in the Permian Period started secreting the signaling molecules like; jasmonates, volicitins to inject it to the SAM, affecting the secretion of phytohormone, Gibberellins at the tip of the host plants (Rothwell and Lev-Yadun, 2005; Adam et al., 2007; Busov et al., 2008).

Eventually, these signaling molecules controlling the secretion of other important phytohormones and activating the function of homeotic genes to form bi-sexual cones, the progenitors of flower. In turn the Paleozoic vascular, seed plants produced the signaling biochemicals, known as "Phytoecdysones" inside (by mechanical intake or secretion) the resident insects, signaling the production of hemocyanin respiratory and hexamerin storage proteins in juvenile stages to stop the growth in order to make the smaller body sizes reside in the "shrub lifeboats" that helps survive the adverse periods of time (Thummel and Chory, 2002; Truman and Riddiford, 2002; Coen et al., 2004). It is quite convincible, that all living organisms have had inherently damaged building properties, the Permian seed plants have not found any exceptions so the repairing of damages of SAM was immediately responded by anatomical and structural damage building responses; the structural modifications of the cone formation to the formation of bisexual cones, then setting the flowers is supposed to be the ultimate response to paleoclimatic conditions (Miller, 2008).

According to Miller (2008), the time period of coevolution of flowering plants (angiosperms) and insects and natural hybridization of wide ranges of Gigantopterids and Vojnovskyalians (considered as progenitors of angiosperms) of the Permian Period were supposed to be around 300 million years old, which indicates the polyphyletic origin of angiosperms, rather than a monophyletic origin.

It was found quite uniquely that the advancement of the features of sexual reproduction of the Permian and Lower Triassic seed plants were related to gene duplications leading to the evolution of protein dimers, tetramers, which took place between 200-300 m.y.a., necessary for vegetative module (Sinha, 1999; Aida and Tasaka, 2006) and reproductive module (Rothwell and Stockey, 2002; Friedman and Williams, 2004) identities; concomitantly, the evolution of floral dimers and quarters in the seed plants (also recognized as shrub lifeboats) of the same time were induced by gene duplications (Doyle and Endress, 2000; Kwiatkowska, 2006; Miller, 2008). Eventually, a second surge as well as a 2nd time molecular evolution of hexamerin storage proteins were witnessed in a group of resident insects in the shrub-life boats leading to the divergence of bees (Coleoptera), flies (Diptera), bees and wasps (Hymenoptera) and moths (Lepidoptera) (Labandiera, 2006; Miller, 2008), which indicate that the coevolution of flowering plants and insects should be considered as a much more fusion type of evolution, “ecological evolution” (as it is the ultimate result of mutualism), based on the mutual relationship of plant and animal groups rather than stereotypical evidences of phyletic evolution, transcending through a specific streamline.

Chapter 15

Evolution of Angiosperms and the Challenges of Evolutionary Taxonomy

"The flower remains ill-defined and its mode (or modes) of origin remain hotly disrupted; some definitions and hypothesis of evolutionary relationships preclude a role for the flower in delimiting the angiosperms." – Bateman et al. 2006

Whatever we see we consider that as the absolute truth, but not necessarily all visible objects represent the true archetype. There is no doubt that all angiosperms are seed plants but all flowering plants are not angiosperms; hence considering flowering plants as typical in the place of angiosperms is misleading as flowers or flower-like reproductive organs have also been noticed among the members of *Welwitschia*, Benettitalian and Gnetalian groups of gymnosperms.

In the "Seeds and Sex" chapter of a unique book on plant evolution, "Tree –A Life Story," Suzuki and Grandy (2004), mention the great synthesis of Aristotelian principles and the Christian doctrine, where nature was defined as "the Book of God's work." Production as well as overproduction of the seeds have been defined as a natural cause or "Nature's plenitude" as well as "the sign of divine providence," a simple example of douglas fir's ecosystem was referred to here; where it was noticed that around 65% of the annual production of Douglas Fir's seeds were consumed by birds, squirrels, mice, chipmunks, voles consumed most of the seeds of remaining part and one tenth of the remaining seeds touched the ground, able to cover a vast area of forest habitat.

The well-established prey-predator relationships in ecology as well as the food-trophic system of science has some earlier theological metaphors as well as biblical doctrines "All flesh is grass," which indicate towards food chain systems like producers (e.g., grass), eaten up by grasshoppers (e.g., of a primary consumer), the toad (e.g., of the secondary consumer); the snake (e.g., of the tertiary consumer). The earlier concept of plant science, which was reconciled with the theological doctrine, considers that production of enough seeds by seed plants was primarily meant for feeding most of the animals to meet the criterion of evolution which is the "diversity of life on Earth." John Ray, the great Catholic priest and founder of the British natural history in the sixteenth century, started working on botany and like his other co-workers, involved himself in the system of the taxonomy of plants to organize the plants according to the consistent characteristics of roots, fruits, and seeds etc. Though the characteristics of fruits and seeds were taken into consideration, Ray being a Catholic priest of the sixteenth century, was not very convenient with plant sexuality. Rather this issue was considered as "a Scandalous Notion in Puritan England" but it became fashionable in Europe. The progress on this apparently forbidden topic, "existence of sexuality of plants," was carried forward by a great French naturalist, Sebastian Vaillant, when he delivered his observations on a pistachio plant, grown in the garden of the Museum d'histoire naturelle in France on this forbidden topic one early morning in a fully packed auditorium, adjacent to the Museum in 1717. It was published as his posthumous publication in 1722. The posthumous article of Valliant made a huge change in the thought process of a young student of the University of Uppsala, Sweeden, Carolus Linnaeus, considered as the father of taxonomic studies. It does not necessarily mean that no botanist has considered the sexual identities or characteristics of plants whenever they engaged in taxonomic classification of the plant kingdom in sixteenth and seventeenth centuries.

According to Suzuki and Grandy (2004), around 300 taxonomic studies and reviews by different plant taxonomists in the sixteenth and seventeenth centuries went through the labyrinth of complex crissed-crossed lines of outcome for considering inconsistent characteristics (e.g., size, shape, fragrance and colour of flowers, fruits etc.) to help reach a taxonomic conclusion.

The observations of Sebastian Vaillant about certain characteristics of plant's sex organs or floral features (e.g., Stamen, Carpel etc.) were found consistent in a number of species to be grouped in a certain sequence, as was done by Linnaeus (Vaillant, 1717) and it helped undertake an artificial system of classification of plants.

According to the initiation of the artificial system of classification of organisms, the foundation was laid by the great philosopher and naturalist Theophrastus as identification of the specimen that was conducted at the level of genus and species, which was taken to the next level by Linnaeus by adding two more groups: Class and order, to assign them in their respective slots. The class of any plant was considered on the basis of the number and arrangement of stamens (the male sexual organ) and the order of any plant was considered on the basis of the number and arrangement of carpels (the female sexual organ). Still, Linnaeus's system of taxonomic classification (1753) was considered as the main reference of all taxonomic classifications, but if we consider it from an evolutionary perspective, this artificial system of classification did not help define any organism and did not give any clue about any phylogenetic lineages to track the predecessors or successors. According to Liannaeus' system of classification (1753), when we consider the classification patterns of

Douglas Fir (*Pseudotsuga menziesii*), it was classified under the domain of Eukaryota; kingdom Plantae; phylum: Coniferphyta; Class: Pinopsida; Order: Pinales; Family: Pinaceae: Genus: *Pseudotsuga*; Species: *menziesii*. At the same time if we consider the classification of human beings (*Homo sapiens*), it was found under the Domain: Eukaryota; Kingdom: Animalia; Phylum: Chordata; Class: Mammalia; Order: Primates; Family: Hominidae; Genus: *Homo*; Species: *sapiens* (Suzuki and Grandy, 2004). If we look at the evolutionary history, the two classifications are linked by the Domain: Eukaryota, but it did not help predict that human beings would supposedly evolved between 5.2-23.2 million years ago in the Miocene Period, whereas the Douglas fir supposedly evolved between 280-345 million years ago in the Carboniferous or late Devonian Period, and the phylogenetic lineages were not missing, did not define the organism, proper differentiation or relationship between these two kingdoms. Referring to Linnaeus' system of classification, Darwin mentioned in "The Origin of Species" that "The ingenuity and utility of this system is indisputable." During his visit to pay tribute to the shrine as well as visiting the tiny backyard garden of Linnaeus at Uppsala, Sweden, the eminent writer John Fowles, expressed his feelings about the potential of the work of the pioneer taxonomist Linnaeus; "the place where an intellectual seed landed and is now grown to a tree that shadows the entire globe."

According to the writer John Fowles, the main flaw of Linneaus' taxonomic classification is that it is an artificial approach of individuation of plants, as well as reduction of natural phenomena to a certain class under a certain order, is an expression of an anthropocentric view of nature through the viewfinder of a camera, without experiencing the nature through our own vision.

It is quite obvious that the expectation towards phylogenetic classifications to dig out the evolutionary history of any species or a group of plants would not be achieved in the formative template of the artificial system of classification and Fowles expressed his disappointment "…this is the bitter fruit from the tree of Uppasalan knowledge" (Suzuki and Grandy, 2004).

Since the seventeenth to the early part of the nineteenth century, the morphological taxonomists basically consider the morphological features of similarities and differences for systematic analysis and that is the typical shadowing practice of the Linnaean system of classification. Broadly categorized as Phenetics or Taximetrics (proposed by Michel Adenson, 1763), it was an effort to classify the organism based on overall similarity (either morphological or structural), without considering the phylogenetic as well as evolutionary relationships of that organism or group of organisms. The ultimate goal of systematics is narrating the evolutionary relationships (a graphical presentation of the tree of life) of a group of organisms and segregation of taxa is necessary for that and in order to achieve that, both phenetics and cladistics methods were initially considered.

Apparently, the advanced cladistics method beat the primitive phenetics method as contemporary researchers related the evolutionary aspects, they put absolute importance on the evolutionary relationships among species but cladistics analysis was not found to be friendly when a diverse group of organisms with subtle levels of differences from one another, were taken into consideration and phenetics analyses extended the numerical tools to give better opportunities to the researchers to make better taxonomic judgements to segregate a species from the rest of its group members. Some advance phenetics methodology like DNA-DNA hybridization (used in Sibley and Ahlquist taxonomy for birds, 1990) was found reasonably close to phylogeny when the advanced cladistics method like the "Bayesian inference" was found to be computationally too complex to narrate the phylogenetic outcome. Sometimes modern phenetics analysis was less erroneous than DNA sequencing of cladistic analyses when contemporary analyses related to plant genomics were concerned.

Between 1950 and 1960 most molecular biologists supported the "Point Mutation Hypothesis" of protein-encoding genes as an evolutionary process and they were in favor of the Darwinian doctrine of gradual evolutionary changes of life. The orthodox supporter of cladistics taxonomy, Willi Hennig, passed away in 1966 but his suggestions to prioritize the cladistics analysis and the refined results from cladistics analysis, in the background of evolution of a new generation of cladistics algorithm, new generation of software and molecular technologies, proved to be superior than and helped recognize it as a unique evolutionary taxonomy of the 21st century. The role of plant genomics particularly the advancement of research on mitochondrial DNA, chloroplast DNA, multi-chromosomal nucleus, and histones play an important role in the advancement of cladistic analysis as well as evolutionary taxonomy to draw a tree of life from unicellular algae to angiosperm plants (Lynn Margulis, 1970).

Another challenge of the Linnaean taxonomy or phenetics and evolution of cladistics, is the "punctuated equilibrium" (Eldredge and Gould, 1972) against the Darwinian "phyletic gradualism." And it narrated further that evolution gradually occurs as increments of small changes, smoothly and continuously. Such evolutionary action transformed the whole lineage that was designated as anagenesis; whereas the "punctuated equilibrium" was considered rare but increments of change is huge in geologically short periods of time and this process of speciation was considered as cladogenesis (a species split into two species instead of gradually transforming into another one).

According to the punctuated equilibrium, once the species was found in the fossil form (also known as "Stasis") there was hardly any further evolutionary change and it remained stable throughout its geological life span. The mosaic of evolution is an incomplete form of punctuated equilibrium, where the different rates of evolutionary changes occurred in different parts of the body or in a group of organisms instead of a complete evolutionary change. According to Eldredge and Gould (1972), the prevalence of punctuated equilibrium was noticed in fossil records in the clade or family of diversified angiosperm groups of (with diversified wood anatomy, diversified leaves, floral patterns and seed architectures, etc.) in the Cretaceous Period.

The visible fossilized transitional form or "footprints on stone" between birds and reptiles like "Archaeopteryx" was not available between angiosperms and its predecessors but the "molecular foot-prints" of plants indicated that factors like the "punctuated equilibrium" or "punctuated evolution," and "mosaic evolution" have played a key role in instant evolution in small isolated populations of the Earth. The short-term changes of genetic configurations of any living organism played a cardinal role in the evolutionary radiations and that's why "Stasis" exists anywhere and everywhere in the fossil histories of the biological entity and it could conveniently be interpreted under the principle of "punctuated equilibrium," rather than the non-existent principle of "phyletic gradualism" of the great evolutionist, Charles Darwin. The understanding and acceptance of the principle of punctuated evolution as the driving force of evolution and diversification of the plant kingdom and the consideration of the latest generation of computers, mathematical algorithms, molecular and biological innovations, make the young group of molecular taxonomists and researchers lean towards the Cladistics of the 21st century rather than the erstwhile phenetics of 17th century.

According to contemporary molecular taxonomists and phylogeneticists, the advancement of researchers in the area of plant genomics between 1966-1970 changed the conventional Darwinian concept of evolution though those might not bring any radical change in the evolutionary taxonomy, and some of the important changes are like identifying the location of the promoter and regulatory components of genes responsible for genetic expression. This is responsible for organismic evolution and some factors like RNA interference, chromosomal aberrations, polyploidy, genome duplications, etc. (Cornet, 2008) play an important role in the organismic evolutions and most likely the consideration of any particular taxonomic rank of organisms or group of organisms should always be considered in the background of spatiotemporal perspectives to have an idea about its phylogenetic status or evolutionary appearance. Apparently, Linnaean taxonomy, as well as phenetics, is obsolete today in terms of phylogenetic systematic studies and research but an eminent supporter of Cladistics, Bruce Cornet (2002) raised his concern by referring to two genera *Euryle* and *Victoria* and mentioned that for the last 200 years the general perception among plant lovers and botanists about these two genera and another genus *Nymphaea* (around 50 species), belonging to the order Nymphaeales, and the segregation of these genera is mainly based on the sizes, ecological diversities, thorns, and etc. On the basis of contemporary cladistics analysis, the Linnaean status of *Euryle* and *Victoria* is dismantled and merges those two in a newly formed clade, creating a big vacuum in our perception. Thus it is necessary to not completely eliminate Phenetics even when we prioritize cladistics, though we cannot ignore the importance of DNA footprints if we are really inquisitive about the phylogeny of any living organism.

Contrary to the "Punctuated equilibrium," "Quantum speciation" (Simpson, 1944) was found to be more convenient in understanding the sudden radiation of angiosperms in the Upper Cretaceous Period. According to Quantum speciation, the new and different types of daughter species evolved from a "semi-isolated peripheral population of an ancestral species in a cross-fertilizing organism" (Grant, 1971). It was observed that continuous gene flow in a stable population of cross-fertilizing species has a "swamping effect" on the new variations (Grant, 1971). When comparing the "Geographical speciation" with the slow, conservative and gradual process of divergence, the dynamics of "Quantum speciation" is instant. This is faster and radical as reflected in the genotypic configuration and phenotypic expressions, hence it has also been designated the "Genetic Evolution" (Mayr, 1954). The phylogenetic investigations of the members of *Clarkia* (belonging to the family Onagraceae) established the relationship of *C. biloba* and *C.lingulata* as parents and offspring (Lewis and Roberts, 1956) and the offspring has always been found to be different from the parental species in terms of morphological outlooks, ecological interactions, and genetical configurations and has occupied the small geographical area overlapping to the adjacent larger areas of distribution, occupied by the parental species (Lewis and Raven, 1958). Lewis (1962) observed that peripheral populations in the unfavorable conditions in smaller habitats might be the easy object of extinction and it has always been followed by regeneration of new populations from the few alive species of genetically aberrant configurations leading to the instantaneous formation of offspring species with unique genetic configurations (Lewis, 1962). Owing to the fast pace of speciation, an abrupt fluctuation in the size of the newly formed population was witnessed, hence it was recognized as the "Catastrophic evolutions" by Lewis (1962), but Grant (1971) considered this as the combination of "selection and drift." The preliminary surge of angiosperms in the Upper Cretaceous Period, which has also been recognized as the "Radiation of angiosperms" witnessed by the appearance of a huge number of cross-pollinated, closed seeded, vascular, flowering plants and its widespread mode of diversification in a short period of time, most likely occured as a result of 'Quantum Speciation" rather than the process of "Phyletic gradualism."

Apparently, it seems that the Linnaean system of classification no longer has any phylogenetic value as this artificial system could not give any idea of the evolution of a group or its phylogenetic relationship with other groups. Thus the Linnaean system of classification was declared a dead end, but it was later found that as a starting point in classifying groups of organisms in a certain order (from simple to complex), it was an effort to get in touch with the "true phylogenetic system of classification"; more convincingly, it's the legacy carried out by the molecular systematists and molecular botanists of the nineteenth century to concentrate their studies on cladistics analysis on the basis of DNA sequencing, and etc.

The contemporary effort to isolate and comparatively analyse DNA with precise accuracy from living organisms to determine evolutionary lineages and phylogenetic classifications, should be considered as fulfilling the dreams of Linnaeus, Vaillant, and Ray in the Quaternary Period. Sequencing stamens and carpels to group and regroup a number of plants, apparently looking identical is the continuation of the legacy to end up with DNA sequencing to adjudge the extent of relations even between apparently non-identical species. The hidden treasure behind the success of DNA is the four structural bases of DNA, known as: Adenine or A, Guanine or G, Thymine or T and Cytosine or C. The four bases form the linear array of molecular strings and two strings are spirally twisted with each other to form a double helix, where the Adenine of one string is paired with the Thymine of anther string, likewise the Guanine of one string is paired with the Cytosine of the other string. The sequential arrangements of bases on a string of a genetic message formed of a series of three-letter words known as triplet codons which encrypt nucleic acids and a series of nucleic acids together in a particular sequence to form proteins or enzymes to perform gene expressions. The application of heat to a solution of DNA molecules break up the bond between the four bases and the free bases can be found in an isolated condition (denatured condition) in the heated suspension of DNA and gradual cooling reverses the whole thing to construct the double-stranded DNA once again.

Technically, if the suspended DNA solution of one species in a heated (denatured) condition mixed with the suspended DNA solution in the heated condition of another species, those base pairs mix together and then the gradual cooling of the two suspensions together may cause a strand of DNA from one species to find a sequence of DNA from the other species, identical to its own. Thus two DNA strands could merge together to form a hybrid DNA double strand. Then the hybrid DNA strand could be compared to the proportionate part of the two separate parental DNA strands (Suzuki and Grandy, 2004).

If a high proportion of DNA of each species is able to form the hybrid of the other, it can supposedly possess a large number of identical sequences and would be considered to be closely related, though it hardly happens in real life. So the determination of affinities in the earlier days between two or more taxonomic groups on the basis of morphological, anatomical, embryological, palynological, and biochemical resemblances in Linnaean taxonomic legacy till early nineteenth centuries went through a shift of paradigm in the twenty-first century where contemporary cladistic analysis, as well as DNA analysis, was applied to determine the phylogenetic relationships between the species, looking apparently not related, or judging the established relationships between the morphologically identical species, predicted under the classical Linnaean system of classifications.

Chapter 16

EVOLUTION OF ANGIOSPERMS – PALEOBOTANICAL, MOLECULAR FOOT-PRINTS, THE UNIVERSAL TREE OF LIFE AND THE ANITA GROUP OF BASAL ANGIOSPERMS

While the presence of a wide range of structural differences and the absence of transitional forms between morphological entities were found to be the key impediments for recognition by botanists; to consider *Amborella* and the water lily members of Nymphaeaceae as the same phylogenetic complex. However, critical review of the molecular configurations and cladistics analysis of these taxa were found helpful for the unequivocal consideration of these plants as the ancient (basal) predecessors of extant angiosperms.

Even, the supporters of the "Cretaceous Origin Hypothesis" could not establish paleobotanically that the evolution of basal angiosperms might have taken place earlier than the Jurassic Period. However, they instead prioritized Cretaceous angiosperms (the outcome of relict type of evolution) being the few survivors and the descendants of the "Triassic-Jurassic mass extinction" which was a natural massacre which happened during the Jurassic period.

Referring to the existence of *Sanmiguella,* the other group of evolutionary biologists considered the primitive members of Cretaceous angiosperms as the outcomes of "*de novo*" (new) evolution of the "pre-angiosperms," that existed during the Triassic period; a group of unusual and rare types of angiosperms, apparently having no structural resemblances among the extant angiosperms. According to Fowell et al. (1994), the transitional forms between *Amborella* and Nymphaeaceae basal angiosperm complexes, supposedly became extinct during the Triassic-Jurassic mass-extinction events.

According to the critical molecular botanical reviews as well as the cladistics analysis of the extant group of angiosperms, it was observed that around eight primitive (might be that more than eight branches existed but those don't exist in the living form) branches formed the primitive, basal angiosperm phylogenetic tree (popularly known as ANITA) they were supposedly represented by their evolutionary successors (Cornet, 2002):

1) *Amborella* (least derived as well as more primitive).
2) the Nymphaeaceae
3) the Hydatellaceae (the sister group of Nymphaeales)
4) *the ITA (Illicium,* Trimeniaceae, and *Austrobaileya*)
5) the Chloranthaceae
6) the Eudicots (Buxaceae, Trochodendraceae, Neumbo, Platanaceae. Proteaceae etc.)
7) the Monocotyledonae + Piperales
8) the Magnoliales + Winterales and
9) Laurales (least primitive as well as most derived).

On the basis of the observations of a number of horizontally transferred genes in *Amborella*, and the evidence of branching off *Amborella* alone at the base of the crown of angiosperms, by referring to the personal communication of the eminent molecular systematist, Ken Kinman, (Cornet, 2002) who expressed his concern about probably splitting off the *Amborella* clade at the base crown group Magnoliophyta. He was referring to the splitting off Nymphaeales at the base crown of Magnoliophyta, as revealed in 1990s, Ken Kinman also expressed his concern about *Sanmiguelia* splitting off from other angiosperms during the Permian Period (Cornet, 2002).

According to the critical observations of molecular evidences of angiosperms as well as the cladistics tree of Doyle and Endress (2000), (Figure 14) combined with the critical studies of angiosperm fossils (e.g., Crinopolles group, Liliacidites, *Magnolia* group, *Sanmiguelia* etc.) help highlight the unique characteristic features of *Sanmiguelia,* a basal angiosperm of the late Triassic period, that had the composite characteristics of monocots and dicots. Contemporary, palaeobotanical evidence of *Amborella*-like flowering plants from the Upper Cretaceous period criticize the considerations of *Amborella* as basal angiosperms (Krassilov and Golovneva, 2004). Based on the observations of molecular analysis of 59 low-copy nuclear genes along with the critical review of 110 evolutionary morphological features of a number of flowering plants, Zheng et al. (2014) corroborated the phylogenetic status of ANITA by recognizing its five clades: eudicots, monocots, magnoliids, Chloranthaceae and Ceratophyllaceae under the supra-clade-Mesangiospermae, which supposedly diverged between the Middle to Upper Jurassic period. However, the critiques of the "ANITA basal angiosperm" support group raised some concerns:

> "With both *Nymphaea* and *Amborella* at the base of the latest angiosperm tree, a wider range of characteristics might be expected in the archetypical angiosperms or angiosperm sister groups than implied by the previous consensus on *Amborella* alone. What of the contemporary projections derived from the analysis of seed plant relationships? Do these analysis point towards likely archetypal angiosperms or angiosperm ancestors? No, not when we incorporate the recent analyses challenging the validity of the anthophytes. "Instead, angiosperms become more distantly related to all existing seed plants, leaving a gap populated only by the extinct taxa that may or may not be represented in the fossil record."—Crepet (2000)

Regarding the formation of the basal clade of angiosperms, Saarela et al., (2007) and Friedman (2008) mentioned their observations on the flower group belonging to Hydatellaceae that forms the basal clade of the angiosperm group, ANITA. This inconspicuous group of sub-aquatic plans from Asia-Oceania region (Australia, New Zealand, India), was morphologically and anatomically found to be linked to gymnosperms and monocots but the re-examination of morphological, ultrastructural data in combination with molecular data showed that it was the basal sister clade of Nymphaeales. So, instead of ANITA (primarily comprised of *Amborella*, Nymphaeales, *Illicium*, Trimeniaceae, and *Austrobaileya*), the name of the basal angiosperm clade should be "ANHITA," the "Mother Goddess of all the Flowers" (Miller, 2008). Due to the simple floral morphology of the members belonging to Hydatellaceae, it was considered as the "Prefloral" stage of well developed "bisexual" flowering forms and the flower architecture of other members of the basal clade ANITA or "ANHITA" "could hardly be more simple" (Friis et al., 2011).

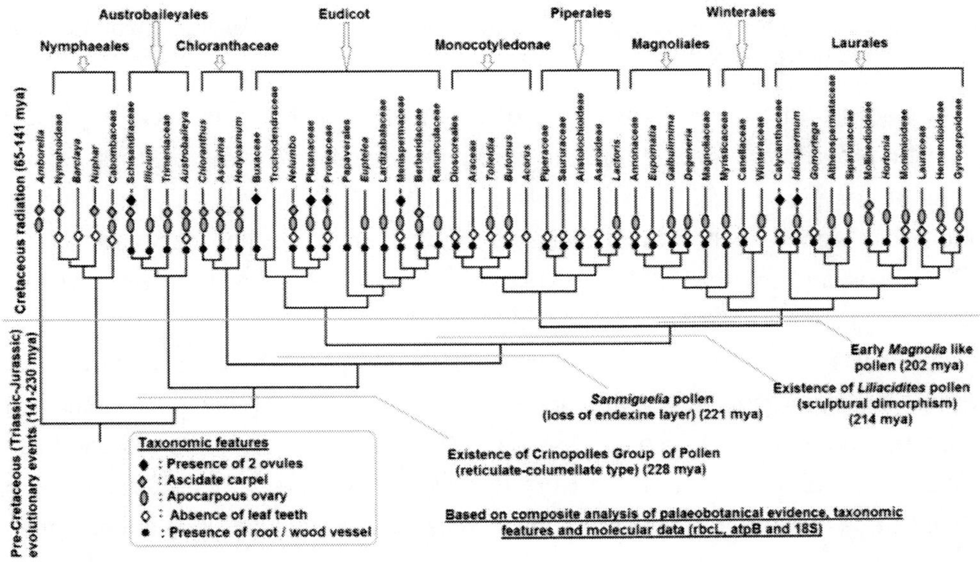

Figure 14. The divergence of primitive flowering plants on the Earth (modified atter Doyle and Endless, 2000).

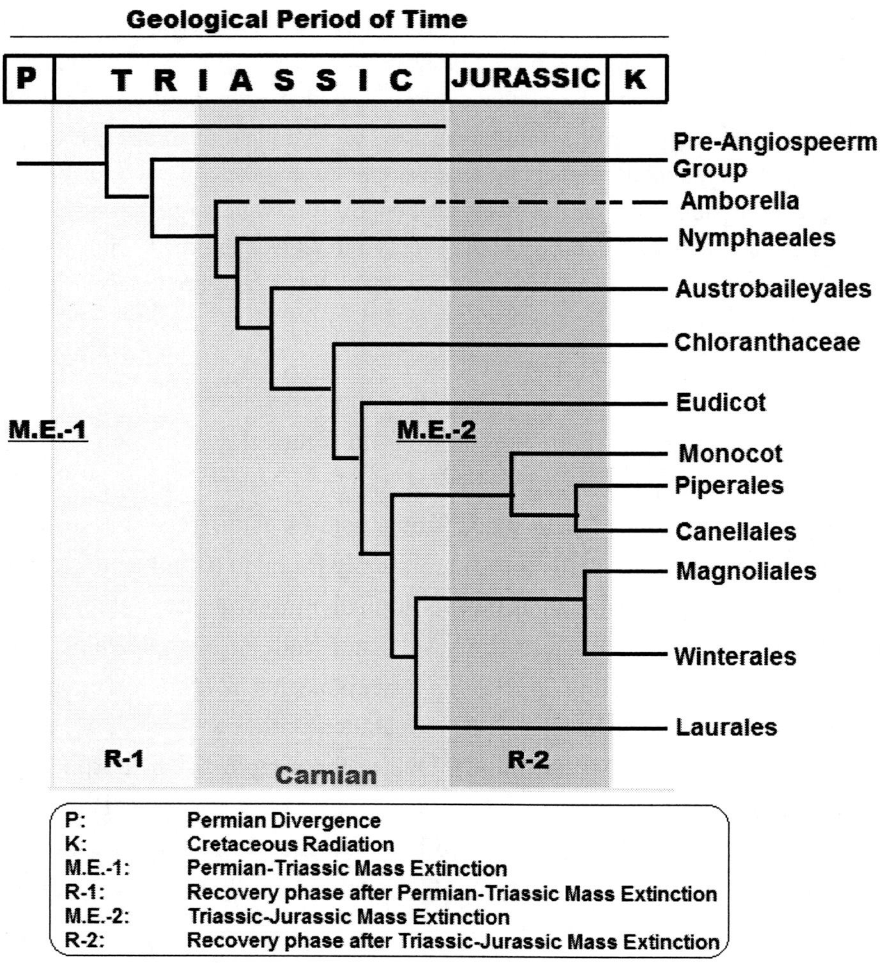

Figure 15. The recovery of primitive angiosperms went through two mass extinctions (modified after Figure 37, Bruce Cornet, 2008).

This cladistics tree gives an overview of the relationship of extant basal angiosperms but failed to provide any specific information about the extent of evolution pressure or rate of evolution (might be varied from one clade to another). However, it ensured one thing, that more than eight basal branches of angiosperms (Figure 15), supposedly passed through the bottle-neck and existing till the end of the Triassic Period or the beginning of the Jurassic-Triassic mass extinctions.

On the basis of the critical observations on the angiosperm phylogenetic tree based on the composite analysis of taxonomic and molecular data, shows that basal branching of angiosperm took place in the Triassic period, rather than the Jurassic Period and that this diversification took place during the early Cretaceous Period (Cornet, 2002). Moreover, the 18S data has helped a group of phylogeneticists (Soltis et al., 2000) to conclude that allopatric speciation induced genetic drifting occurred during the early Jurassic Period and no substantial evidence of the occurrence of the branching of basal angiosperms and diversifications of angiosperms in the early Cretaceous were found responsible for the coevolution of insect pollinators, whereas, another group of phylogenetists (Labandeira et al., 1994; Labandeira, 2006) cited the paleobiological evidence of 97 million years of "angiosperm-insect association" at the molecular level as "coevolution," which is thought to play akey role as well as put selective pressure on the diversification and radiation of angiosperms in the Triassic Period.

According to a group of evolutionary biologists, chronological ambiguity might affect the inappropriate indication of timing of evolution of basal angiosperms. Like violations of sequential occurrence of events, like the appearance of organ features (e.g., sequences of leaf rank) of angiosperms in the Cretaceous Period (Hickey and Doyle, 1977) which has rather mistakenly been depicted as the evolutionary occurrence at different periods of time due (e.g., *Pannaulika*) to disrupted sequential expressions of evolutionary features (Cornet, 1993). Palynological evidence created some ambiguity to predict the actual time of evolution of any particular type of angiosperms, like the palynological features of some of the extant angiosperms were considered to be derived, found geologically primitive, e.g., the polyplicate pollen types of derived ones were found in the species of *Sanchezia*, belonging to the Acanthaceae family and *Spathiphyllum*, belonging to the family of Araceae, were reported in the fossil sites of the Permian Period and the Tertiary Period (Cornet, 1989; Cornet and Habib, 1992). Hochuli and Feist-Burkhardt (2004) reported the unique angiosperm-like extinct plant *Retisulicites*, collected from the Middle Triassic (Ladinian) marine deposits of the Norwegian Arctic, Barents ocean, having eight pollen-grain taxa (e.g., aperture types vary from trisulcate, operculate, syncolpate, monosulcate etc.) also considered the maiden angiosperm. They went through the Permian-Triassic mass extinction events. Critical observation of its pollen type-A shows affinity to *Polycolpopollis* spp., collected from the Richmond basin from the Late Triassic (Carnian) period, whereas, pollen type-N has a resemblance to *Penetecrinopollis* spp., collected from Richmond basin, of the Late Triassic Period and most surprisingly, the occurrence of Middle Triassic taxa outside of the tropics as well as warm-temperate climates is most unlikely.

The global hunt for fossil evidence of ancient and extinct life in the middle of the 1960s, initiated another shift of paradigm, to look into the molecular foot-prints of organisms to construct a universal tree of life, to establish the evolutionary relationships among the extant organisms and this unique approach is mainly based on the following principles (Schopf, 2001):

1) All living organisms are able to produce proteins as well as enzymes, controlling the metabolism process and all these proteins were produced by the globular organelles, called ribosomes.
2) All these ribosomes or protein processing units are chiefly made up of RNA (Ribonucleic acids) a single-stranded biomolecule comprising of ribose sugar-

phosphate backbone with the rib-like attachment of linearly arrayed nitrogenous bases Adenine (A), Guanine (G), Cytosine (C) and Uracil (U).

3) As we know that as part of central dogma process, mRNA has formed from DNA template by the complementary sequencing of nitrogenous bases as a result of transcription and any mutational change of DNA molecule of any organism could change the configuration of mRNA configuration. The mutagenesis of rRNA or ribosomal RNA could also be triggered in the same way.
4) According to the molecular systematics, molecular biologists, paleo chronologists, and cladistics analysts, the DNA configurations of the closely related living organisms are identical. Hence, the rRNA configurations of the closely related members of different living organisms are identical. Whereas, the rRNA configurations of distantly related organisms are not identical at all.

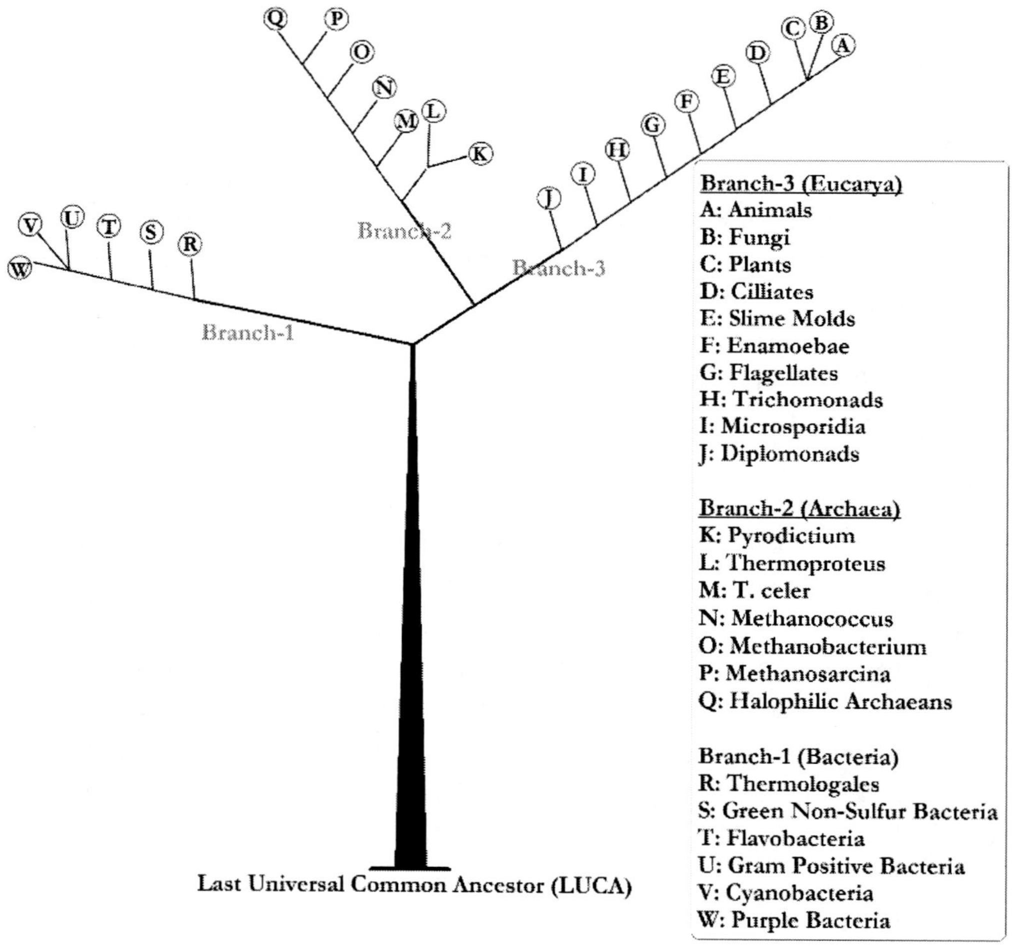

Figure 16. The diagramatic sketch of Universal Tree of Life (Schopf, 2001).

So, the computational analysis of sequences of nitrogenous bases of rRNA (a particular type "16S" type) of a diverse group of living organisms have given an impression about the close or distant relationships among the members of that group of living organisms. In the "Universal Tree of Life," the vertical length of each branch represents the change of a number of nitrogen base sequences from the nearest node, it splits off from the closest relative. The graphical presentation of the "Universal Tree of Life" (Figure 16.) has presented the cladistics relationships among the different groups of living organisms (Schopf, 2001).

1) From an evolutionary viewpoint, the plants and animals taxonomically belong to the Eukaryote group; but there are more than twenty evolutionary branches between these apparently close relatives.
2) In the Universal Tree of life, most of the living organisms are microscopic forms. There are three groups of living organisms (plants, animals, and fungi) having predominantly macroscopic members and some microscopic members. Rest of the living members, belonging to the twenty branches (approx.) are microscopic.
3) The eukaryotes were found to be the primitive predecessor of all extant living organisms and these groups were found closely related to Archaea group.
4) The root of the Universal Tree of Life, as well as the earliest predecessor, is closely related to the bacteria followed by *Archaea* and *Eucarya.*
5) The Universal Tree of life shows that all living members are grouped under three groups of organisms:
    a. Bacteria: This group represents all cyanobacteria (producing oxygen) and all other bacteria.
    b. Archaea: This group of non-nucleated, methane gas producing organisms survived in hyper-acidic, high-temperature conditions that's why these group of organisms have ecologically been recognized as "Acidophilic-thermophiles" or "extremophiles."
    c. Eucarya or Eucaryotes: This group of organisms have chromosomes, organized in the sac-like nucleus.

While branching pattern of the Universal Tree of Life adheres to the earlier observations on the fossil records on Archaea and Bacteria, the groups supposedly appeared earlier than Eucarya but the rRNA trees could not give any idea about the time of branching of different groups as these trees were structurally formed on the basis of 16S analysis of all extant members. According to the ideal graphical presentation, all branch-tips are supposed to extend to the same height, if those represent the right evolutionary time then the top of the tree is supposed to be flat instead of the zigzag patterns of bouncing type (which have given indication of occurrence of different rates of evolution in different group of organisms). As an example, it could be inferred that Diplomonads, Microsporidia, and Trichomonads of the Eucarya domain evolved faster than the groups having shorter branches like animals, fungi, and plants. Though the conventional age-dating process of ancient or extinct representatives of living organisms is using the radioactive isotopes (popularly known as carbon or uranium dating process). The Universal Tree of Life pattern was not helpful in terms of looking back at the evolutionary clock, as the rRNA trees or the molecular footprints of the evolving protein did not yield any satisfactory results. Even the fossil evidence of all organisms were not found due to the complex process of fossilization, which left some gaps to predict the evolutionary time of those group of organisms; even the presence of fossil evidence could help to decipher the occurrence of a biological group of organisms rather than its actual appearance on this Earth.

In order to understand the evolution and diversification of angiosperms in general, the "land plant developmental toolkit" was developed to work on the functional aspects of homeotic genes of Palaeozoic seed plants by Floyd and Bowman (2007). Baum and Hileman (2006) worked on the functional aspects of homeotic genes on spermatophytes to understand the mode of evolution of a flower from bi-sexual cones. Floyd and Bowman (2007) also carried out molecular phylogenetic analysis on a diverse group of plants starting from algae to angiosperms e.g. *Chara* (algae), *Physcomitella* (bryophyte), *Selaginella* (pteridophyte), *Picea* (gymnosperm), *Arabidopsis* (eudicot angiosperm), *Antirrhinum* (asteroid angiosperm), *Oryza* (monocot angiosperm) etc. They found that a number of homeotic genes, integrator genes, meristem identity, phytohormones, with micro-RNAs have made up the "developmental toolkit" of land plants, vascular plants, and angiosperms to better understand the mode of evolution and the diversification process at the molecular level. It was found by Floyd and Bowman (2007) that a number of homeotic genes of the "developmental tool kit of the land plant," could decipher the tentative time of appearance of the flower on Earth: Class III HD-Zip, APETALA, AP2, ARF, DELLA, MIKC-type, MADS-box, PIN, GRAS, KNOX, LEAFY, YABBY, TIR genes. Every approach, technique, process, and effort has its limitations and the "Tree of Life" is not an exception; even the sudden appearance of enormous paleobotanical evidence and critical review of that palaeobiologic evidence could give no more than a predictive age of the branches of the "Tree of Life" (Schopf, 2001).

Chapter 17

ANGIOSPERMS – ON TEMPORAL SPECS: WALKING THROUGH HYPOTHETICAL ODDMENTS

The evolution of contemporary angiosperms is not properly conclusive due to the lack of straightforward evidence or interpretations of the few scattered foot-prints in the spatiotemporal background, having no clue of its beginning or end. Since the beginning of 18th Century, the evolution of flowering plants is an "An Abominable Mystery" as this phylogenetic jigsaw puzzle has not been solved by plant taxonomists, phyto-morphologists, plant anatomists, palaeobotanists and molecular biologists either individually or collectively. Since, Charles Darwin's "Voyage of Beagle," the plant phylogenetists of different regimes construed various hypotheses from time to time. In order to understand and explain the origin and diversity of flowering plants or angiosperms or anthophytes (taxonomically recognized by the plants as having predominant sporophytic generations and distinguishing features like the presence of flowers, having ovaries or fruits, bearing ovules or seeds etc.) on this Earth, different scientists have involved themselves in unearthing the evolutionary history of angiosperms; basically it's a philosophical quest rather than a scientific effort to look for the earliest ancestors of the diversified group of contemporary angiosperms (Arber and Parkin, 1907; Wettstein, 1924; Zimmermann, 1930; Dahlgren, 1938; Hutchinson, 1969; Takhtajan, 1969; Stebbins, 1974; Martin et al., 1993; Margulis and Schwartz, 1998; Soltis et al., 2000; Doyle, 2001; Byng, 2014). In the absence of availability of the complete form of paleo-botanical epigraphy for determining the ancestry of primitive angiosperms, most of the paleobotanists tend to consider the gymnosperms as the mother-stock of the primitive angiosperms and some of the hypotheses are in favor of the monophyletic origin, whereas other hypotheses have supported the polyphyletic origin of angiosperms, which are as follows:

1) Hemi-angiospermous Hypothesis, Euanthial Hypothesis or Magnoliid Hypothesis: Scientists duo Arber and Parkin (1907) advocated in favor of a hypothetical

missing link of "hemi-angiospermae" that maintained a transitional bridge between the unique gymnosperm group of plants, Cycadeoideae, and the primitive angiosperms. The contrived reproductive parts of hemi-angiospermae have a close resemblance to the cycadeoid bi-sexual strobilus-like flower bearing cylindrical axis in the central position, encircled by spiral rows of perianths, a number of androecium and a number of gynoecium having sporangia in its margins. Arber and Parkin (1907) suggested further that the angiosperm flower of Magnoliaceae family had close identical features to this hypothetical floral structure of hemi-angiospermae and the possible extension of this hypothesis indicates that primitive angiosperms evolved from Ranalian mother-stock and dicotyledonous plants evolved earlier than monocotyledonous plants. According to the Euanthial Hypothesis, the extant and ancient group of flowering plants: Magnoliaceae build the evolutionary bridge between flowering plants (phanerogams) and non-flowering (cryptogams) (Martin et al., 1993; Doyle, 2001). The Euanthial Hypothesis suggests that basal lineages were noticed by small trees with the slower life cycle. This hypothesis has further elucidated the monophyletic origin (that is from a particular mother-stock) of early angiosperms, but lack of supportive fossil evidence and lack of supportive views of the renowned phylogenetists like Bessey (1915), Cronquist (1968), Hutchinson (1969) etc. have made it unacceptable.

2) Pseudanthial Hypothesis or Hamelididae-Piperales Hypothesis: According to this hypothesis the angiosperms originated from the plants having unisexual naked seeded gymnospermous plant carrying tiny, simple unisexual flowers also found in the members of Hamelididae and Piperales. According to Stuessy (2004); Doyle (2012), the progenitors of angiosperms belonged to Coniferales, Gnetales, and Casuarinales.
3) Paleoherb Hypothesis: The Paleoherb Hypothesis of the origin of angiosperms was found to be remarkably different from Psedanthial and Euanthial Hypothesis as the basal lineages of the group members predominantly comprised of herbaceous and fast-growing plants. Burger (1981) proposed that primitive angiosperms evolved as monocotyledons, having the following features:

i. large and complex form of angiosperms evolved from small, simple, primitive forms.
ii. dispersed vascular bundles got organized in the tubular form in the stem in the due course of time.
iii. simple aerial branching gave rise to a complex form of branching.
iv. a complex form of leaves with lamina and petioles originated from undifferentiated leaves.
v. complex venation pattern of leaves evolved from leaves with single or no veins.

According to Dahlgren and Clifford (1982), the progenitors of angiosperms were identified as shrubby monocots with subterranean rhizomatous stem (coping with a shift of wet and dry period of the year) which has given rise to the aerial stem in course of stable environmental conditions.

This hypothesis has also further elucidated the monophyletic origin (that is from a particular mother-stock) of early angiosperms but lack of supportive fossil evidence made it unacceptable.

4) Anthophyte Hypothesis, Englerian Hypothesis or Gnetalian Stock Hypothesis: Another monophyletic origin of angiosperms was clarified by the scientist Wettstein (1924), who suggested that the monosporangiate, primitive angiosperms evolved from the Gnetalian flowering stock. In order to substantiate, his view, Wettstein (1924) considered the *Casuarina* species as the living predecessors of ancient angiosperms and the existence of bisexual angiospermous flowers was construed as the formation of pistillate flowers on the staminate inflorescence stock. The extant group of Gnetales was considered as the evolutionary bridge between flowering (phanerogams) and non-flowering (cryptogams) plants (Markgraf, 1930; Barkmann et al., 2000; Doyle, 2012). Based on the unique double fertilization pattern, a typical embryological developmental pattern of *Ephedra* sp. (an ancient representative of Gnetales group) and angiosperm group of plants, Friedman (1992) has considered these two groups as closely related in terms of their phylogenetical identity. Eminent molecular systematists, Mathews and Kramer (2012) reported that a major group of gymnosperms (e.g., Gnetales) evolved parallely with angiosperm group of plants from a common mother-stock. The contemporary researches on molecular phylogeny (on the basis of comparative analysis of two sequence datasets of mitochondrial genes) of Gnetales and angiosperms ensured that these groups evolved independently.
   The critical analysis of morphological and molecular phylogenetic data failed to establish a relationship between extant seed plants and angiosperms and plausible fossil evidence is required to explain the evolution of angiosperms, hence this hypothesis was discarded (Doyle, 2001).
5) Polyphyletic Proangiospermous precursor Hypothesis: Deviating the monophyletic origin of angiosperms from their gymnosperm precursor, scientist Krassilov (1975, 1996) recommended that from a common ancestral stock an advanced (structurally and functionally) group of gymnosperms like *Ephedra, Gnetum, Welwitschia* etc. evolved, whereas, another lineage given rise to the precursor of the angiosperms of today, known as "Proangiosperms" (the progenitors of ancient angiosperms) in polyphyletic terms. While another plant phylogenetic, Anderson (1934) did not agreed with this Polyphyletic Proangiospermous precursor Hypothesis, he supported the polyphyletic ancestry

of angiosperms from the diverse form (combination of primitive and advanced forms) of gymnospermous stock, earlier. On the basis of the appearance of characteristic features of the reproductive organs of gymnosperms in ancient angiosperms helped Anderson (1934) recommend that primitive angiosperms evolved polyphyletically, as a result of the hybridization of widely related gymnosperms in the palaeoclimatic environment. According to the great plant phylogeneticist, Cambell (1928) discarded the hypothesis of the ancestry of angiosperms from the gymnospermous stock, particularly from the Cycadeoideace groups and recommended the evolution of primitive angiosperms from the Pteridophytean root-stock.

6) Biome/Phytogeographical origin Hypothesis: In the "Uplands Hypothesis," Axelord (1952) has postulated that the origin and evolution of flowering plants occurred in a particular topographical area. Later, on the basis of extensive review of paleobotanical evidences, secondary information on the center of origin of angiosperms, different hypothesis on evolution of angiosperms and plant-insect coevolution theory, Axelord (1970) come to the conclusion that angiosperms evolved in the Triassic/Jurassic boundary of the tropical upland in Gondwana and subsequently migrated to identical climatic zones and he has further suggested that primitive angiosperms and insect pollinators dispersed over semi-arid forests in low-lands and radiated in subtropical lowland forest areas in the lower Cretaceous period. Similarly, Retallack and Dilcher (1981) the identified coastal sea-shore as the center of origin of flowering plants in their "Coastal Hypothesis." Stebbins (1974, 1984) located the alpine biome of the northern latitude as the center of origin of flowering plants; whereas, G. Sun et al. (2001) consider Eastern Asia as the cradle of speciation of flowering plants in their "Eastern Asian Centre Hypothesis." On the basis of critical studies on palynomorphs, collected from the boreholes of the sea-bed, Hochuli and Fiest-Berkhardt (2004) considered that radiation of angiosperms happened in the "Boreal cradle'. On the basis of ecological data, place of occurrence on the upland habitat in the Pacific island, subtropical-temperate forests, along with data from cladistics analysis on the deep-rooted clade ANITA, Feild et al. (2004) have considered *Amborella* as basal angiosperms but there are no equivocal assertions from the scientists and phyllogenetists, globally (Crepet, 2000; Rudal et al. 2007).
7) Protangiospermous ancestral lineage hypothesis: The great pioneer in plant phylogenetics, Diel (1936), rejected the earlier views of the origin of earlier angiosperms from the Gymnosperm mother stock, either "Cycadeoidean descent" or 'Gnetalian descent." Rather, Diel (1936) postulated the existence of a hypothetical ancestral-stock in the Mesozoic Era, known as "Protangiosperms," giving rise to the primitive angiosperms. Furthermore, he stated that the protangiosperms stock supposedly originated from the eusporangiate

Pteridophytean ancestor, like *Ophioglossum* and protangisperms act as the progenitors of monocotyledonous and dicotyledonous lineages in the course of evolution. The hypothetical protangiosperms might have possessed bisporangiate flowers, either rudimentary perianths or absolutely achlamydeous.

8) Stachylosporous Hypothesis: The great paleobotanists, Sahni (1920) and Lam (1948) emphasized the origin of seed plants from two different mother groups, known as a. Phyllospermae and b. Stachyospermae, in the "Stachyosporous Hypothesis." Furthermore, this hypothesis has attributed phyllospermae to give rise to dicotyledons and their derivatives bearing megasporangia, enclosed by leafy carpels in the apocarpous ovary. Whereas, Stachyosporae was attributed to giving rise to monocotyledons, monochlamydeae and a sympetalae group of plants designated by pioneers duo in plant taxonomy Bentham and Hooker (1862-1883), bearing megasporangia, covered by some sterile organs.
9) Mass mutagenic Hypothesis: The great plant phylogenetist, Goldschimdt (1940) postulated a unique form of mutagenesis on the earlier plants in a paleoclimatic environment in his "Mass mutagenic Hypothesis" to narrate the origin of primitive angiosperms. Though this hypothesis is a new line of thought to the plant phylogenetists to consider the inter-disciplinary research related to pollination biology, population genetics, cytogenetics, molecular biology etc. of the early group of plants to analyze and establish the phylogenetic relationships between an ancient group of the predecessors of angiosperms of the present day. However, the lack of evidence in support of mass mutagenesis among the progenitors of the present day angiosperms, make this hypothesis an unreliable format of explanation to unfurl the evolution of angiosperms.
10) Chloranthoid Hypothesis: According to this hypothesis, the angiosperms evolved from the primitive chloritoid ancestral stock which was structurally different from the contemporary members, belonging to Chloranthaceae (Leroy, 1983). According to recent data on cladistics analysis, the species *Hedyosmum* of Chloranthaceae was located at the clade basal of all angiosperms (Stuessy, 2004), hence it was considered as the mother-stock of all angiosperms.
11) Nymphaean, Nymphaean-Amborella-Magnoliid or ANITA Hypothesis: On the basis of developmental patterns and genetic evidences, and cladistics analysis, the molecular phylogenetists have considered the ancient, extant group of plants belonging to Nymphaeales (water lilies) and established an evolutionary link between phanerogams (flowering plants) and cryptogams (non-flowering plants) (Doyle and Donoghue, 1986). According to the recent findings on molecular systematics of Angiosperm Phylogeny Group (APG), Barkman et al., (2000) have considered that the members of Nymphaeales and *Amborella* (the member of Laurales and closely related to the family Magnoliaceae) and together form the rootstock of early angiosperms. The interesting primitive characteristic features of

Amborella were noticed by the absence of petals in its flower; whereas the primitive feature in the Nymphaean members was evidenced by the unique transitional floral structure, like a transitional form of stamens from the petaloid structure. On the basis of recent phylogenetic investigations on the diversified group of angiosperms in three consecutive levels (i.e., APG-I, APG-II and APG-III), Soltis et al. (2005), came to the conclusion that the members of Nymphaeales, *Amborella* (belonging to Laurales) and the three clades of Magnoliaceae (belonging to Magnoliales) polyphyletically formed the basal stock of angiosperms.

The embryological investigations on female gametophyte of angiosperms show that most of the female gametophytes are seven-celled and eight nucleate whereas critical observations on female gametophytes of *Amborella*, revealed that it is eight called and nine nucleate and embryologists expressed their concern about basal angiosperms as recent investigations reveal that the embryological profile of those primitive plants are remarkably different from other angiosperms (Friedman, 2006).

After traversing through the labyrinth of hypothetical phylogenetic lineages of ideas and interpretations of systematic botanists, plant anatomists, phyto-morphologists, paleobotanists, embryologists, cytologists, evolutionary biologists, phytogeographers and molecular systematists, the evolution of flowering plants has still been an enigmatic issue as no particular hypothesis of any of the nine (discussed above) resolved this phylogenetic puzzle. Basically, the enigmatic quest for ancestral lineages of the earliest flowering plants supposedly originated from either of these two distinct hypothetical basal lineages of plants:

i. Paleoherb form: The plants with herbaceous habit and faster (apparently) life cycle.
Or
ii. Magnoliid form: The plants with arborescent habit with the slower (apparently) life cycle.

A number of plants of a diverse group of plants, comprising of Aristolochiaceae, Piperales, Nymphaeales, Monocotyledonae etc. supposedly perform composite assemblages of the paleoherbs, which were earlier considered as dicots but recently been related to monocots (Reveal, 2012). The paleobotanical evidence ensured that the primitive form of reproductive organs of paleoherbs was characterized by a number of small flowers, growing around the floral axis. Age dating process of fossilized flowers of Piperales was evolved around 120 million years ago. So far as the vegetative features are concerned, the fossil form of leaves of Nymphaeaceae were discovered as the earliest form of fossil evidence in flowering plants.

The vegetative and reproductive features, phytogeographical observations and paleobotanical evidence ascertained that Magnoliid ancestry (i.e., the ancestor of all extant flowering plants should have an arborescent habit with slower life cycle) is more tenable (Hutchinson, 1969; Takhtajan, 1970, 1991; Stebbins, 1974). Takhtajan (1991) has favored on the reproductive features over vegetative features of Magnoliaceae to narrate the ancestral lineage of flowering plants as he has noticed that vegetative characteristics are reversible whereas the reproductive features are non-reversible. So far as vegetative characteristics are concerned, the arborescent habit of Magnoliids, were considered as relict forms and representing the predecessors of the primitive group of flowering plants in comparison with the contemporary, diversified, herbaceous successors of advanced groups of flowering plants and this evolutionary attainment of advanced forms from the primitive stock has presumably transpired as a result of the structural and functional weakening of the cambium tissues of the arborescent form of plants (Takhtajan, 1991). It was hypothetically conceived that the evolution of closed forms of megasporophyll (carpel), a distinct feature of female reproductive organs of Magnoliaceae, the progenitors of all flowering plants, supposedly rose from the strobilus of non-flowering, Cycadeoid plant by a series of transitions like: differentiation, oligomerization[25] and fixation (Bessey, 1915; Hutchinson, 1969; Stebbins, 1974; Takhtajan, 1991). The elongated axis of receptacle in Magnoliaceae was considered as resembling the hypothetical elongated axis of the receptacle of an ancient flower; though it was explained as an evolutionary reversal rather than retention of the relict form of plant life and spiral arrangements of floral parts on the axis of the receptacle of the Magnoliid flower was recognized as the relict feature of the angiosperms in comparison to the cyclic and whorled arrangement of flowers on the axis of the receptacle of diversified and advance groups of angiosperms (Bessey, 1915; Takhtajan, 1991). According to the clarifications of the petaloid structure of primitive stamens (which is devoid of filament or anther) of Magnoliaceae is supposed to be laminar in origin having three vascular traces and the microsporangia presumably embedded in the stamen in the early stages. The appearance of filamentous appendages from stamens were integrally related to the co-evolution of pollinators (a group of insects), strategically enhanced the pollination of an advanced group of contemporary flowering plants (Takhtajan, 1991). In support of the Magnoliid ancestry of flowering plants. After analyzing the chromosomal configurations of Magnoliaceae, Stebbins (1974) construed the chromosomal evolution of angiosperms in following steps:

i. aneuploid modifications of the basic, low number of chromosomal configurations of n=6 to n=8.
ii. polyploid modifications of a low number of chromosomal configurations to the numbers of n=19 to n=21. And

---

[25] Oligomerization: Its an incident of fusion of floral parts (e.g., sporophylls) of flowering plants and leaves of non-flowering plants.

iii. The structural diversifications in generic level with the inherent tantamount level of polyploidy etc.

Paleobotanically, it was evidenced that the disappearance of gymnosperms and appearance of angiosperms took place during the Cretaceous Period. The marshy, wet low-land terrestrial habitat of gymnosperms gradually receded with the disappearance of predominant, diverse groups of gymnosperms by the early groups of angiosperms in the changed climatic conditions on the modified habitat. The few representatives of flowering plants were reported in the upland areas as most of the primitive flowering plants have presumably grown in the lowland habitat, not so congenial for fossiliferous process, whereas the upland habitats and were not found to be a suitable abode of flowering plants due to the acclimatization process (Stebbins, 1974; Sporne, 1991). The high rate of extinction of gymnosperms in the post-Jurassic Period and rapid radiation (appearance of a large number of angiosperm family members in short period of time) of angiosperm families in the Cretaceous Period have made the transitional fossil evidences inconclusive to establish a direct transitional lineage between flowering and non-flowering plant groups (Takhtajan, 1991; Judd et al., 1994). Naturally, Magnoliaceae was recognized as a relict group as well as a "living fossil" to build an evolutionary mother-stock of all diversified, flowering plants (Stevens, 1980; Takhtajan, 1991). The age-dating process of fossiliferous Magnoliid flowers ascertained that members of Magnoliaceae appeared around 90 million years ago. Hence, the preferential claim of the ancestry of Piperales as the progenitor of the modern group of angiosperms was found to be more tenable when a comparison was made between Paleoherbs and Magnoliids. According to the contemporary DNA sequencing analysis or cladistics analysis, done by Martin et al., (1993), Magnoliaceae appeared as a branch of the Liliidae group. As it was not found in the basal root positions to be considered as the primitive ancestor of all extant angiosperms. Naturally, the direct hypothetical interpretations of "strobiloid ancestry of carpel" (closed megasporophyll) was not found valid anymore. Rather Martin et al., (1993) postulated that an ancestral basal rootstock originally separated from Magnoliids and that transitional mother-stock went through the gradual stages of independent reduction and process of simplification to form a diverse array of contemporary angiosperms.

12) Developmental and Genetical Hypothesis: Basically, it is a composite hypothesis where the evolution of angiosperms was elucidated by combining the evidence on organogenesis, development, and diversification of reproductive organs of flowering plants along with the consideration of its genetic properties, related to its ontogeny. The following hypothesis was taken into consideration to explain ontogeny of vascular flowering plants:

i. Growth retardation theory of angiosperms-proposed by Asama (1960, 1982)

ii. Gamoheterotophic Hypothesis- proposed by Meyen (1988)
iii. Mostly Male theory proposed- by Frohlich and Parker (2000)
iv. Out of male and out of female Hypothesis-proposed by Becker and TheiBen (2003); Theiben and Meizer (2007)
v. Evolution of the flower by the elimination of LFY gene paralog Hypothesis-proposed by Albert et al. (2002).
vi. Gene duplication and polyploids induced evolution of angiosperms-proposed by DeBodt et al. (2005).

Among all these hypotheses the Mostly-Male theory of Frohlich and Parker (2000) and De Bodt et al. (2005)'s genome duplication hypothesis need to be mentioned.

In their preliminary observations to postulate the "Mostly male theory," Frohlich and Parker (2000) stated that for testing their hypothesis they would have to test modern extant vascular seed plants without accumulating data on fossil primitive angiosperms. According to this hypothesis, the early-acting gene controls the evolution of flowers were evidenced in male reproductive organs of gymnosperms rather than in female reproductive organs. Later, on the basis of gene expression in the evolution of flowers, Frohlich (2002) projected Corystospermales as the earliest flowering plants, that evolved during lower Mesozoic Era. In the following year, Frohlich (2003) postulated that ectopic ovules on the abaxial leaf surface of gymnosperms attracted insect pollinators for expediting pollination and fertilization. Paleobotanical evidence of angiosperms and Corystosperms from same fossil site of Tasmania, Australia lead to discarding the Mostly male theory (Mcloughlin et al., 2008).

As per "Gene duplication and polyploids induced evolution of angiosperms Hypothesis" (DeBodt, et al., 2005), during the gene duplication module, the signalling genes, regulatory genes and the genes for organ development play a key role in the phenotypic variation and increase biological complexity. The consideration of the earlier polyploidy is used to elucidate biological evolutionary incidents like the evolution of flowering plants with certainty.

13) "Sneaky-Herb Hypothesis": The contemporary findings on molecular systematics, which have neither fully supported the Paleoherb Hypothesis nor Magnoliid Hypothesis to establish the ancestral basal lineage of a diversified group of angiosperms, have helped conceptualize a composite hypothesis, known as the "Sneaky-Herb Hypothesis." Reveal (2012) claimed that the ancient group of angiosperms was divided into two groups: i. the predominant, herbaceous, Piperoid group of "forest floor dwellers" and ii. Rare, arborescent, Magnoliid group of "marginal forest dwellers." It was further clarified that that period's forms have retained the relict form of herbaceous habit and diversified further, concomitantly, this group evolved aquatic plants too. In the due course of

evolution, the terrestrial herbaceous habit gave rise to dicots and monocots; whereas the aquatic habit was found to be an evolutionary bottle-neck for angiosperms. Reveal (2012) stated that the mass extinction at the juncture of Jurassic/Cretaceous Period and the post-Cretaceous climatic changes and changes of habitat lead to the partial extinction and limitation of expansion of arborescent the habit; whereas plants with herbaceous habit flourished due to better reproductive evolutionary advantages (e.g., double fertilization, better seed structure, seed dormancy etc.). The co-evolutionary relationship between plants and pollinators, the evolution of better defense mechanisms among flowering plants and pollination strategies etc. helped diversify the modern angiosperms (Reveal, 2012). It was also predicted that the evolution of Magnoliids of the arborescent habit, in the Cretaceous period would be another "evolutionary bottleneck" rather than considered to be an "evolutionary mother-stock" of diversified angiosperms (Reveal, 2012).

14) Gonophyll Hypothesis: According to Melville (1969), flowering plants evolved from Glossopterids of Gondwana in the Mesozoic Era and it was further elucidated that Glossopterids along with Gigantopterids and conifers formed the arid-subtropical Pangaean vegetation during the Permian Period and mostly disappeared when they through yhe Permian-Triassic mega extinction (around 250 m.y.a.) and survived by some living fossils belonging to glossopterids and conifers. Retallack and Dilcher (1981) presented a graphical analysis where they suggested the structural resemblance of Glossopterid fruits to the primitive angiosperm fruits. According to the non-cladistic hypothesis by White (1986), the Glossopterids micro-fructi were found as basal group upon Caytoniales, angiosperms, *Cycas*, Podocarpaceae, Araucariaceae, Casuarinaceae etc. sometimes branching and reticulate line of evolutions have also been found with those groups; whereas the Glossopterids mega fruictification was found to be the second basal group upon Ranalian angiosperms, monocotyledonous plants including *Pandanus*, Bennettitalean member *Williamsonia,* some cycads evolved.

15) Seed-fern Hypothesis: According to this hypothesis, angiosperms evolved from the Paleozoic and Mesozoic Seed-fern mother stock belonging to primitive gymnosperms (Arnold, 1947; Andrew, 1961; Doyle, 1978, 2006; Frohlich, 2002, 2003). According to Doyle (1994, 2006), cupulate seed fern, *Caytonia* was recognized as the ancestral stock of angiosperms as structural homology of fruit-bearing rachis of seed ferns were found to be functionally the same as fruits, carpels and seeds of angiosperms, whereas the morpho-anatomical analysis of both these groups are yet to be studied (Doyle, 2006). Further studies on Caytoniales by Arnold (1947) ascertained the true identity of Caytoniales as pteridosperms, evolved in the Mesozoic Era. Recent efforts were made to study the comparative analysis of reproductive organs between Glossopterids and Caytoniales to

establish the evolutionary lineage that lead to primitive angiosperms. While, reproductive features have not been critically analyzed, the permineralized reproductive organs of Vojnovskyalians, Gigantopterids of the Permian-Carboniferous Period but leaf venation and advance levels of paleobotanical analysis (chromatographic analysis of fossilized geo-chemicals "Oleanane," used for petroleum exploration as a biomarker), and leaf venation analysis to establish an evolutionary link with angiosperms (Doyle, 2006). The fossilized triterpenoid: Oleonone triterpenes, popularly known as "Oleananes, was found in the leaves of Gigantopterids, Bennettitatles, and flowering angiosperms but this important geochemical has not been found in other seed plants (Taylor et al., 2006).

According to Zhang (1995), and G. Sun et al., (2001) angiosperms evolved from Gigantopterids in 200-250 m.y.a. in four consecutive phases:

i. Pregnant stage (phyllospermous): from Permian-Triassic Gigantopterids and Sagenopteris.
ii. Adaptation stage (folded lamina in conduplicate carpels): In Triassic-Jurassic anthophytes (e.g., *Sanmiguellia* and *Schimeissneria*).
iii. Expanding stage: Vicariance of Pangaean seed-plant stock and dispersal from tropical area to Antarctic area in Middle Jurassic-Cretaceous Period.
iv. Flourishing stage: Radiations of angiosperms in the current biome in the Cretaceous-Neogene Period.

This hypothesis was categorically supported by Mathews and Donghue, 1999; Stuessy, 2004; Soltis et al., 2005; Miller, 2008, but due to a lack of enough paleontological evidence it could not be introduced as a theory to explain the evolution of angiosperms (Darrah, 1939).

16) Permo-Triassic: Thigmo-morphogenic Hypothesis: The critical reviews of the paleobotanical, palynological and molecular-botanical evidence failed to affirm "Cretaceous evolution and primary radiation of angiosperm Hypothesis." The theory was supported in part by observations and in part by assumptions, like the presence of primitive leaf fossils and considerably low angiosperm pollen species diversity in the Neocomian-Aptian (i.e., Lower Cretaceous) regime rock deposit and apparently little or no paleobotanical evidence in any rock deposits older than the Lower Cretaceous Period (Brenner, 1996). The eminent scientists on evolutionary biology like Grimaldi (1999), Crepet (2000) postulated that the evolutionary success of origin, evolution, and diversification of angiosperms in short period of time must not be related to evolution of the protective layer around the seeds (encasement of seeds inside the ovary) as a group of naked seeded plants or gymnosperms (e.g., Araucariaceae, Cheirolepidaceae and Ephedraceae)

evolved and diversified in the same conditions, rather the evolution and diversification of angiosperms is the ultimate result of symbiotic relationships between plants and pollinators as well as angiosperms and insects and floral evolution was induced by an insect-driven selection pressure (plant species specific pollinator) and floral evolution has carried out pollinator evolution analogously. The floral architecture helps portray the mode of pollination and indicate the taxonomic affinity of particular plant groups towards particular pollinators and the detailed fossil evidence of insect pollinators provide brief information on plant-pollinator symbiosis. Furthermore, it has focused on issues like angiosperm radiation and insect diversification in the backdrop of a temporal dimension (Grimaldi, 1999). In nature, plants have sustained different survival challenges and insect attacks where the touch of a pollinators is one of the important factors. In the course of the evolution of plants, a sensitive mechanism was evolved in the plant body to accept that stimulus and respond accordingly, known as "Thigmomorphogenesis" (Chehab et al., 2009). A number of signaling molecules and phytohormones are involved in this stimulus induced signal transduction process (e.g., Ethylene, Auxin, Nitric oxide etc.). Genes are involved in encoding proteins to induce a response system accordingly. When the touch of herbivorous plants could induce calcium sentencing or cell wall modification, the touch of pollinators induces optimization of cross-fertilization. Evolutionary biologists consider that angiosperm radiation is linearly proportional to anthophilous radiation which indicates that the evolution of derived floral characteristics and evolution of an advance group of anthophilous insects strictly linked to each other, which ascertains a "Co-evolution Hypothesis". According to Crepet (2000), "the rate of floral innovation/million years and the angiosperm diversifications during Cenomanian/Turonian (i.e., late Cretaceous) interval coincide with the evolution of a group of derived pollinators. Cornet and Habib (1992) moved a step further to narrate that in the pre-Oxfordian (i.e., the Late Jurrasic) regime, the low or non-existing "plant species specific pollinator" condition might not boost-up evolutionary pressure on angiosperms to evolve and diversify (rapid radiation) in a short time and they refer to the role (like domestication, agriculture, horticulture etc.) of different members of mammals like canines, bovines, human beings etc. in sub-speciation, variation etc. in the Pliocene to Holocene Epoch in the Tertiary and Quaternary Period. According to Labandeira, et al. (2002) there are four groups of insects involved in angiosperm pollination and according to the functional ability of the pollinators of these four groups as well as the rate of optimization of pollination among the different flowering plants, they are (in descending order) as follows:

1. Bees or Hymenoptera
2. Butterflies and moths or Lepidoptera

3. Flies or Diptera
4. Beetles or Coleoptera.

The presence of the primitive members of these groups in the Triassic Period and further through the diversification of each group in the Late Triassic (Climate optimum: Carnian), Early Jurassic (post mass extinction: Hettangian-Sinemurian) and Late Jurassic periods (floral recovery: Oxfordian Kimmeridgian), prior to the Early Cretaceous (Aptian-Albian) Period, the fossil evidence of angiosperms indicate the common appearance of angiosperms on Earth and the template of insect diversification has given a better presentation on origin, evolution, and diversification of angiosperms than the Cretaceous Hypothesis (Miller, 2008).

Chapter 18

IN THE QUEST FOR PRE-CRETACEOUS ANGIOSPERMS: FOLLOWING PALEO-PALYNOLOGICAL FOOT-PRINTS OF CIRCUMPOLLES AND CRINOPOLLES

From one end, the footprints of a tiger are clearly distinguishable from an elephant's; from the other end, the critical observations of footprints of two different species could ensure the existence of two different species and age-dating processes further ensure their chronological presence in a certain geological regime. Likewise, the analysis of paleo-palynological and palynological sample could primarily ensure the identity and existence of certain species in a particular period of time and further investigations could help determine the transitional forms in between the ancient/extinct line of origin to reach the extant form and focus on evolutionary radiations in the course of spatiotemporal change. Regarding interpretation of evolutionary radiations, Darwin's "phyletic gradualism" could be tested as a less functional principle after being challenged by "punctuated equilibrium." The initiation of evolutionary journey was mostly kicked off from the simple form to get to the complex form over the course of time; convincingly, the evolution of advanced forms of distinguishing "angiosperm pollen" should be initiated with the earlier appearance of a precursor "proto-angiosperm pollen" form, known as "Crinopolles group" (e.g., *Liliacidites, Monocriopolles, Nymphaeales, Polycolpopollis, Ricclisporites, Sanmiguelia, Tricrinopolites* etc.) showing affinity to the "basal angiosperms" or ANITA type, the basal angiosperms according to the recent cladistic analysis (Cornet, 1977, 1993; Brugman, 1983) and *Monocrinopolles* was found the dominant type among the Crinopolles group. Whereas, the gymnosperm pollens were found to be the descendants of "pro-gymnosperm pollen group' or "Circumpolles group" (e.g., *Chasmatosporites, Classopolis, Granuloperculatopllis*, *Praecirculina* etc.) showing affinity towards the Cheriolepidaceae group of gymnosperms also recognized as the "Partitisporites group" and *Classopolis* was found to be the most dominant conifer pollen, representing 90% palynoflora of the Rhaetian-Lower Jurassic border in Europe (Venkatachala and Goczan, 1964; Fischer, 1972; Scheuring, 1974; Brugman, 1983). Both of these groups coevolved in the Ladilian of the Middle Triassic period, supposedly in Laurasia but in different habitats, having gone through the phase of diversification through the Carnian, Norian, and Rhaetian Periods of the Upper Triassic age till the Triassic-Jurassic border of mass extinction (Dunay and Fischer, 1974; Litwin et al., 1991).

According to Milankovitch-type climatic (lacustrine) cycle, the evolution of the advanced form of a reticulte-columellate-monosulcate type of pollen from the simple form of the auriculate-polyplicate type was attained after 100,000 generations of plants (Cornet, 2002). Till the middle of the 19th century, the preponderance of paleobotanical evidence provoked the scientists to be blind supporters of the "Cretaceous angiosperm radiation theory"; but the missing link of this hypothesis, the transitional form of fossils with common features of gymnosperm predecessor and angiosperm successor has yet to be discovered, which is the primary flaw in this hypothesis. In order to narrow down this evolutionary gap, scientists have chosen an alternative path, hunting of paleo-palynological evidence such as palynomorphs (the pollens and spores) which were found to be collected from any geological strata of any spatiotemporal extent in an unaltered condition and the magical ingredient was "sporopollenin." The "sporopollenin" is a biopolymer, found in the walls of palynomorph which can protect the spore and pollen from being unaltered through the ages to sustain any extreme physio-chemical variables, so the identity of the pollen species remains the same forever, even after being released from the rocky substrate by using strong acid solutions. The advantage of working with paleo-pollens and the pollens of extant species is establishing the transitional phases in chronological order, the ancient plant groups, forming and reforming of flora and vegetation in response to variable palaeoclimatic, edaphic, environmental, geographical and geological changes that went through the temporal scale. The palynological methods of the 18th Century, which started under the compound microscope went through transitional phases of advanced technologies of TEM (transmission electron microscopy) and SEM (scanning electron microscopy) and advanced methodologies in sample preparations in the latter half of 19th Century and this progress was popularly interpreted as a 'shift of paradigm in palaeontology from 10X to ultrastructure' and contemporary palynological analysis is trying to go through the experimental threshold of 'nanotechnology'.

The ultra-structural palynological differences of angiosperms and gymnosperms are distinguished by features of angiosperm pollens as well as descendants of Crinopolles, like i. bi-layered pollen with the outer layer (ectexine) separated from the inner layer (endexine) by tubules or columellae; hence these are designated as a columellate type. Whereas the ectexine of pollens of gymnosperms as well as the descendants of Circumpolles, were found identical to angiosperms but the endexine of these pollens were found having multi-layered laminate (alternate stack of light and dark-coloured lamina) or laminated whereas the endexine of angiosperms was non-laminated and these types of pollens were designated as a granular type (Doyle, 1978). The presence of columella in angiosperm pollens and absence of non-laminated endexine has not helped in establishing the theory of the gymnospermous origin of angiosperms, which lead the phyllogenetists to look for an alternative path of origin of angiosperms. Another type of pollen was also noticed in the primitive angiosperms, apparently looking like granular pollens like gymnosperms but with non-laminated endexine (Cornet, 1989; 2002).

The graphical interpretation of a critical review of the Jurassic palynofloras in Laurasia have helped understand the dynamics of Crinopolles in comparison to Circumpolles and focus on the certainty of the origin of basal angiosperms in the Triassic–Jurassic Period. Though Crinopolles and Cricumpolles originated during the Middle Triassic Period, the dynamics of these were found to behighly influenced by the dynamics of the paleoclimatic regime in the course of time till the Triassic–Jurassic boundary which was presented by the extent of relative abundance in the different Triassic palynozones of the world. If the dynamics of the Crinopolles was graphically presented on a chart, the overall mode of representation of this group was looking like a gradual, converging radiant type (Figure 17a) having maximum representation in Carnian, going through the bottle-neck of Norian, to reach a minimal representation at the penultimate level at Rhaetian, before reaching the Triassic-Jurassic extinction borderline whereas the Circumpolles group was initiated with minimum representation in Carnian, getting higher representation in Norian; subsequently getting higher representation in Rhaetian to reach the Triassic-Jurassic extinction borderline in the shape of a gradual, diverging radiant type (Figure 17b).

The abundance of basal angiosperms producing Crinopolles group pollen types remarkably decreased in the Lower Triassic Period due to low adaptation and shifting of the paleoclimatic regime from moist tropics to an arid subtropical regime in Pangaea whereas the abundance of Classopolis type of pollen of the Circumpollis group of pollen produced by the Conifers, the members of Cheriolepidaceae, were found to be remarkably increased due to the adaptation to favorable climatic conditions. The dominance of Cheriolepidaceae and surviving struggle of a minor group of basal angiosperms since the Carnian of the Upper Triassic to the Albian of the Lower Cretaceous periods, it should better be interpreted as (in the boundary of Triassic–Cretaceous) the abundance of Circumpolles, as inversely proportional to Crinopolles, possibly due to better adaptation and sudden surge of diversity of angiosperms and the disappearance of Cheriolepidaceae could be the result of the combined effects of:

i. Reversal of the favorable paleo-climatic regime for Crinopolles
ii. The extinction of members of a cheriolepidaceae group of advanced gymnosperms in the struggle for existence during Triassic-Jurassic mega extinction calamities.
iii. The radiation of depleting members, belonging to basal angiosperms after the Triassic- Jurassic mega-extinction and the immediate occupying of vacant niches, left by Cheriolepidaceae.

The continental drifting induced movement of Pangaea in the middle of the Panthalsa during the Carnian as well as Lower Triassic periods has conceivably gone through a change of palaeoclimatic regimes. Scientists reported that around 226-230 m.y.a., the progenitors of angiosperms evolved in short periods of time in the paleo-equatorial as well as tropical rift basin and those primitive floristic elements or palaeotropical vegetations were represented by the palynofloras, having an identical affinity to the angiosperm pollens, also known as "Crinopolles" or lily-like pollens (e.g., some of the Crinopolles looking like polyplicate type of pollens, found in the extant members of Araceae) (Cornet, 1989a). Eventually, the distributional index of crinopolles in paleo-palynoflora was estimated at 1-2%. The age dating process of these crinopolles helped determine the Triassic period of evolution of these progenitors of angiosperms. Naturally, the evidence of the existence of angiosperms or angiosperm-allies was found as a fact against the biased claims of the "Cretaceous school."

The appearance of crinopolles hold an immense evolutionary importance to elucidate the justification of the "punctuated equilibrium" and chronological deciphering of a particular period of time for "angiosperm radiations" and understanding of the following features are as follows:

i. It seems that pollen sculpture, outer morphology (particularly exine) and biochemistry plays a key role in the successful execution of sexual reproduction by maneuvering the way of its dispersal, germination, and formation of the pollen tube, mode of fertilization. The factors like successful execution of reproduction and radiation of angiosperms as well as the evolutionary progress of any living organism depend upon structural evolutionary changes of pollen.
ii. Pollen is responsible for 50% execution of sexual reproduction. The male gametes, as well as pollens, are repositories of all genes and could change genetic information, causing mutations for short-term and any irreversible changes for the long run to yield a new form of organism with unique features, might be recognized as a new species in future. So, the appearance of crinopolles in the Upper Triassic Period has played a key role in the evolution of angiosperm pollen.
iii. The variation of organs and irreversible changes from one generation on the way to achieve the objective of formation of a species (with new features) were found as a cumulative or interactive results of multi-gene that's why the chance of

evolutionary progress of this kind of macroevolution is higher compared with one or few number of genes responsible for variation of pollen architecture and wall structure (particularly the exine). This is an ideal instance of microevolution where pollen morphology or ultrastructure remained unaltered for a long time.

iv. According to the contemporary level of palynological investigations of around 12 crinopolles regarding its origin, evolution and extinction were determined to occur in one complete cycle of the equinox (around 100,000 years) (Olsen et al., 1996; Kent and Olsen, 2000), with a surge of its diversity in a couple of precision sub-cycles of the equinox (around 50,000 years). The cladistics analysis has indicated that the microevolution of crinopolles is the combined effect of less functional, phyletic gradualism and punctuated equilibrium.

v. According to Wolfe et al., (1989), the appearance and evolution of crinopolles, which supposedly took place around 220-240 m.y.a., and the features like the germination of pollen on stigma, progress of pollen tube to reach at encased ovule, formation of viable seeds etc. clearly point out the period of origin and evolution of angiosperms.

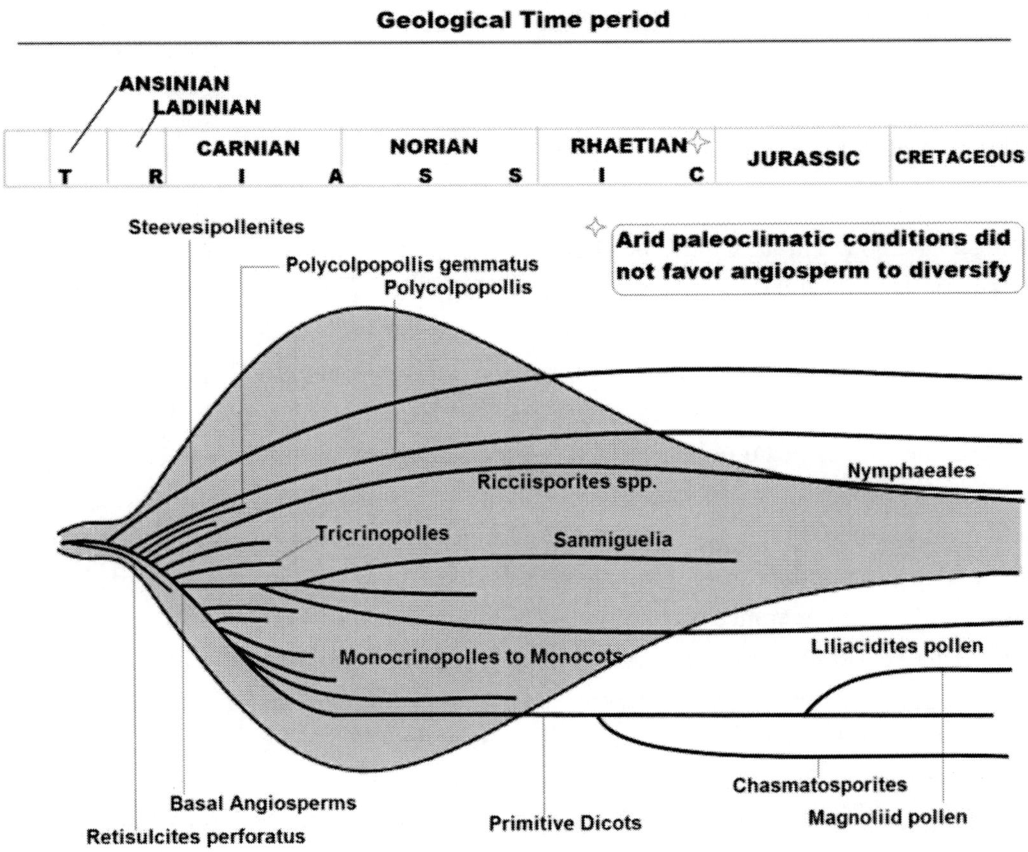

Figure 17a. The divergence and diversification of Crinopolles group during Triassic (modified after Figure 71 of Bruce Cornet, 2008).

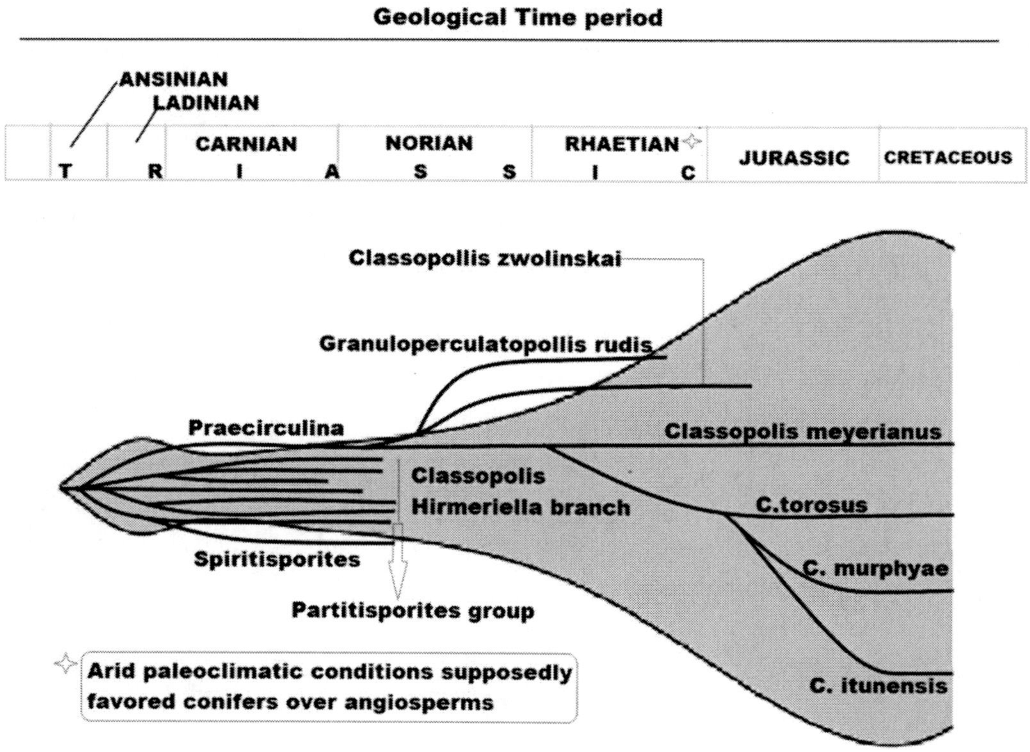

Figure 17b. The divergence and diversification of Circumpolles group during Triassic (modified after Figure 71 of Bruce Cornet, 2008).

According to the Cretaceous school, multi-aperturate angiosperm pollen evolved from monosulcate pollen in the Cretaceous Period but Cornet (2002), argued that the appearance of crinopolles took place in the late Triassic Period (way before Cretaceous) and went through a struggle for existence with various numbers of pollen apertures to cope in different palaeoclimatic regimes for 100 million years. Cornet (2002) further opined that in the course of time, the species bearing multi-aperturate pollen gradually disappeared, whereas the monosulcate pollen bearing species survived and he explained furthermore that the evolution of monosulcate pollen should be considered as a result of merging of a number of distal apertures (as a result of coalescence) (found in monocotyledon group of angiosperms) rather than an evolution from gymnosperm pollen type with a single aperture. Based on the preponderance of columellate palynomorphs as well as pollen species in the Cretaceous Period, the phylogenetists of the early nineties have come to the conclusion of "Cretaceous radiation of angiosperms" but the existence of columellate pollens and granular but non-laminated pollens in the pre-Cretaceous age in the late nineties discarded the earlier version of morphology-based phylogenetists claim of the "Cretaceous theory."

The contemporary palynological researchers on Crinopolles and advancement of the researchers of molecular botanists to find out the "period of appearance of angiosperms" on the basis of molecular evidences and analysis in a multi-disciplinary approach, helps consider the period between the Lower Jurassic and Upper Triassic Periods should be considered as the time of birth of basal angiosperms rather than Cretaceous Period. The early angiosperm pollens in Lower Jurassic from Sweeden and United States (Nilsson, 1958; Cornet and Traverse, 1975) have indicated the appearance of angiosperms in Laurasia before the Cretaceous period. The presence of megafossils and angiosperm pollens from North America in the Triassic period (Cornet, 1989b, Fowell et al., 1994; Hochuli and Feist-Burkhardt, 2004) and angiosperm pollens from West Europe in Jurassic period (Cornet and Habib, 1992) have indicated the evolution of basal angiosperms in Laurasia and some of those basal angiosperms supposedly moved northward in order to avoid the arid climatic regime as well as moist tropical climatic regime across the dry desert-like palaeoequator IAL zone (Yang and Lin, 2005). The migration of basal angiosperms from Lauratia to Gondwana was restricted by arid, climatic barrier in pre-Cretaceous Period, which was changed by the reversal of moist-subtropical regime in the Lower Cretaceous Period and it helped the early angiosperms migrate southward across the Strait of Gibraltar and moving forward to cross the rift basin of Gabon in between Africa and South America to reach Gondwana from Laurasia during the Cretaceous Period.

The paleobotanical and paleo-palynological evidence indicates that a group of angiosperms migrated to Gondwana, colonized the north-eastern coastal region of Gondwana e.g., *Liliacidites, Clavatipollenites* etc. (Brenner, 1996). Recent palynological investigations ensured that monosulcate-reticulate angiosperm pollens were recorded from Namibia, Southern Africa in the Carnian Period of the Upper Triassic Period, which indicates that even some of the basal angiosperms were able to migrate before a climatic barrier set in (Cornet, 2002). The disappearance of climatic barriers in the Berriasian Period of the Lower Cretaceous Period opened up the rifts as the sea barrier widened up, stunting the chance of angiosperm migration from Laurasia to Gondwana again in the Hauterivian of the Lower Cretaceous Period, leading to the diversification of certain angiosperms in Laurasia, belonging to the family Chloranthaceae and diversifications of angiosperms, belonging to Winteraceae in Barremian of the Lower Cretaceous Period (Doyle and Endress, 2000; Friis et al., 2000).

Like a number of scattered finger-prints in support of the "Pre-Cretaceous origin of angiosperms, "Auriculae," the sculptural witness of the pollen grains played an important role in the "palynological litmus test" to determine the tentative divergence of angiosperms. Under the ultramicroscopy, "Auriculae," is a tuberous outgrowth, found on the outer wall (exine) of the apical region of pollen grains; might also be dispersed all-over the exine (known as clave) (Cornet, 2002). The existence of these typical sculptural fingerprints was found in the polyplicate pollens of seed plants in the Mesozoic (Triassic, Jurassic, Cretaceous) Era, Tertiary Period and are still found in the extant monocot pollens in Quaternary Period (e.g., Crinopolles of *Pentecrinopollis gemmatus*, *Ricciisporites unbonatus* in Carnian-Norian of Upper Triassic; crinopolles of *R. tuberculatus* in Rhaetian of Upper Triassic; palaeo-pollen of *Ricciisporaties* sp. in Albian of Lower Cretaceous; *Elaterocolpites castelainii* in Cenomanian of Upper Cretaceous; *Bomarea lycina* in Tertiary and operculate-clavate- monosulcate pollens in Quaternary) (Lundblad, 1954; Felix and Burbridge, 1977). According to Cornet (2002), "Auriculae" supposedly the "synapomorphic for angiophytes (i.e., stem-group + crown–group angiosperms), representing expressed or non-expressed (archived) genes that survived in some crown group and recent angiosperms." Based on the presence of "aureculae," the morphotypic, palynological resemblances has indicated that extant pollen of *Nymphaea tetragona* (of the family Nymphaeaceae), found a member of the basal angiosperm complex, inherit the "auriculae synapomorphies" (Cornet, 2002).

On the basis of molecular data analysis, Martin et al. (1993), have considered the members belonging to Nelumbonaceae (*Nelumbo nucifera*, popularly known as Oriental lotus and *N. lutea*, known as American lotus) under order Nymphaeales, have considered as basal angiosperms, phylogenetically segregated from rest of the angiosperms (including Magnoliidae). Based on the contemporary molecular data analysis of Friis et al. (2001) and Doyle (2006) have put the Nymphaeales at the bottom clade in their phylogenetic tree, also known as "Cladogenesis of Seed Plants." Among the order of Nymphaeales, members of the genus- *Nymphaea* as well as Nymphaeaceae are noticeably different from the genus- *Nelumbo* or Nelumbonaceae and a palynological (exine sculpture), transitional evidence was noticed between these two family members. The sculptural pattern ascertained that *Nelumbo* possesses the palynological features of typical angiosperm like reticulate, columellate pollen whereas *Nymphaea* possesses granular to echinae to clavate pollen (e.g., *Pentecrinopollis gemmatus*), apparently found among gymnosperms. The typical angiosperm exine evolution- "reticulate-columellate" type of pollen was considered as the derived feature of *Nelumbo* by contemporary phylogenetists (Cornet, 2002), which has indicated towards two hypothetical options: either pre-Cretaceous existence of pre-Crinopolles (pre-tectate form of pollen; either monosulcate or zonasulcate) in *Nymphaea* lost the eectate exine sculpture in the course of time. The phylogeneticists strongly opposed the gymnospermous origin of angiosperms, sometimes referred to for their clear structural differences, between the orientation of saccate pollen of *Caytonia* and monosulcate pollens of basal angiosperms, but stratigraphic interpolations indicate that both groups have supposedly originated from the common ancestor and in in the due course of evolution both groups have lost sac-like structures and it was assumed during the investigation of the ovular structure of *Callistophyton* (Doyle, 2010). The position of "Crinopolles" in the Doyle's (1998) model of "Cladogenesis of Seed plants," was found just above of the clade "Nymphaeales" helped to establish synapomorphic relationship between the members of Nymphaeales and Amborallaceae and Crinopolles on the basis of "Pre-Reticulate – Columellate" exine sculpture of pollens.

Later, a variation of pollen aperture in *Nelumbo* was witnessed, ranges from trisulcate to tricolpate (Kuprianova, 1979) and the cladistics node of *Nelumbo* was placed above

Choloranthaceae but below the Eudicot in the revised work of molecular phylogenetic analysis of Doyle and Endress (2000). According to Kuparinova (1979), the palynological transition starts from monosulcate exine sculpture to evolve to Tricolpate dicot pollen. But palynological studies of (Cornet, 1989a), based on morphological and geological evidence indicate that monosulcate is the derived one that has discarded the "Monosulcate progenitor of Tricolpate Hypothesis" of Kuparinova (1979) where the trisulcate condition of *Nelumbo* is vestige in this line of evolution. Furthermore, Cornet (2002) has argued that tetrad formation (either radial or bilateral) was noticed in the Triassic paleo-pollen of Tricrinopollis and the extant member of *Nelumbo*. Hence, both morphotypes could be considered as the progenitor of Tricolpate pollen. The structural resemblance of *Nelumbo* and *Tricrinopollis* have indicated further that Nelumbo pollen supposed to be the "Crinopolles survivors" (where the expression of genes were suppressed and archived, and later reactivated) of Triassic/Jurassic mass extinction and radiated in Cretaceous (Doyle, 1978; Cornet, 2002).

The basal angiosperms should be considered as survivors of Crinopolles descent and stoically addressed issue like "in quest of pre-Cretaceous origin of angiosperms" is a shift of paradigm to come out of biased, stereotyped ideas to look for truth, hidden somewhere where following the palaeo-palynological foot-steps could help to solve this mystery: "Time of birth and radiations of green angiosperms on this blue planet."

Chapter 19

CLADOGENESIS OF SEED PLANTS – IN SPECIAL REFERENCE TO DIVERGENCE AND RADIATION OF FLOWERING PLANTS

The mode of organic evolution on Earth has typically been categorized into two categories- Anagenesis and Cladogenesis. While "Phyletic gradualism" supported the mode initial phase of organic evolution in accordance with "Anagenesis," a slow, gradual, continuous mode of an evolutionary pattern where parental form remains unsplit and gone through gradual changes to diverge a new form in course of long time span. In contrary, "Cladogenesis," is a short-term evolution, where a parental form splits to form two new, distinct forms (from supra-species level onwards) to be recognized as "Clade" (literally its dichotomous or branching formation of new clades from parental clade). Genetically, cladogenesis is the splitting of an ancient gene pool to two new gene pools, could give rise to new species in course of evolution (in short span of time). In order to have an idea of the origin and evolution of a particular group of an organism like seed plants and angiosperms and to have feasible explanations of the critical issues like- "radiation of flowering plants" in Cretaceous, a simulative phylogenetic analysis is necessary for which cladistic analysis through the computer-aided algorithm is the best option. In this regard, the contemporary phylogenetic studies of Smith and Goldberg (2015) on macroevolution of flower color should be referred for better understanding of phylogenies of floral radiations; they have tested the gain or loss of anthocyanin pigmentation on four clades of flowering plants and it was observed that the mode of color shifting in different clades were found in "gain" mode over "loss" mode, controlled by genetic, ecological and physiological factors and this macro evolution of color of flower was evolved through cladogenetic mode of evolution rather than anagenetic mode of evolution in course of co-evolutionary divergence of plant-pollinator relationship.

On the basis of a set of comparable features of the seed plants, belonging to different orders, Crane (1985), the pioneer worker on "cladogenesis of seed plants," has set up a cladistics frame-work to reconstruct the evolutionary history of seed plants, emphasizing on the character polarity, homology, stratigraphic history and evolutionary relationships and this was considered as maiden approach to deal with applied aspects of cladogenesis. On the basis of initial observations of morphological and ultrastructural features, palaeobotanical observations, critical analysis of molecular data of the seed plants, compiled with the previous cladistics analysis done by Crane (1985); Crane et al. (2004); Doyle (2006, 2008, 2010) has attempted to reconstruct the evolutionary history of seed ferns (gymnosperms) and flowering plants and divergence of basal angiosperms by forming an advanced cladistics framework. On the basis of molecular, cladistics analysis of Crane (1985), the validity of "Anthophyte Hypothesis" of origin of angiosperms was discarded (Qui et al., 1999; Donoghue and Doyle, 2000). Based on the DNA sequences of the major genes, involved in photosynthesis actions, Chase et al., (1993) have published the phylogenetic works on 500 angiosperms, on the basis of computational analysis of large molecular data. Surprisingly, the result of DNA sequencing has presented that all monocotyledonous plants belonging to a group, unlike the dioicts like – magnolias, roses, daisies, don't appear as a group; apparently, some of the distant groups of plants like *Nelumbo* and *Platanus* were found phylogenetically close related.

According to the contemporary reviews on plant species on the Earth, so far 374,000 species were identified; out of the total plant species, there are 44,000 simple Thallophytes or algae; followed by 21, 925 species of archegoniate Thallophytes or bryophytes (comprised of 9,000 liverworts, 225 hornworts and 12,700 mosses); followed by 11,850 embryophytes or pteridophytes (comprised of 1.290 lycopods and 10,560 ferns); followed by 1079 naked seeded plant species or gymnosperms and rest of the vascular plant species (Christenhusz and Byng, 2016). There are 308,312 vascular plants were recognized, having 295,383 flowering species (comprised of 74, 273 monocots, 210,008 eudicots and a number of dicots) (Christenhusz and Byng, 2016). Based on the comparative morphological, anatomical and embryological resemblances of vascular plants, Coniferales and Gnetales were recognized as phylogenetically distant groups in flowering seed-plants earlier, but cladogenesis of seed plants on the basis of molecular data ensured that these two groups are phylogenetically closely related (Qiu et al., 1999); Magallon and Sanderson, 2002; Burleigh and Mathews, 2004).

Based on the morphological, embryological features, phylogenetically, the close proximity of Gnetophytes were established earlier than the angiosperms were later found to be quite distant after analyzing the molecular data and reconstructing the cladogram (Donoghue and Doyle, 2000). In his cladistics analysis, "Cladogenesis of seed plants," Doyle (2008, 2012) has dealt with 36 clades of seed plants; out of which 22 seed plants are seed gymnosperms and rest of the clades are belonging to the flowering plants

(Figure 18). Legacy of Crane (1985) has further been carried forward by a big crowd of contemporary phylogenetic investigators like – Chase et al., (1993); Crane et al., (1995); Goremykin et al. (1997); Nandi et al. (1998); Magallon et al. (1999); Barkman et al. (2000); Sanderson and Doyle (2001); Hilu et al. (2003); Chaw et al. (2004); Jansen and Brenner, (2004); Sanderson et al. (2004); Magallon and Sanderson (2005); Doyle (2006); Friis et al., (2006); Moore et al. (2007); Zavada (2007); Kramer, (2009); Doyle, (2012) and finally the molecular phylogeny has able to draw a cumulative, "Bigger Picture" to resolve the evolutionary mystery "origin of seed plants" by Soltis et al. (2000) and Doyle (2008), in the monumental works "Cladogenesis of Flowering plants on the composite analysis and consideration of - molecular data, phylogenetic systematics, palaeobotanical evidence and stratigraphic history."

Palaeobotanical research on the well-preserved fossil remnants of reproductive organs of flowering plants during the Cretaceous period have helped to elucidate the radiation of angiosperms (Friis et al., 2006; Magallon and Castillo, 2009). On the other hand, the advancement of computer algorithms and methodologies to analyze a wider array of molecular data have helped to track the floral evolution, phylogenetic relationships among different group of angiosperms (Irish, 2006; Bell et al., 2010; Endress, 2011) as well as diversification of flowering plants as well in last decade.

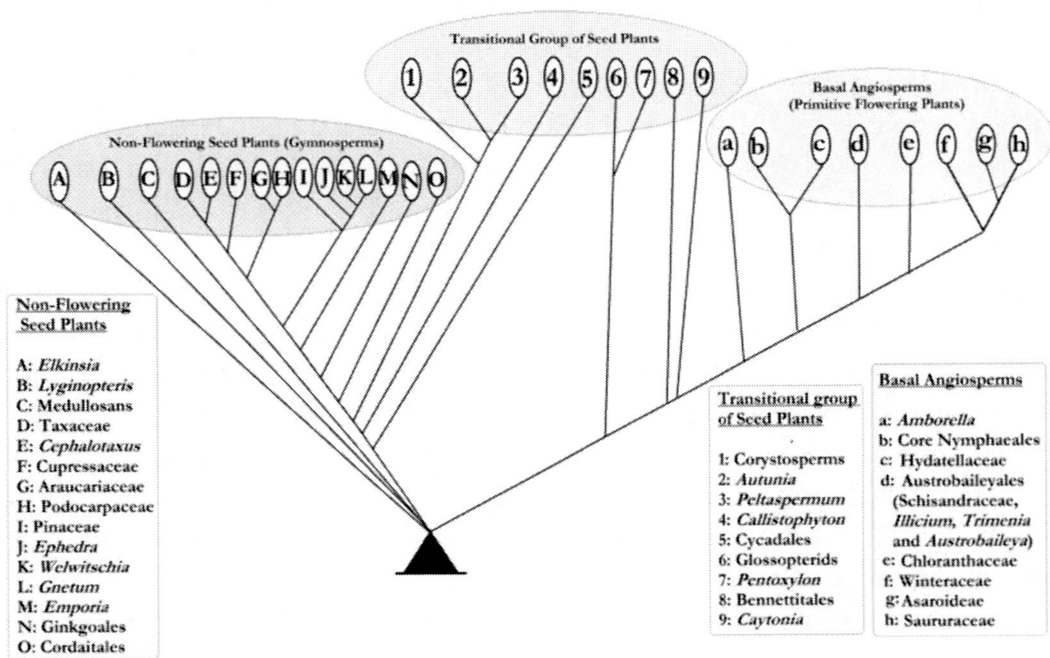

Figure 18. The phylogeny of early seed bearing plants (modified after Doyle, 2008).

Table 2. Palaeobotanical existence of the transitional, non-flowering seed plants through Carboniferous to Paleogene

Name of the Order	Carboniferous	Permian	Triassic	Jurassic	Cretaceous	Palaeogene
Bennettitales	-	-	+	+	+	-
Caytoniales	-	-	+	+	+	-
Corystospermales	-	+	+	+	+	+
Glossopteridales	-	+	+	+	-	-
Nilssoniales	-	-	+	+	-	-
Pentoxylales	-	-	-	+	+	-
Petriellales	-	-	+	-	-	-

[Based on primary palaeobotanical data (Cornet, 2008)], (-): Absence (+): Presence.

Based on the cladistics analysis, phylogenetically, the cladograms of a number of seed plants (comprising of non-flowering seed plants or gymnosperms and flowering seed plants or angiosperms) was reconstructed in vertical arrays (Doyle, 2002). The vertical arrays of cladogram of seed plants has broadly been dichotomized in two clusters, apparently correlated as polyphyletic groups, on the horizontal plane; the upper cluster comprised of 20 clades of non-flowering seed plants or gymnosperms (starting from *Elkinsia* and end in Cycadales) and the lower cluster was comprised of 16 clades (starting from Glossopterids and end in Saururaceae), comprised of 12 basal angiosperms (*Amborella*, Austrobaileyales, Hydatellaceae and Nymphaeaceae) or ANITA complex (the predecessor of most of the extant flowering plants) the from bottom part of lower cluster (Mathews and Donoghue, 1999; Wikstrom et al., 2001; Zanis et al., 2003; Leebens-Mack et al., 2005; Qiu et al., 2006; Saarela et al., 2007; Doyle, 2008) and interspersing 4 non-flowering seed plants or gymnosperms (comprised of Glossopterids, *Pentoxylon*, Bennettitales and *Caytonia*) on top of basal angiosperms (Frohlich, 2002; Soltis et al., 2005; Doyle, 2006) as paraphyletic group. Among four members of non-flowering seed plants, *Pentoxylon* has maintained a monophyletic relation to Glossopterids and *Caytonia* was found to be the sister clade of basal angiosperms "ANITA" complex (Soltis et al., 2005; Doyle, 2008) in the contemporary phylogenetic tree reconstructions, done by molecular botanists. According to the classical and contemporary phylogenetic reconstruction works, a group of evolutionary biologists considers "Corystospermales" as progenitor of angiosperms (Frohlich, 2002), but contemporary reconstruction works on cladogenesis of seed plants assigned this group of non-flowering seed plants in the upper cluster among other non-flowering seed plants like *Peltaspermum*, Ginkgoales, *Gnetum*, Pinaceae etc. rather than closer to basal angiosperm cluster (Doyle, 2008; Soltis et al., 2009). The critical flaws of these contemporary, phylogenetic tree reconstructions of seed plants that most of the models had not considered inclusion of "Gigantopterids" seed-bearing plants in Permian (Moldowan and Jacobsen, 2003). Due to lack of critical analysis of fossiliferous reproductive organs due to lack of preservation of fossil or complex process of permineralization (Li and Yao, 1983), though most important biomarker "Oleanane" of flowering plants have also been extracted from Bennettitales and Gigantopterids which gives an indication of origin of angiosperm in Permian rather than the Cretaceous.

According to most of the contemporary phylogenetic trees (on the basis of molecular data), the transitional group of seed plants (comprising of Bennettitales, Caytoniales, Corystopermales, Glossopteridales, Peltaspermales, Pentoxylales and Petrietales etc.) in between upper cluster of non-flowering seed plants and basal angiosperm (i.e., primitive flowering plants) of bottom cluster were found as a transitional node of evolution seed plants from non-flowering stage or "naked seed" stage to flowering stage or "closed seed" stage. The palaeobotanical footprints (Cornet, 2008), corroborated the molecular phylogenetic evidence (Table 2). The critical observations on the palaeobotanical history of these transitional group of non-flowering, seed plants and co-evolutionary evidence (referred in the Figure 13 earlier) from Permian to Cretaceous (Cornet, 2002; Zheng et al., 2014) have substantiated that the divergence of flowering plants occurred much earlier than the Cretaceous Period.

The critical reviews on the fossil history of the transitional group of seed plants, mentioned above, have indicated that those plants (including Gigantopterids) have either appeared or existed in Permian or little earlier and disappeared during early Cretaceous which is concomitantly overlapping with the geological period of origin and divergence of flowering plants. According to the contemporary phylogenetic analysis on age determination of basal angiosperms by adopting the "Penalized likelihood" cladistic methodologies, it was determined that primitive angiosperms evolved around 251.7 million years ago on the Earth (Schneider et al., 2004) and the big surprise is that the "Penalized likelihood" cladistics methodologies has also determined that concurrent occurrence of "EPE" (End Permian Extinction), overlapped with evolution of basal angiosperm. This has helped to hypothesize the primitive angiosperms are the luckiest survivor of End Permian Extinctions; indirectly, it has attributed more strength to the "Permian- Triassic origin of basal angiosperms" rather than so-called "Cretaceous Hypothesis." Phylogenetically, there is a clear distinction between "Crown group" and "Stem group" of angiosperms. The crown group of angiosperms is the monophyletic origin, comprised of extant members and the immediate ancestors of any particular clade, whereas the stem group was recognized by the members starting below the crown group of that clade to end up to the node, split up from the sister group. As an example, if the family Proteaceae is considered, the crown group of Proteaceae is comprised of the extant taxa along with few fossil taxa, so age determination of evolution of crown group of any particular taxonomic rank is considerably easier, whereas determination of time of evolution of the stem group of Proteaceae family is quite difficult as the gap between the node of splitting from sister group family Platanaceae to the base of crown group of Proteaceae is quite long. It may be 100mya and the ancestral lineage of this stem group of angiosperms might have some structural resemblance to gymnosperm. The phylogenetic investigations, based on analysis of molecular datasets of flowering plants by Bell et al. (2010); Smith et al. (2010) and Magallon et al. (2015), ascertained the age of angiosperms in between Paleozoic to Cretaceous and diversification of angiosperms supposedly took place in early Cretaceous. The phylogenetic investigations of Bell et al. (2010) determined that diversification of a crown group of angiosperms were taken place around 132 mya, supposedly in between Lower Jurassic and Lower Cretaceous. In order to reconfirm the "Permian-Triassic origin of basal angiosperms," cladistic analysis of molecular data on "crown group" of angiosperms of Thornne and Kiishino, (2002) and Bell et al. (2005; 2010) should be referred here where molecular phylogeneticists have used Bayesian "Relaxed Clock" methodologies and them able to determine that "crown group" of angiosperms have evolved between 106.1 - 229.4 million years ago. Bateman et al. (2006) focussed on applied aspects of cladistics analysis to estimate the age of divergence of the "stem group" and the "crown group" of angiosperms.

In the 17th century, the journey of classification of plants has started with artificial system of classifications in "Species Plantarum" of Linnaeus (1753) has lasted for two and half centuries, followed by most acceptable natural system of classification. Popularly known as "Bentham and Hooker's system of classification" in 19th century; subsequently followed by so-called "phylogenetic" (which is refined form of natural system of classification) system of classifications of "Cronquist and Takhtajan system of classification" in the 20th century. All these systems of classifications were basically the result of individual efforts of taxonomists and the framework of classification had regional attribution. The beginning of 1990, initiated the new paradigm shift from classical taxonomic classification pattern to the evolutionary exposition of molecular phylogeny and which was effectuated by unique marriage between "DNA sequencing technology" and "Computer algorithm." Though, the preliminary phase of "cladistics studies" there was the individualistic approach of "molecular taxonomists" like their earlier predecessors "classical taxonomists" and thematically, the ultimate objective of their findings would not overcome the influence of regional perspective and this trend has continued until the end of the 1990s. Since 1998, the investigative approaches of molecular taxonomy or molecular botany oriented research radically changed from individual effort to collective efforts of a group of scientists from a number of countries; naturally the perspective of phylogenetic investigations also changed from regional level to global level and the birth of APG or Angiosperm Phylogeny Group in 1998 with their maiden approach on "phylogenetic classification of angiosperms" was initiated. This began a new era to decipher the evolutionary divergence of flowering plants. The phylogenetic classification of diversified group of flowering plants of APG (1998) has clearly established the monophyly of different groups above the rank of family. In this supra-familial phylogenetic system of classification, around 462 families, under 40 monophyletic orders and some monophyletic informal higher groups evolve the unique branching pattern of the dendrogram; in 2016, which was presented with required modifications (APG-IV) (Figure 19). Noticeably, a number of families have unique phylogenetic flexibility either recognized as "Bracketed families" (e.g., Asparagaceae has 7 bracketed families under it which have all independent familial status as well) comprised of a number of families under it (*Sensu Stricto*) or all those families could be recognized as all independent families (*Sensu Lato*) under a certain order. The major orders of the flowering plants were assigned under supra-ordinal groups or clades- monocots, commelinids, eudicots, core eudicots, rosids (having 2 subgroups of eurosids- I and eurosids-II) and asterids (having 2 subgroups of euasterids-I and euasterids-II). A number of families were assigned under informal groups, as these families have not been assigned under specific orders and a separate list of families were appended, those don't have enough molecular data to be assigned in the informal groups. The jackknife value (>50%) was mentioned on each branch; the first jackknife value was derived from analyzing of 545 sequences of genes (*atp*B, *rbc*L, and 18S rDNA), whereas the second jackknife value was derived from 2538 *rbc*L gene sequence (APG, 1998). Despite, enjoying it advantageous benefits, the downside of this phylogenetic system of classification of angiosperms, are assigning a number of families that were assigned under the supra-ordinal groups (e.g., Dilleniaceae, Gunneraceae, Vitaceae under core Eudicot clade) but have not been assigned under any specific orders. The list of families (i.e., 25 families starting from Balanophraceae to Tepuianthaceae), show no phylogenetic interrelationship to assign in any order or supra ordinal group and the position of the basal clades (also known as basal angiosperms)[26].

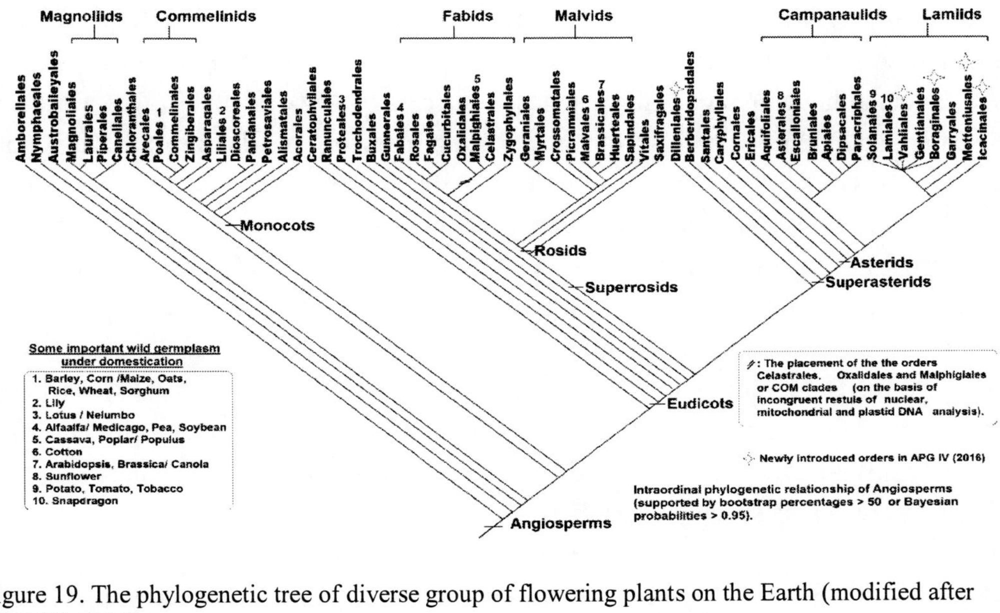

Figure 19. The phylogenetic tree of diverse group of flowering plants on the Earth (modified after APG-IV, 2016).

[26] Evolutionary view point the root-stock of all angiosperms, maintained synapomorphic relationships with other clades is not well defined.

The revised version of APG 1998 classification was published in 2003, known as APG-II (2003) phylogenetic classification of flowering plants, where earlier framework of APG (1998) was remained unchanged like main cladistics formats, the flexible status of 55 bracketed families etc. with some major modifications like- i. assignments of most of the families under supra-ordinal groups without any specific orders to assign in different orders with certainty (on the basis of critical analysis of newly accumulated molecular data), ii. Restructuring of the families, previously assigned in different clades, iii. An effort was made to the unassigned group of families to accommodate them in existing supra-ordinal frameworks, iv. Reorientation of those families under basal clades under specific orders etc. finally, APG-II (2003) has accommodated 457 families under 45 orders and the major success of this classification is bringing down the number of un-assigned families from 25 to 3; whereas 36 families were assigned in supra-ordinal groups or clades without assigning these families in any existing orders and it has left an area to explore and add more orders in the existing cladistics framework.

In APG-III (2009), the phylogenetic liberalization was witnessed in the revised version of APG-II (2003), where main cladistics framework was remaining unaltered with major resolutions on the bracketed families, to be recognized as inclusive forms (*Sensu Stricto*) instead of abolishing it early privilege to be recognized as all independent families to clear-up infra-familial ambiguity in molecular taxonomy of flowering plants. At the same time 20 new families were included in the APG-III (2009), not included earlier. As a result of restructuring, the number of families has gone down from 457 to 415 and the number of families under supra-ordinal groups, without any orders have gone down from 36 to 10 in APG-III (2009); whereas two families like – Apodanthacaceae and Cyanomoriaceae were taken out of consideration to include in this supra-ordinal system of classification, although temporarily. The major success of APG-III (2009) was witnessed in the emerging number of orders from 45 to 59 in respect to APG-II. One of the important findings of APG–III (2009) is coming out of "Monophyletic evolutionary bias" started since APG-I (1998) (as molecular data analysis of the orders: Diascoreales and Santanales don't comply this hypothesis). So, the success of development of phylogenetic system of classification of flowering plants have succeeded due to the factors like- gradual increasing trends of orders, decreasing trend of family numbers, better orientation of basal clades, decreasing trends of family numbers in the supra-ordinal groups or clades (without inclusion under the specific orders) and the decreasing number of families with uncertain positions in APG-III (2009). APG-III (2009) has created an opportunity to understand the mode of transcending synapomorphies between basal clades and the distant, branched clades related to it, phylogenetic interrelationships of different orders and families in the standard cladistic framework, under a supra-ordinal classification of flowering plants.

Apparently, the satisfactory portrayal of phylogenetic framework of APG-III (2009) on flowering plants was in the crossroad of evolutionary bottleneck, on the parallel phylogenetic observations of Doyle (2006, 2008, 2010) on "Cladogenesis of seed plants," an effort to narrate the evolutionary transitions of seed plants from non-flowering seed plants to flowering seed plants and it lead to a further revision of APG-III (2009). As a result, another, more liberal form of supra-ordinal classification of flowering plants has appeared, globally recognized as APG-IV (2016). In APG-IV (2016), the number of orders was increased from 59 to 64 and the number of families was increased from 415 to 416 as a result of the thorough restructuring of families in this classification. For accommodating the new orders like- Boraginales, Dilleniales, Icacinales etc., APG-IV (2016) have initiated the major changes like addition and alteration of main cladistics framework, and followed since APG-I (1998), by introducing two informal clades: Superrosids and Superasterids, related to existing major clades: rosids and asterids, respectively. The families like Apodanthaceae and Cynomoriaceae, considered earlier as the out of scope issues in APG-III (2009), were rehabilitated in APG-IV (2016) by assigning the family Apodanthaceae to the order: Cucurbitales and the family Cynomoriaceae to the order: Saxifragales. The basal clade of the modified cladistic model of APG-IV (2016), popularly known as "ANA" grade or "ANITA" grade or "ANAHITA," was recognized as the 'operational clade" to initiate radiation of angiosperms. In order to stop ambiguous recognition of different genera or family members, "bracketed familial" systems were technically accepted in favor of inclusive principle as well as "broader circumscriptions" but APG-IV (2016) has left an option for further consideration of "narrow circumscriptions" if necessary, to explain the phylogenetic issues like " tracing out trail of synapomorphies" or " mode of divergence" or " molecular affinity" to establish intra or inter-cladistic relationships and apparently the unique findings on "cladogenesis of flowering plants" is not finished yet, as the comparative assessment of "narrow circumscriptions" versus "broader circumscription" is an ongoing issue, critically reviewed by eminent group of plant phylogeneticists (Wearn et al., 2013; Christenhusz et al., 2015).

Though, Article 16 of ICBN (International Code of Botanical Nomenclature) has clarified the visible morphological and embryological distinctions of cotyledonary leaves have attributed the seed-bearing flowering plants with two seeds dicotyledons or "dicotyledonae" and the seed-bearing flowering plants with one seed as monocotyleadons or "monocotyledonae." Though, molecular botanists and evolutionary biologists don't consider this apparently, ephemeral features like -number of embryonic leaves could not be the principal criterion to distinguish the plants into two broadly categorized groups, unless the molecular data analysis recognizes the monocot as monophyletic and dicots as a paraphyletic group. So, it's not incorrect to consider the "plant with single cotyledonary leaf" might either be recognized as "monocotyledon" or "dicotyledon" and the "plant with two cotyledonary leaves" might be recognized as "monocotyledons" or "dicotyledons."

Contemporary taxonomic reviews and floristic enumerations revealed that species diversity of flowering plants on this blue planet ranges from 250,000 to 400,000 (Wilson, 1992; Scotland and Wortley, 2003; Pimm and Joppa, 2015; Christenhusz and Byng, 2016) and the extent of species diversity in different supra-ordinal groups varies from one clade to another clade; as the species diversity in eudicot clade was observed 75% (in highest extent) followed by monocot clade: 23% followed by was magnoliid clade: 2% and rest of the clades altogether: less than 0.1%. Though all over species diversity in dicots are comparatively higher than monocotyledons but the extent of species diversity in family level was found to be highly diversified; as it was found in to be highest in Asteraceae (23,600 nos) followed by Orchidaceae (22,075 nos), followed by Fabaceae (19,400nos), subsequently followed by Rubiaceae (13,150nos) followed by Poaceae (10,035nos). Out of these five top families, three families are belonging to dicots and rest of the two families (Orchidaceae and Poaceae) are belonging to monocots (APG-I, 1998: APG-II, 2003; APG-III, 2009; APG-IV, 2016). Apparently, species diversity in different families doesn't give any trendsetting results, from which it's difficult to conclude that phylogenetic pressure drives the mode of diversification of flowering plants. But the overall dominance of species diversity in dicots over monocots indicates that molecular evolution of flowering plants were bestowed with a better transcending advantage in the paraphyletic phylogenetic inheritance rather than the monophyletic phylogenetic inheritance.

Chapter 20

THE MYSTERIES OF THE MESOZOIC SINCE THE PERMIAN AND THE BIRTH OF BASAL ANGIOSPERMS: "ANA" OR "ANITA" (ALIAS "ANHITA")

The history of evolution and radiation of primitive angiosperms have gone through the biased explanations of Phylogenetic Hypothesis and paleobotanical evidence since 18th centuries. The pioneer naturalist and father of theories of organic/biological evolution, Darwin, has proposed the "theory of phyletic gradualism" (i.e., observation gradual evolutionary changes in a single line rather branching out) has a sharp contradiction to the contemporary "Rapid evolution Hypothesis." In the light of his theory of phyletic gradualism, he has considered the origin and evolution of angiosperms as "an abominable mystery," a sudden appearance and full-blown radiation of angiosperms occurred during Albian of middle of the Cretaceous Period (Hickey and Doyle, 1977). Naturally, the availability of fossil evidence, the diverse research analyses of that evidence and limitations lead to a bias, over-whelming explanations of "Cretaceous radiation of angiosperms" have masked the "pre-Cretaceous radiation of angiosperms," which was initiated in the 18th century by Darwin (Axelrod, 1961). In order to come out of any biased observations and ideas about the tentative period of time of origin and evolution of basal angiosperms, an intensive review of various hypothesis and observations on origin and evolution of primitive angiosperms needs to be considered. The observations and analyses of the contemporary multidisciplinary researchers in molecular botany, Ultramicroscopy, molecular systematics, palaeobotany, palynology, geology on the palaeobotanical evidence, particularly the angiosperm-like fossils have overturned the earlier Phylogenetic Hypothesis of angiosperms and it disheartens both groups of "angiosperm phylogeneticists" of the two opposing schools, the contender of "Cretaceous Hypothesis" and "Anthophyte Hypothesis."

Apparently, scanty and fragile nature of early palaeobotanical evidences in Pre-Cretaceous Period (either from Triassic or Jurassic); the factors like lack of evidence to explain incomplete evolution and radiation of angiosperm plant group above the rank of family level and absence of evidence of angiosperm radiation in Gondwana during pre-Cretaceous, could not provide acceptable clarifications on radiation of primitive angiosperms in pre-Cretaceous Period but recent, collective evidence on molecular phylogenetic analysis, palaeoclimatic and palaeogeographic simulative reconstructions have ascertained that factors like mass extinction and tectonic plate movement could help to explain logically that evolutionary evidence on radiation of angiosperms in Jurassic period or even earlier have mostly considered "in transit" status in the middle of mass extinction and "resurrection of epibiotic," whereas the angiosperm have not migrated and invaded Gondwana till Neocomian of Lower Cretaceous Period. It seems difficult to swim against the tide; till the early of 20th centuries, it has apparently been difficult to go against the biased observations and widely accepted theories (in the 1970s the hypothesis of Cretaceous radiation of angiosperms was elevated to the status of theory from the status of hypothesis) to raise the concern about over-simplified version of major group of evolutionary biologists and botanists: "Cretaceous radiation of angiosperms is the ultimate result of initial evolution."

In the late 20th century, the cumulative investigations and researches on phylogenetic research on angiosperms, particularly on pre-Cretaceous origin of basal angiosperms have well been supported by the recent cumulative analysis of molecular botany, palaeobotany, palaeo-palynology, molecular taxonomy, geochronology along with ultra-microscopic and morphological research on primitive angiosperm fossils, extinct and extant members have revealed a unique facts like existence of angiosperm pollens (e.g., *Afropolis, Clavatipollenites, Stellatopollis* etc.) have been witnessed in Western Europe and North America from the Jurassic-Triassic Period (Vigran and Thusu, 1975; Cornet, 1989b; Cornet and Habib, 1992; Hochuli and Feist-Burkhardt, 2013) and the discovery of fossilized flower *of Euanthus panii* (having features of typical angiosperm flower like sepal, petal, androecium with tetrasporangiate anther, gynoecium with close ovule) from Middle and Upper Jurassic of Liaoning, China (Liu and Wang, 2016). The presence of the "basal-crown-group of angiosperm" in the tropical part of North America in diversification mode and the disjunctive existence of Triassic angiosperm fossils and palynofloras in North America have helped to convince the contemporary phylogeneticists Laurasia as the evolutionary matrix of basal angiosperms.

One of the firm supporters of "Plant-insect Co-evolution Hypothesis" of decrypting the phylogeny of radiation of angiosperms, Crepet (2000) has claimed that the identical resemblance of the rate of floral innovation/million years to the rate of angiosperm diversification during the interval of Cenomanian-Turonian has subsequently resembled the appearance of derived insect pollinators. The extinction of relict form of the members belonging to Cheirolepidaceae during Cenomanian, has paved the way for sudden evolution of modern group of angiosperms with unique floral innovations (coping with insect pollinations) in Turonian (around 90 mya) (Vakhrameev, 1970; Crepet, 2000) which contradicts the initial ideas of evolution "theory of phyletic gradualism," suggested by Darwin, who has considered Albian is the time of ultimate time of magical appearance and radiation of angiosperms (Crepet, 2000).

One basis of abundance and diversity of Circumpolles pollen types adjacent to the areas of the Tethys Sea (e.g., Moroccan Triassic basins) and palaeoclimatic investigations of Upper Triassic, it could be predicted that the physically (semi-arid tropical and subtropical) and physiologically arid (low altitude subtropical saline habitat close to ocean) regions were dominated by tall, arborescent coniferous gymnosperms with microphyllous leaves belonging to the family Cheriolepidaceae; whereas the shady, apparently moist, understories habitat of theses Cheriolepidaceae members dominated forest ecosystems (Vakhrameev, 1970), supposed to be occupied by minority group of small shrubby, small-sized, "basal angiosperms." Apparently, the favourable palaeo-climatic shift towards Cheriolepidaceae helped establish its dominance by increasing its abundance and diversity in Rhaetian of Upper Triassic, when its symbiotic relationship with pollinator insects (e.g., Hymenoptera) and herbivorous group of dinosaurs boosted up its dominance till Triassic–Jurassic border of mass extinction and extended it up to end of Jurassic. The extinct members of Cheriolepidaceae of the Triassic period have an identical resemblance to extant members of Araucariaceae, having seeds, encased by scale-bracts (Vakhrameev, 1970). Till Jurassic, sauropod group of herbivorous dinosaurs supposed to consume the fruits with well-protected seeds along with green tender shoots of the crown of the members, belonging to Cheriolepidace. Eventually, it getting the seed coats digested and the seeds of diverse form of Cheriolepidaceae, getting protected in the moist excreta. And the moist excreta acting as the nursery of Hymenoptera beetle larva; the entire organic nutrients enriched microenvironment supported germination and expediting abundance of Cheriolepidaceae till end of Jurassic and continued till Lower Cretaceous. Simultaneously, Hymenoptera, the insect pollinators have played important role by inducing pollinating entomophilous members of Cheriolepidaceae (e.g., fossilized Classopolis pollen of Cheriolepidaceae, elytra beetle of Hymenoptera group and *Conanthus*, a derived gymnosperm and similar to Magnoliaceaous reproductive organs were reported from the same sediment in Upper Jurassic-Lower Cretaceous period. The change of food habit of herbivorous dinosaurs from treetop tender-shoot eaters to understory shrub grazers have given an edge to the "basal angiosperms" in Lower Cretaceous for widespread dissemination of its seeds in the same way the members of Cheriolepidaceae enjoyed once; simultaneously Hymenoptera have switched its insect-pollination activities from entomophilous gymnosperms to entomophilous "basal angiosperms' Bakker, 1978).

The habit and shape of the predominant Cheridoleiaceaeous members have started changing since Triassic-Jurassic border of mega-extinction event (apparently it's induced by hitting extremely hot mass of asteroids on earth surface); hence the dominant, arborescent tall Cheridolepiaceous members were gradually replaced by smaller, shrubby, xerophytic members (e.g *Frenelopsis*) till Albian of Lower Cretaceous and those have formed horse-tail like vegetation in coastal areas (Cornet, 2002). The reversal of palaeoclimatic changes started at Triassic-Jurassic border, have not found been favorable for maintaining abundance of diversity, and dominance of Cheriolepidaceae in coastal regions but the restricted form of a minority group of basal angiosperms were favored by these palaeoclimatic changes from arid subtropical profile to moist subtropical profile (Doyle and Hickey, 1977). The gradual disappearance of Cheriolepidaceae was expedited lead to its total extinction (evidenced by the disappearance of Circumpolles in Cenomanian) in Upper Cretaceous, resulting in vacant slots of its niches in coastal regions, which were occupied by the basal angiosperms in Upper Cretaceous (Lupia et al., 1999).

Simultaneously, the critics of "Cretaceous theory of angiosperms" (Brenner, 1996) expressed their concern that little palaeobotanical evidence (like fossil pollens) of primitive angiosperms in Neocomian-Aptian could be availability and fossilization issues rather than over-simplified observation like "Cretaceous radiation of angiosperms." According to the conventional morphology-based phylogenetic analyses, the early diversification of basal angiosperms in Cretaceous were attempted by the families of Chloranthaceae and Winteraceae but the recent molecular botanical reviews, as well as cladistic analysis, has ascertained that earliest diversification attempt of basal angiosperms (including ANITA grade, Magnoliids, Chloranthaceae, and Ceratophyllaceae), happened among the Eudicot group followed by Magnoliid group (Crane et al., 1995; Endress, 2010) in flowering plants. The contemporary research on palynological (mainly presence of *Clavatipollenites, Liliacidites*) and palaeobotanical evidence with molecular evidence of cladistics analysis reveal that:

i. members of Nymphaeaceae and *Amborella,* the basal angiosperms were reported along with Winteraceae, Chloranthaceae, ITA, *Liliacidites*, Piprales in Upper Triassic - Lower Cretaceous period in different locations in Lauratia and sequential position of Winteraceae on the top of other basal angiosperm group ANITA, has convincingly designated it as Cretaceous angiosperms whereas; the other members have reportedly evolved before Cretaceous in Laurasia and this has happened before migration of angiosperms to Gondwana from Lauratia (Friis et al., 2000; Thien et al., 2009; Endress, 2010).
ii. critical analysis and age dating process has confirmed in Bathonian of Middle Jurassic from the Western part of Norway and the evolution of basal angiosperms below Monocotyledonae and Piperales were taking place in Callovian of Middle Jurassic (Vigrans and Thsu, 1975; Cornet, 2002).
iii. existence of Liliacidites and *Clavatipopollenites* in the Northern part of France in Oxfordian of Upper Jurassic were noticed (Cornet and Habib, 1992).

iv. age dating process has confirmed the fossilized insect pollinators, belonging to Hymenoptera in Kimmeridgian of Upper Jurassic the existence of angiosperms, prior to Upper Jurassic has validated the theory of successful radiation of entomophilous angiosperms in the post-Jurassic Period (Labandeira, 2002).

v. The palaeo-climatic simulation model has confirmed that reversal of climatic regime from arid –subtropical to a moist subtropical condition in Berriasian, in Lower Cretaceous has lead the migration of angiosperms from Lauratia to Gondwana (Cornet, 2002).

vi. Discovery of *Sanmiguelia*, the "basal crown-group" of angiosperm, having unique features (sharing the vegetative characteristics of Monocotyledonae with the reproductive characteristics of dicotyledonae), reported from North America in Carnian-Norian of Upper Triassic has proven the evolution and diversity of basal angiosperm has taken place in Laurasia during Triassic, much earlier than Cretaceous (Cornet and Traverse, 1975; Ash, 1976; Cornet, 1989b;).

vii. The confirmation of megafossil and palaeo-palynological evidence of basal angiosperms in Ladinian of MiddleTriassic from palynofloras at Barents sea (Cornet, 1989a, 1989b; Cornet and Waanders, 2006), North America has confirmed evolution of basal angiosperms in Laurasia in Triassic; existence of Crinopolles in Carnian of Upper Triassic from Eastern Tosha stratigraphic deposit, Namibia (Cornet, 2002), which has confirmed the migration of basal angiosperms in Gondwana before "climatic barrier" stopped migration of angiosperms from Laurasia to Gondwana and apparently it created a confusion that migration of angiosperms did not start before Lower Cretaceous.

viii. Based on the existence of "oleanane triterpenes" from the leaves of Gigantopterids, Bennettitales and flowering angiosperms, it was ascertained that evolutionary lineage was established to even Permian-carboniferous Gigantopterids (Taylor et al., 2006, Miller, 2008) and most likely Gigantopterids maintained the evolutionary descendency by passing its genome to primitive angiosperms on or before mega-extinction event during Permian/Triassic Period.

ix. According to the contemporary cladistics analysis ("Bayesian Relaxed clock" methodology), the origin and evolution of a crown group of angiosperms were taken place between 106.1- 229.4 m.y.a. (Thorne and Kishino, 2002; Bell et al., 2005).

x. The recent molecular investigations based on computational algorithm (penalized-likelihood cladistics method), primitive angiosperms were evolved around 251.77 million years ago (Schneider et al., 2004). Coincidentally, the molecular clock also determined the age of EPE (End Permian Extinction) and were found to be the same, which indicates that most probably at the end of EPE, maiden appearance of primitive angiosperms were taken place.

Based on the above findings, reaffirmation of evolution of basal angiosperms in Jurassic Period of Mesozoic Era, as well as earlier than Cretaceous in Laurasia, discarded the earlier "Cretaceous origin of angiosperms"; hence it has confirmed two important facts that the location of evolution of basal angiosperms was identified as Lauratia (as, climatic barrier played the key role to stop migration of early angiosperms from Lauratia to Gondwana, the chance of finding fossil evidence of basal angiosperms from Gondana before Lower-Cretaceous is less likely) and the period of origin and evolution of basal angiosperms took place before Cretaceous Period and which has further been determined in between Triassic-Jurassic Period or even further down in Permian. Hence, there is one important information missing here- whether, the evolution of angiosperms happened at the same time or it happened at different point of times?

While it is not so clear whether the primitive form of angiosperms have undoubtedly evolved during Permian or late Triassic but molecular phylogenetic analysis of angiosperms has ascertained that primitive flowering, vascular, seed-bearing plants were evolved before Cretaceous (Wang, 2010). The fossil remnant of leaf and female reproductive organ of a seed-bearing plant was collected from Haifanggou Formation, Western Liaoning of China from Middle Jurassic (Wang et al., 2007), known as *Schmeissneria microstachys*. Its ultrastructural investigations revealed that the female reproductive organs were found in pairs, attached to the short peduncle and each female organ was comprised of a central unit of two locules, separated by a vertical septum and the central locular unit was encapsulated by longitudinal ribs (Wang et al., 2007) and apical region of the central locale was found closed. Although *Schmeissneria* has structural resemblance to the members of Ginkgoales but closed apex of the central locular unit of this species has striking resemblance to the closed carpel of angiosperms (Wang et al., 2007) and which has given opportunity to the molecular phylogeneticists to consider it either as angiosperm, evolved in the Triassic–Jurassic boundary or a missing link, the new seed plant, parallel to angiosperms and other vascular, spermatophytes in Mesozoic.

A group of phylogeneticists (Sun et al., 2002) have reported two extinct species of angiosperms (*Archaefructus liaoningensis* and *A. sinensis*) belong to the genus *Archaefructus,* under subclass: Archaemagnoliidae and the family: Archaefructaceae from the Jianshangou bed (fresh-water lake deposit), China in Tithonian of Upper Jurassic (around 144mya) and critical analysis on its age-dating process has helped to redetermine its age (125 mya) and it was considered as basal angiosperm. According to Ji et al., (2004) *Archaefructus* was considered as an aquatic, submersed angiosperm with branched shoots and dissected leaves and lack of sepals and petals in the flower stock (considered as primitive features) has indicated its tentative evolution long before insect-plant co-evolution regime but nature of dissected leaves and presence of pair of stamens in each flower indicating its advanced features; so far *Archaefructus* is one of the basal angiosperm, likely to be evolved during early Cretaceous (Sun et al., 2002).

The modern phylogenetic reviews ascertain that diversified angiosperms in post-Cretaceous are the phylogenetic descendants of basal angiosperms and five predecessor lineages of Chloranthaceae, magnoliids, monocots, *Ceratophyllum* and eudicots. It's quite unique that dominance of flowering plants in the terrestrial ecosystem was dominated by eudicots (comprising of basal and core eudicots). The discovery of *Leefructus*, the earliest basal eudicot (evolved between 122-127 million years ago) paleopollens and leaf remnant from the Yixian Formation of China, from Lower Cretaceous (between Barremian to Lower Aptian) has indicated that basal eudicots are geologically contemporary to *Archaefructus* (Sun et al., 2011).

Very recently, the mythical first flower of aquatic habitat was re-discovered from the limestone deposit of Pyrenees of Spain, known as "*Montsechia vidalii,*" cladistically found sister group of *Ceratophyllum* clade from Barremian age of Lower Cretaceous, indicating geologically evolved 125-130 million years ago, as well as contemporary to the group of dinosaurs belonging to Brachiosaurus and Iguanodon (Gomez et al., 2015). It could be retained as a matter of debate to the phylogeneticists to determine the "mythical first flower" in a metaphoric "photo-finish of a 100 m. phylogenetic sprint," as analytical reviews among these possible contenders, the earliest representatives of angiosperms of favorable aquatic habitat (though few of the members were reported from mountain ridges of terrestrial origin): *Amborella, Archaefructus, Euanthus, Leefructus, Montechia, Sanmiguelia, and Schmeissneria*. From the palaeobiological evidence, it is quite convincible that the evolution of angiosperms has supposedly taken place in aquatic habitat and such prehistoric transmigration has played a key role in the diversification of early angiosperms on this blue planet.

In order to decipher the mode of divergence of angiosperms, particularly the basal angiosperms, contemporary analyses on molecular genomics have played the key roles, where *rbcL* and 18S genes were bootstrapped and this critical analysis was performed on the basis of the concatenation of sequencing of genes (Delsuc et al., 2005). The result of this experiment is quite thrilling as this simulative model shows that in last 300 million years span of time (at the average rate of mutation for this period of simulation), angiosperms have evolved at least three intervals of times on earth (Doyle, 2001, 2006), according to the geological time-scale. The earliest evolution of angiosperms, took place in a particular interval of time, started at the Middle Triassic and finished at the Lower Jurassic as per 2nd position of *rbcL.* Moreover, allied results indicate towards Upper Triassic as the highest rate of divergence in this span of time; preliminary effort of colonization was witnessed among the small shrub-like angiosperms, evolved during this phase (Sanderson and Doyle, 2001). The most recent evolution of angiosperms, were taken place in the interval of time, started from Cenomanian and finished at Turonian of Upper Cretaceous as per 3rd position of *rbcL.* Diversification, as well as modernization (in terms of structural modifications, better cytogenetical configurations, better physiological profile) of Cretaceous flora, reportedly has taken place during this time (Sander and Doyle, 2001). The next one (in between recent and earliest) incident of evolution of angiosperms, were taken place in the interval of time, at the boundary of Lower and Middle Jurassic, as per position of 18S and evolutionary biologists have preferred this second incident of divergence of angiosperms as diversity and abundance of angiosperms (predominance of arborescent, crown group of angiosperms) took place in the phase of evolution (Doyle, 2006).

Meiotic cell division is an integral part of cross-pollination induced sexual reproduction of bisexual flowering species, (mainly during intra-specific crosses). The pairing of two normal set chromosomes (each set from each of the parents) took place usually. But, there is an inbuilt screening system in the chromosome, called "MSUD" or "Meiotic Silencing by Unpaired DNA" detect any unwanted gene pairing by screening out the odd one (in terms of size or length) with deviating from the original parental type (Shiu et al., 2002). As MSUD could not detect any alteration of DNA sequence (when the size of the gene remained same at the time of pairing), point mutation has bypassed this screening system. During interspecific crosses, the silencing of MSUD was noticed (Shiu et al., 2002). Hence, it seems that "point mutation" supposedly played an important role along with factors like plant-pollinator co-evolution, climatic reversal in post-Triassic/Jurassic mass extinction etc. regime paved the way for "radiation of angiosperms" in Cretaceous.

So, it is convincible that mass extinction events like extreme climatic conditions, desertification etc. and mainly the mass extinction event at Triassic/Jurassic boundary play the key roles in slowing down the process of radiation of angiosperms (Fowell et al., 1994). According to Tanner et al. (2003), the initial attempt of radiation of angiosperms during Norian to Hettangian of Jurassic was aborted furthermore that mass extinction had wiped out the members of three families of flowering plants belonging to Glossopteridaceae, Peltaspermaceae, and Corystospermaceae (Ash, 1986; Traverse, 1988). On the other hand, the molecular biologists and evolutionary biologists claimed that climatic pressure slowed down the floral diversity and radiation process by a triggering point mutation on the 18S gene location rather than *rbcL* position. Then, there is an indication that reversal of climatic change in post Triassic/Jurassic boundary might have helped recovering suppression of floral diversity as well as expediting the radiation of angiosperms along with reversing the earlier occurrence of point mutatio at the same time molecular evolution of plant-insect relationship has expedited the completion of "radiation of angiosperms" in Cretaceous and it is little-delayed event supposed to be taken place during Triassic. Geologically, it is quite unique to notice that little gap between the two events"origin of angiosperm" in Triassic and "radiation of angiosperms" in Cretaceous concomitantly related to breaking-up of "Pangaea" and formation of "rift basins," the time in between (Fowell et al., 1994; Rees, 2002).

The different mode of pollination pattern like waterborne and wind-borne pollination were witnessed in Lower Cretaceous but the execution of successful outcrossing and radiation of angiosperms have not been witnessed in that time. The evolution and radiation of angiosperms were found to be exclusively related to co-evolution of insect pollinations as successful outcrossing of colourful, showy, bisexual flowers would not be possible without evolution and diversification of insect pollinators (conducting outcrossing and dispersal of fruits and seeds) and the expression of evolution and radiation of angiosperms (Dilcher, 1995) have gone through the node of evolution of angiosperms in Paleocene and Eocene, characterized by bursts of evolution of reproductive organs like stony, entomophilous fruits and viable seeds. So, success of radiation of angiosperms should be considered as an evolutionary success in the Upper Cretaceous which was related to recognize and maximize the mutualistic ecological services of plant and animals tended to each other and ultimate success of genetics of outcrossing or successful execution of sexual reproduction process (Endress, 1994) and it has helped in Upper Cretaceous Period onwards for diversification, dispersal and better adaptation in different part of the biome throughout the world as well as on the Earth, the biodiversity, climatic, edaphic, topographic and environmental factors were controlled in different ways by anthropogenic actions in Holocene.

Apparently, natural calamities like sand-storm in the semi-arid and desert areas cause soil degradation by continuously degrading the top soil layer and exposing the sub-soil layer exposed to heat and light, making it moisture free and unfit for the growth of the root-stock of the existing plant communities. On the other hand, anthropogenic interferences like man-made deforestations have affected the natural habitat by degrading the natural landscape lead to make the natural landscape barren forever to be occupied by any new plant species. Superficially, both issues have no direct relationship but it was noticed quite different in real time. The Gurban Tonggut desert of North West China was originally occupied by 70,000 sq.km of semi-desert type *Haloxylon* spp. in this typical forest and it used to support camel ranching in that desert ecosystem. The unique *Haloxylon* forest was found severely degraded in 1950 and shrunk to 38.6% (i.e., 27,000 sq.km) after few decades due to the anthropogenic exploitation of *Haloxylon* fuelwood by local ethnic groups. At the same time, the parasitic root system of *Haloxylon* helped grow the *Cistanche desert cold* Maemend, the important ethnomedicine of North Western China, which was harvested annually around 150 tons (Gensheng et al., 2001). It was observed that bi-fold exploitation of *Haloxylon* forests make expedite the soil degradation and enhance the soil erosion to make it absolutely barren for any plant species to grow further. As a result of severe degradation of top-layer of sand and exposition of sub-soil sand particles, the cycles of sand-dust storms were enhanced from 6 times a year to 22 times a year between the year of 1950 to 1980. The result of exposition of the forest habitat by the desert ecosystems due to enhanced sand-storms and adjacent ecosystems have a huge collateral damage in expansion and stabilization of the existing plant communities, naturally the diversities of wild angiosperm plant species have faced a huge challenge in the regime of Anthropocene not only from human beings but also from natural disasters.

In the post-Cretaceous Period, diversification of angiosperms was initiated since the Oligocene of the Tertiary Period; diversity of herbaceous angiosperms were initiated during Miocene of Tertiary and expansion of grassland or savanna type of forest ecosystem expanded all over the world, which have supported the increasing populations of herbivorous grazing animals. During mass extinction event of Pleistocene of the Quaternary Period, arborescent angiosperms have mostly been disappeared as those have mostly been lost in the struggle for existence. During Holocene of Quaternary, the appearance of human beings (*Homo sapiens sapiens*) were taken place; concomitantly, a number of herbaceous and few arborescent angiosperms able to survive Pleistocene mass extinction. Since Holocene, the earth has started being dominated by the single mammal species *Homo sapiens sapiens* and the factors like the formation of vegetation, diversification of angiosperms, migration, colonization, endemism, and extinction of angiosperms were highly influenced by human beings. Hence, this current phase of Holocene has also been recognized as Anthropocene.

Chapter 21

PHYTOGEOGRAPHY: DISPERSAL AND MIGRATION IN RELATION TO THE GATEWAY OF ANGIOSPERMY

> "Every feature of the general distribution of plants over the world is due to the combination, in varying pattern, of the separate individual distributions of all kinds of plants." - J. D. Hooker

Phytogeography or plant geography is a branch of botany that helps identify and designate different parts of the world in a certain order, characterized by different assemblages of plants distributed over the surface of the world. The main aim of phytogeography is to interpret plant migrations in time and space and visualize through factual records of plant distribution (contributed through standard, comprehensive taxonomic and systematic work), then transferring these records to a map, which represent tracks of origin, migration dispersal etc. The three gates of angiosperms have been, (i) African (ii) Magellanian and (iii) West Polynesian, evidenced through plant fossils in different parts of the would from different geological periods, which help draw an outline of the origin of the angiosperm centers, the routes of their migration and dispersal through the geological time scale (Croizat, 1952). Hence phytogeographical studies should not be considered to be a collaborative study between Botany and Geography, but rather a multidisciplinary approach of different branches of Science-Geology, Meteorology, Zoology, Biodiversity, Anthropology, Statistics etc. as well as other branches related to applied Botany (Schouw, 1823; Engler, 1879; Ridley, 1923; Willis, 1923; Herzog, 1926; Good, 1947; Turrill, 1964; Gleason and Cronquist, 1965; Takhtajan, 1969, 1974; Heywood, 1978; Wilson, 1992; Chapman and Reiss, 1995; Christopherson, 2011).

The distribution of plants over six continental landmasses and numerous islands is mainly influenced by inherent geographic factors. According to Good (1947), the inherent factors are mainly genetical, concerned with the evolution of an individual, whereas geographic factors are topographical and ecological. The geographical factor is subdivided into several subfactors - (i) climatic, (ii) edaphic (iii) biotic and (iv) barrier factor (Turrill, 1964; Kellman, 1975). The Barrier is the most significant to the spreading or dispersal of any species. The nature of the barrier may be physical (obstacles of mountains, deserts, oceans) or ecological, such as the plants of that belong to a habitat with different geographical locations, e.g., the mangrove forest community in Florida is remarkably different from the mangrove forest community in Sunderbans.

Ridley (1923) critically analyzed the observations about the flowering plants distributed over large areas across the world's surface and categorized such cosmopolitan plant groups into four categories:

i. Weeds, which are dispersed by anthropogenic vectors (wind dispersal helps it sometimes)
ii. Plants dispersed by waterfowl.
iii. Plants dispersed by sea currents.
iv. A few plants thriving in temperate and tropical regions descended along the mountain chains in the South of the equator.

Willis (1918) explained the distribution of plants in his "Age and Area Hypothesis" in which he opines that the distribution of plants is directly linked to their age, although this view was ultimately proven false; e.g., Members of Cycadaceae which are aged enough to occupy a wide area were found confined in limited areas whereas the diversification of angiosperms (mostly eudicots) were found to cover a large expanse over a very short period of time in the Upper Cretaceous Period.

There are 20-30 species, occupying half of the earth's surface and 0.001 percent of the total flowering plants of the world have occupied one-quarter of the earth's surface (Wulff, 1943). DeCandolle (1855) pointed out that there are no true cosmopolites (i.e., ubiquitous species, inhabiting the entire globe from the south to the north pole across the equator, which is found to be nonchalant to environmental variations). DeCandolle introduced a new term; "endemic area," to designate the habitat of any species with limited distribution.

Wulff (1943) mentioned that the "Vicarious area" was the most part mutually exclusive and belonging to closely related species, differing only in a few specific characters (e.g., subspecies) and linked by the fact that they derived their origin from one initial form. According to the eminent ecologist, Korovin (1934), the spread of a species from one ecological region to another, preserving its chief characteristics due to its availability and likelihood of ecological variations, is known as "successive migration," which sharply contradicts the "autonomous migration," when a species is spread over from one ecological region to another, accompanied by the formation of vicarious species. The stunting of "successive migration," or initiation of "autonomous migration" leads to the formation of an "endemic area," dominated by "Paleo" or "Neo-endemic" plants.

Chapter 22

THE ORIGIN AND DISPERSAL OF FLOWERING PLANTS IN RELATION TO GEOLOGICAL GATEWAYS

In course of time, the earliest and most simple plant life on earth gradually changed and developed in the process of evolution. The knowledge of the earliest plant life come from the 'record of rock', or fossils contained by sedimentary rock, through entrapment of organic remains. The earliest record of plant life has been traced out permanently in the rocks of the Proterozoic Era (which is also known as the age of Thallophytes). The Proterozoic Era representatives are followed by the Paleozoic Era ferns, club-mosses and simple type seed plants. The Paleozoic Era was followed by the Mesozoic Era, represented by ferns, tree ferns, and gymnosperms. The Mesozoic Era was followed by Cainozoic Era, known as the era of flowering plants or angiosperms. The Mesozoic Era is subdivided into the Triassic, Jurassic and Cretaceous Periods. The fossil evidence of the earliest flowering plants on Jurassic rocks prove that angiosperms first appeared on Earth during this period (Wulff, 1943; Cornet, 2002; Endress, 2011) or a little earlier, rather than the biased observations of the "Cretaceous Origin" theory.

According to Croizat (1952) the geologic revolution began in the Late Cretaceous Period and was more or less accomplished by the end of the Tertiary Period and is the master key to an understanding of the history of modern plant life. It is evident that maps were radically altered in due course of time through the geological timescale. Based on the correlation of phytogeography, ecology, taxonomy, constructive morphology, paleobotany and geology, Croizat's (1952), the "Phytogeographic Hypothesis" postulate that the main angiosperms forms evolved in pre-Cretaceous continents as the fossilized remnants and footprints of those early angiosperms are to be traced out in modern maps of phylogeneticists. Existing Antarctic lands possibly constitute the largest of these remnants. Geologically, it has been ascertained that the pre-Cretaceous continents that stood as the cradle of modern angiospermy, was a landmass, which occupied the entirety of the modern Indian Ocean. The continent Gondwana was a welded and widespread paleocontinent one, composed of Africa, Asia, Malaysia and Australia, additional lands in the Western Pacific and continuing to South America, Caribbeans and certain sectors of North America and Europe.

Figure 20. The Gates of Angiospermy: 1. African Gate, 2. Magellanian Gate and 3. Western Polynesian Gate (Croizat, 1952).

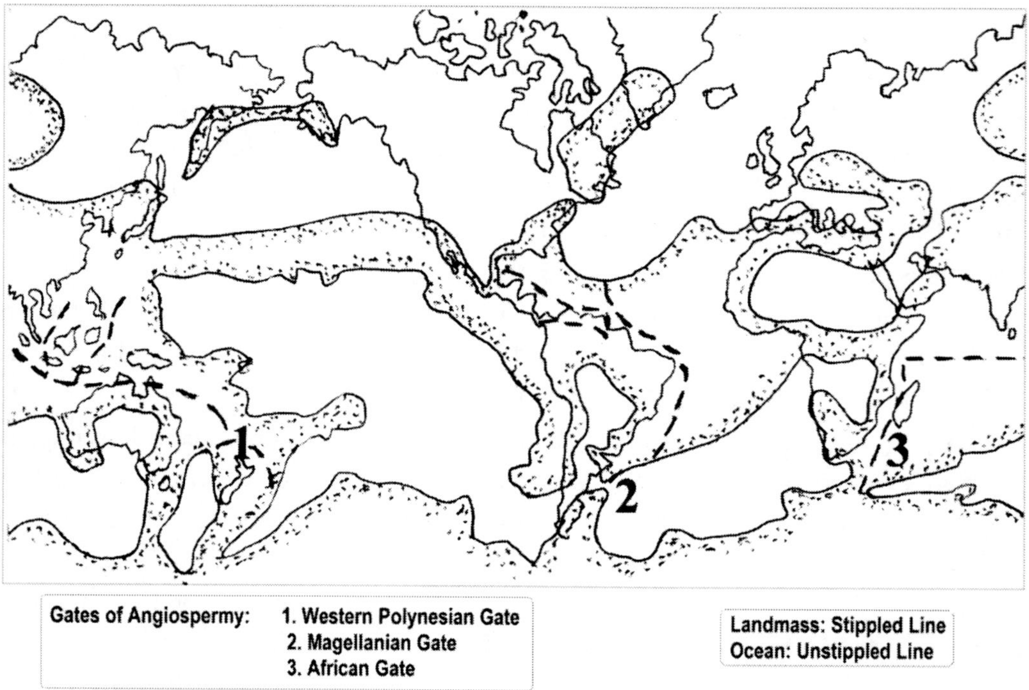

Figure 21. The inception of earlier angiosperm migrations on the Earth (Croizat, 1952).

The footprints of all primitive angiosperms started revealing their existence in certain locations, mostly from the Southern Hemisphere of the present world map of migration, dispersal and early distribution of primitive angiosperms (Raven, 1979). After their origin, these primitive angiosperms uniformly migrated throughout the earlier continental shields to cope with spatiotemporal changes to achieve a unique form of floristic identity of contemporary continental as well as regional form and this initial phase of metaphoric journey of primitive flowering plants was noticed in certain well-defined geographic centers, identified by Croizat (1952) as the "Gates of angiospermy," recognized as - (a) African Gate, (b) Western Polynesian Gate, and (c) Magellanian Gate. The "Gates of angiospermy" and tracks of active migration of primitive angiosperms are presented in Figure 20 and Figure 21, respectively.

a. African Gate:

The "African Gate" can actually be broken down into Madagascar-Mascarenes-Seychelles; Natal-Kerguelen Islands Tristan da Cunha, known as the Gondwanic-triangle and Afroantarctic-triangle, respectively. It has been observed that flora wedded to the lands of warm or tropical climates usually originate in the Gondwanic-triangle, contrariwise the plants tolerant of cold start from the Afroantartic-triangle. The tracts of dispersal that start from the Gondwanic-triangle are:-

1. Madagascar to Peru, through Tanganyika-Cameroon-Amazonian Brazil.
2. Madagascar to Fiji, through Malaysia and the Coral Sea.

Both of these tracks (1) and (2), mentioned above, were found warm, followed by the tracks of tropical nature from Afroantarctic triangle –

3. South Africa to Europe, Possible North America, through East Africa, Northern Africa, and Mediterranean region.
4. Central Asian (with outlines to East Africa, Java, Sumatra, the Himalayas on occasion) to all circumboreal lands by the Altai node.
5. South Africa to Australia, Tasmania, New Zeland, Magellania.

These three tracks are recognized as 'Cold'.

Besides these five tracks, the tracks which cannot be ascribed to either one of the triangles of African Gate are: -

6. Kalahari center to Brazil, ultimately reaching the Caribbean, parts of North and South America.

South Africa, generally, Madagascar and the Mascarenes to Patagonia, Bolivia, and points beyond in the New World.

b. Western Polynesian Gate:

It is broadly broken down into two distinct ranges, namely: -

i. *Neocaledonian Centre:* (composed of the lands of New Guinea, Queensland, Lord Howe Island, Tonga, Samoa and Fiji).
ii. *Macquarian Centre:* Comprised of Tasmania and the Islands of S. New Zealand (Macquarie, Campbell, Auckland, Antipodes). The Neocaledonian and Marquarian center, release "warm" and "cold" flora intermittently northward and Southward, respectively. The basic tracks of dispersal of the 'Western Polynesian Gate' are:
iia. Neocaledonian Centre to the New Worlds (North, Central or South America).
Iib. Neocaledonion Centre to Europe and Northern Africa.

c. Magellanian Gate: Centred around the shattered "insular arc" between South America and Antarctica.

The track of the Magellanian gate, through which the dispersal of angiosperms occurred, is:

10. Central Chile or Bolivia northward to Mexico, the U.S.A. possibly Hawaii, and the Caribbeans, further extended to Uruguay and SouthEastern Brazil.

In addition to these ten fundamental tracks, other channels of migration are as follows:

11. From the Mediterranean to the Caribbeans and Mexico.

This track may grant itself upon the channel described in (3) (4) or (12).

12. From the South Atlantic to the North Atlantic.

This track emerges as a variant of (5) (6) or (7).

13. From West Africa to Burma and Philippines - This channel is a variant of track (1).

The modern angiosperms began active migration through the thirteen tracks mentioned above, chiefly belonging to three gates, at the close of the Jurassic and very Early Cretaceous Periods. These plants had a second period of active migration in an epoch of general submersions and emersons when the ancient continental mass of the Indian Ocean extended its last phase of crumbling and also crumbled Malaysia and the shores of the Pacific most actively. This process of secondary dispersal lasted from the final eras of the Cretaceous to the initial eras of the Tertiary Period.

In addition to these "Gates," certain important centers of secondary migration and secondary evolution, are:

(a) Kalaharian and Nigerian in Africa.
(b) Romanian and Ozarkian - Appalachian in the Americas.

Croizat (1952) has advocated in favor of his hypothesis as explained here. Let us suppose that we have before us a glass disk, and mark upon it certain sectors as the centers of origin of angiosperms hosts. Let us suppose next that we give the disk a sharp rap with a hammer, right on the center. The disk would fly into pieces. Finally, these pieces are rearranged in such a manner that pieces where the angiosperms did not originate come into contact with fragments of the places these plants appeared. Let us rearrange this broken stuff again more or less in a disk-like outline and observe. It can therefore be concluded that (a) All the centers can still be connected with the positions they formally occupied; (b) Each centre may in its new positions may further angiosperm life; (c) Every point is in contact with a more or less extended life with every other point and the centre.

They may be compared with the pre-Cetaceous world, wherein angiosperm prototypes were few and concentrated. The disk next shattered and rearranged represents, on the contrary, the post-Cretaceous World. Tracks in the shape of lines will connect forever the secondary and primary center and all their tracks will come back, in the end, to their original starting points, the gates of angiosperms. The gates may no larger be whole glass and may be disfigured (e.g., Mascarenes, New Caledonia, Juan Fernandez).

A number of previous attempts were made for interpreting the origin and dispersal of flowering plants by – "Occasional means of secondary-distribution," "Polar shifts," "Floating continents," "Landbridges," "Holarctic theory" etc. However, Croizat's (1952) Phytogeographic Hypothesis seems to be the more realistic approach to solve the problem of origin and dispersal of angiosperm with the evidences found in paleobotany, paleogeography, comparative morphology, anatomical morphology, taxonomy, phylogeny, and ecology.

Chapter 23

Long Distance Dispersal and Migration of Flowering Plants Since the Radiation of Angiosperms during the Post-Cretaceous Era

Most of the time dispersal and migration of plant species in general and flowering plants were used synonymously, although there are significant differences between the two; dispersal is a kind of stochastic state of action of plants, driven by mechanical and physical forces, it could be individual or a synergistic effect of overlapping factors like ocean currents, el-Nino actions, impacts of hot desert storms, severe glacial storms, vortex of tornadoes, multitude of avalanches, drifting of tsunamis, effect of Greenhouse gases, expansion of Ozone Hole in stratosphere and melting of polar ice-shields, etc., while the migration of plants is a much more organized process oriented to ecological actions and factors like-

1) exceeding the carrying capacity of the ecosystems,
2) mode of diversifications of species,
3) likelihood of identical climatic, edaphic, topographic, and environmental conditions,
4) biotic factors like suitable habitat for further colonization,
5) the spree for filling-up the vacant ecological niches of any adjacent, fragile ecosystems, to reinforce the concerned food web of any existing ecosystem for better ecological sustainability of any habitat,
6) nature of herbivory (like smaller size of the seeds, consumed by the larger herbivores with a wide range of distribution),
7) maternal inheritance of genes by forest tree species (like the rate of expansion of the European oak species since the latest post-glacial regime has been determined

as 500 m yr $^{-1,}$ induced by the inheritance of maternal genes in new seedlings, which boost-up metapopulation dynamics to expand in a wide range of distribution) etc.

The possible combination of all factors should have played a key role in driving plant migrations in general or orchestrated radiation of angiosperms in the post-Cretaceous regime (Janzen, 1984; LeCorre et al., 1997; Cain et al., 2000). Whether it is deterministic ecological diversification of species or the stochastic expansion of species dynamics in an uncharted territory, both are possible for plants, (Briggs and Walters, 2016) particularly the flowering plants (mainly eudicots) to radiate or diverge in long run. Plant dispersal is kind of a random process which requires them to adapt in a new space, challenged by unique environmental, climatic, edaphic etc. factors; the survival and existence of species richness has been tested in due course of time (Adams, 2009).

The migration of plants has evolutionary as well as biogeographical advantages, as it is the preliminary step of colonization in the process of plant succession on a biogeographical location and it has a better chance of adaptation due to better survival opportunity due to ecological interaction rather than relying on stochastic conditions. No evolutionary journey is a continuous process as evolution, origin, and diversification of any entity has always gone through the tumultuous ride of "boom and bust" cycles in the due course of time and space, and the incidence of "radiation of angiosperms" is no exception. Immediately after traversing the mass extinction of the Tertiary/Cretaceous boundary, the radiation of angiosperms was witnessed in a short period of time in the Paleogene Epoch and this period of the geological time scale has been marked with diversified fossil representatives of conifers and primitive group of angiosperms (Sauer, 1988). During the Paleogene Epoch the tectonic plate driven movement of continental plates to drive the early phase of plant migrations on the Earth's surface and the disappearance of the ocean barrier in Laurasia, the migration of primitive angiosperms was favorable. Meanwhile, the completion of breaking-up Gondwana restricted the dispersal of plants over there (Raven and Axelrod, 1974).

According to Sauer (1988), oceans expedited the migration of angiosperms rather than acting as a barrier to stop it; as an example, Sauer (1988) referred to *Nypa* (a typical mangrove palm of Eastern Asian origin), which migrated through the Tethys sea to reach the Atlantic region of North America and South America. It has been noticed further that ocean-borne fruits and pollen samples were found from vast areas of South East Asia, Australia, West Africa, Europe, North Atlantic area, North American locations etc. in the Paleocene and Eocene Epoch. The dispersal of another mangrove *Rhizophora* disjunctly migrated to the Indonesian (Borneo) and Carribean Islands during the Paleogene Epoch (Guppy, 1917).

The disjunctive migration of a tropical beach plant *Caesalpinia* was confirmed by its paleobotanical evidence in Nigeria and South America during the Oligocene Epoch (Sauer, 1988). The broad-leaf plants, of Arcto-Tertiary Geoflora, e.g., *Sequoiadendron*, *Magnolia,*

Platanus, supposedly been originated before Cretaceous but known from Paleogene; supposedly these migrated from North America to Europe either before the formation of "Atlantic barrier" or through the "Bering land bridges." Sauer (1988) elucidated that the tectonic plates induced the earliest attempt of migration of angiosperms when the Indian plate moved northward and "ferried" some seed plants from Gondwana to Asia.

Podocarpus, the living member in India of African and Madagascar origin, was found in another instance of widespread dispersal to Neogene, migrating from Australia to South Asia via the East Indies (Sauer, 1988). Sauer (1988) pointed out that the ocean adaptation was the prime factor for disjunctive widespread dispersal and distribution of some coastal and mangrove species like *Cocos, Terminalia, Nypa, Barringtonia* to be located in Africa, Madagascar, and South Asia. According to Renvoise (1979), the combination of high wind velocity with ocean currents in the Indian Ocean drove the plant migration from Malesia to the West Indian coastline and the Arabian Peninsula. Even, some occasions like the disjunctive migration or the presence of *Acacia* in Australia and Mauritius was an unsolved mystery as the lack of land bridges between these two locations, existing sea barriers and lack of adaptive features in the reproductive bodies in *Acacia* make the migration of this plant species a little difficult to explain (Bell and Evans, 1978). The early long-range migration of flowering plants was referred by Raven (1979), it encompassed a distance of 5400km from Australia to Madagascar. The history of plant migration from Madagascar to Australia has been evidenced by following the path of migration of *Adansonia gregorii*, also known as "protoboab."

Between the Miocene and Pleistocene Epochs, the paleoclimatic conditions prevailed in Madagascar and Australia were found to be identical (Rangan et al., 2015), which helped the fruits of the "protoboab" to drift away with ocean currents and high air velocity on the route to reach northern Australia from Madagascar (Beard, 1967; Wickens and Lowe, 2008) and it was concluded on the basis of data available on molecular analysis and the paleobotanical evidence. Based on botanical, anthropogenic and linguistic evidence, Blench (2006) proposed in his hypothesis that Austronesian seafarers might have carried the protoboab propagules from Madagascar from Australia but the Miocene-Paleocene Epoch of time in the molecular clock and age dating process of paleobotanical evidence does not conform Blench's (2007) hypothesis (Rangan et al., 2015). According to Lowe (1998), the pods of the baobab were carried to the interiors of fertile zones deep inside the Kimberly coastline of Northern Australia, ferried by the huge cyclonic tidal waves (around 11m) during the wet season. The recent findings of a 2000-year-old fossilized egg of an extinct Madascarian elephant bird (*Aepyornis maximus*) in the Holocene deposits of the Cervantes period (near Perth), Northern Australia (Wickens, 2008). In the South Western Australia, at the river basin of Scott River, Augusta, another fossilized egg of *Aepyornis maximus* and two fossilized eggs of king penguin (*Aptenodytes patagonicus*), the inhabitant of Kerguelen islands in the Indian ocean which is located 2000 km away from Australia were noticed. It suggests further that the "Antarctic Circumpolar oceanic current" carried those eggs to the Australian coast likewise the migration of "protoboab" took place from Madagascar to Australia (Long, 1993; Lowe, 1998; Wickens, 2008). Based on the taxonomic differences between *Adansonia gregorii* and the rest of the members of the Section: Longitubae, it was assumed that it took a long time for "Protoboab"to evolve so the arrival of "protoboab" should not be considered as a recent event (Wickens, 2008) related to the human migration of Holocene.

Another instance of long-range dispersal and long-distance migration is in the case of two mangrove species of *Barringonia* (*B. asiatica* and *B. racemosa*) of Indo-Pacific center of origin, that moved to Madagascar by oceanic dispersal but the distributional pattern of the both the species was found to be random rather than following identical pathways (Payens, 1967). *B. asiatica* was found to colonize Mauritius and Seychelles islands, and failed to colonize the Reunion islands and failed to colonize in the East African coasts; the other species *B. racemosa* had successfully colonized Seychelles islands and reached the coasts of East Africa, Madagascar but failed to colonize Mascarenes islands (Payens, 1967); hence the dispersal pattern for both mangrove species seems alike but successful migration of both species seems to be different in varied ecological conditions, as *B.racemosa* has preferred to grow in the habitat with tidal inundation whereas the other species *B.asiatica* has preferred to colonize the habitat of higher tide level.

The difference between the long-range expansion in comparison to short range of expansion of dispersal type, of any individual or group of plants depends upon the support of external factors and favorable ecological interactions and structural advantages like-

1) smaller size of fruits,
2) decay resistant seeds,
3) adaptation of fruits in aerodynamics etc.

have helped to move and migrate from one place to next one, conveniently. Whether

1) it is driven by the fecal materials of herbivorous dinosaurs in Jurassic or
2) widely dispersed by the avifauna in Paleocene or
3) induced by tectonic plate movements since Triassic Period or
4) earliest instance of anthropogenic migration related plant migrations in Holocene,

the migration of plants has been considered as an important event in the process of evolution of plants. Particularly in the radiations of angiosperms in post-Cretaceous Period and the ultimate state of expansion of angiosperms was witnessed in Holocene Epoch of Quaternary. The Holocene has been marked as the regime of diversifications of flowering plants, domestication of a number of wild plants followed by its upgrading cytogenetical and bio-molecular configurations as 'crop plants" and supplying energy to the growing populations of *Homo sapiens*. So, the issues like migration and expansion of flowering plants in early ages in pre and post Tertiary/Quaternary end-

1) domestication,
2) induced mutation and
3) genetical and molecular development of crop plants

should be considered as the continuation of divergence of flowering plants rather than treated as isolated issues in the history of angiosperms on Earth. It has been found that long-range migration history of angiosperms from Polynesia to Hawaii, which is around 8000 km is the site of recent findings in phytogeographical discoveries. The contemporary radio-carbon analysis has unearthed the history of earliest incident of human colonization in Western Pacific Islands, Samoa in 800 BC, migrated from Polynesia; further migrated from Westen Pacific Island to settled down at Society and Cook islands of the Central Pacific Islands between 1025-1120 AD and migrated further down to Ruma Nui, Hawaii islands and New Zealand between 1190-1290AD (Wilmshurst et al., 2011). The fossil remnants and anthropological evidence ascertained that the earliest human settlers from Polynesia brought around 72 plant species to Hawaii islands out of those 41-45 plants were predominantly found in the central Pacific regions of Polynesia and around 29 plants were predominantly found in Hawaii (Wilmshurst et al., 2011) and those primitive germplasms of the modern cultivated crops (e.g., mountain apple, gourd, banana, bamboo, sugarcane, sweet potatoes, yam, breadfruit, paper mulberry, coconut etc.). The transportation of those primitive angiosperms with economic potential as well as wild germplasms of modern crops was supposedly performed by canoes through oceanic routes, which indicated that ocean voyages in the beginning of the Holocene Epoch played an important role in the migration of angiosperms related to human migrations.

The Recent story of the anthropogenic induced migration of aquatic weed, water hyacinth (*Eichhornia crassipes* of the family Pontideriaceae), is a unique example of pan-tropical migration and colonization in the Holocene Epoch of the Quaternary Period, from South Africa to North America, Asia, Africa and Oceania water bodies, lakes, rivers etc. Though *E. crassipes* is reported to be from Colombia in 1801, its center of origin has been considered as the Amazon basin of Brazil and found to be very restricted in distribution prior to the 1800s (Barrett and Forno, 1982). As a colourful aquatic plant it was introduced in different parts of the world around 1884, it was introduced to Lousiana, USA (Julien, 2001) followed by its introduction to Australia, Egypt, and Japan in 1890, followed by Java (Indonesia) in 1894, and India in 1896 (Gopal, 1987). In 1902, *E.crassipes* was reportedly introduced in China (Xie Yan et al., 2001) and in 1900 to Madagascar (Binggeli, 2003). In the 20th century, the widespread distribution of *E. crassipes* in all aquatic bodies all over the world and its over-dominance to push the native aquatic plants to the brink of extinction, clogging the water bodies, attributed a negative recognition to it as an obnoxious weed, and extensive biological, mechanical and biochemical measures were taken to control the population growth of this species.

Chapter 24

THE CONTEMPORARY FLORISTIC REGIONS OF THE BLUE PLANET (EARTH)

From divergence to diversification it has been a paradigm shift; a process of evolutionary development of all biological entities, through the spatiotemporal evolutionary trail to formulate a unique identity. It has been recognized that in the evolutionary progress of life and plants in general or flowering plants, in particular, is no exception. As long as the flowering plants go through the transitional phases of divergence in the primary phase of evolutionary development, before entering the next phase of diversification. This phase of diversification of the secondary node of evolutionary develop is obviously a complicated phase of reticulate evolution which involved with continental drifting, isolation of continental blocks, changing of paleo-climatic regimes, passing through the transitional phase of transmigration, hibernation through the suffocating phases of mega-extinctions, radiation in the late Cretaceous phases. Earliest efforts to disperse and settle in adaptive geographical locations across the continents, the assisted phase of plant migrations by taking advantage of plant-animal co-evolutionary relationships and availing the advantages of biological, physiological, and climatic factors (e.g., wind current, structural organization of coconut seeds etc.). The contemporary phase of the Anthropocene, started with the history of human settlement, migration, domestication and utilization of wild resources and establishing authority and manipulation or micro-management of natural ecosystems to determine the dynamics of the producer level of any ecosystem in terms of population size of any particular plant species or a group of plant species. In the post-migration phase, plants struggled to settle in particular geographical areas of far and wide inter-continental blocks, across the blue planet. In due course of time, the complex process of evo-devo like sympatric and allopatric speciation helped evolve unique plant populations with less parental morpho-genetic inheritance and more characteristic, unique morphological and genetic features in the newly settled geographical locations. In a nutshell, it is a story of a metaphoric journey of diversifications through ecological and evolutionary times as well as the secondary phase of evo-devo of plants (particularly angiosperms) to emerge from the primitive identity of the geological past in the quest for diversification and surviving with a new identity (in terms of morphological, physiological, anatomical features and genetic configurations) in the new geological regime (Bennett, 1990). The settlement of diversified plant communities (under specific taxonomic or systematic identities like orders, families, genera, species, and sub-species) at the species level (specifically in the identity of endemic species level), in geographical locations, helped identify specific phytogeographical regions or biogeographical centers. The great plant taxonomist, phylogeographer and evolutionist Dr. Armen Takhtajan (1986) proposed the floristic areas of any rank as "Choria' or "Phytocoria." Naturally, the phytochoria or floristic regions helped identify itself as a matrix of homogenous floristic components (comprised of orders or species in the transcending taxonomic hierarchy), in the broader aspect, or phyto-endemism in a narrow perspective (Axelius, 1991). It went through a labyrinth of an evolutionary trail of divergence, convergence etc. to attain the ultimate phase of geographical isolation, recognizes as the ultimate phase of seclusion.

Whether it is a micro or macro level of evolution, there are clearly distinguished nodes that any entity would have gone through to attain the ultimate phase of diversification. This ultimate phase of diversification is the latent phase like a stepping stone to initiate the next evolutionary cycle to carry on the continuity of the journey of evolution either progressively or regressively, depending upon the limiting or inducing factors driving the pace and direction of evolution. Divergence of flowering plants is no exception. Flowering plants took a long time to emerge and radiate due to several unsuccessful attempts (since the Permian Period or even earlier) as they went through the bottleneck of several mass-extinction events in the early Cretaceous Period, we witnessed its preponderance in the terrestrial biomes of this blue planet. After the successful emergence of the threshold of the evolutionary phase from unicellular, prokaryotic, simple forms of life to the huge, complex, multicellular forms of flowering plants, they tried to cope with the diversified ecological, climatic, physiographic, edaphic and biological factors of diversified geographically and geophysically isolated areas. Some plants settled after coping with all the variables of the consecutive area of its population while some became extinct. In this phase of diversification of evolution, settlement of particular groups of plant populations helped designate the concerned habitat or biome to be designated in unique forms of phytogeographical regions as well as floristic regions. The expansion and contraction of floristic regions were always driven by population dynamics of flowering plants influenced by the selection/limiting pressures like changing climate, topography, supply chain of food or energy trophics, habitat disturbance, over-exploitation, environmental pollution etc. In this regime of Anthropocene, the well-established floristic elements of different floristic regions faced anthropocentric manipulations of the all the ecosystems of this blue planet. Hence, the rate of disappearance of species and the identities of certain floristic regions changed rapidly in the micro-level, which was not witnessed in the broad-based biome level. However, it is convincible that all these cumulative factors played a key role in driving the pace and direction of the evolution of flowering plants on Earth in a certain form and direction in the long run. The enigmatic query is whether we are in the latent phase of the next evolution or the divergence of flowering plants would further continue to form and reform the shape of the floristic regions of this blue plant?

The phylogeographers segregated the plant life of the world into three broad latitudinal zones (a) Polar, (b) Temperate and (c) Tropical zones according to the formation of plant communities in distinct geographical locations. The earliest worker Hansen (1920) divided the scientific classification of world's plant life into distinct floristic regions. He has divided the world phytogeographically into eight distinct regions on either side of the equator, which is presented in Table 3.

However, Hansen's (1920) classification is purely geographical, where distribution of plants in particular zone is not considered in terms of its abundance or its identity. Later, different works classified the phytogeographical regions, according to the distributional pattern of most abundant and important plants of those regions. Schouw (1928) has divided the entire world phytogeographically, into 25 Kingdoms, which in turn subdivided under several provinces, named after distinguished and abundantly distributed plants of that zone. This type of classification is purely species-specific, where identity of the plant community of that particular species has been overlooked and geological, geographical data of that region (which is the key factor to form distinct plant community) is not properly taken into consideration. Good (1947) proposed another format of classification of phytogeographical regions, based on floristic analysis.

Table 3. The phytogeographical classification of world (Hansen, 1920)

Name of Zones	Degrees of latitude on either side of equator
a) Equatorial zone	0.0-15.0
b) Tropical Zone	15.0-13.5
c) Sub-Tropical Zone	23.05-34.0
d) Warm Temperate Zone	34.0-45.0
e) Cold Temperate Zone	45.0-58.0
f) Sub-Arctic zone	58.0-66.5
g) Arctic Zone	66.50-72.0
h) Polar Zone	72.0-90.0

This floristic classification (37 phytogeographical regions) is a positive approach to identify a geographical area in term of its floristic identity (where endemism is also taken into consideration) through this system of classification is complex enough to some extent. However, the early effort on phytogeographical classification of Good (1947), were simplified by later workers (Takhtajan, 1986; McLaughlin, 1994). According to the contemporary perception, the phytogeographical or floristic regions would be categorized under three categories on the basis of the size and extent of geographical areas:

a) Micro-regional areas: Involved with the area of floristic elements under a single boundary of restricted geographical area. e.g., The floristic province of Southern Western Ghats in India is the southern part of the entire Western Ghats (phytogeographically, comprising of Southern, Central and Northern Western Ghats) mountainous floristic regions, designated as UNESCO world heritage sites for being enriched in endemic plant resources but Southern Western Ghats, the landmass formed in the Lower and Middle Miocene of the Tertiary Period, were recognized as the best abode as well as the "hotspot" of biodiversity in terms of the highest degree of phyto-endemism (Kundu, 2001).
b) Macro-regional areas: Involved with the area of floristic elements of several floristic provinces of a wide geographical area, sometimes recognized as floristic kingdoms. e.g., The Antarctic floristic Kingdom comprises of vast geopolitical regions 40^0 South of the Southern hemisphere, the entire continental flora of Antarctica, Patagonian floristic regions (comprised of southern Argentina and Southern Chile), Sub Antarctic floristic elements of islands of New Zealand and the floristic regions of the islands of the Southern Oceans (e.g., Falkland, Kerguelen Gough Islands etc.). The floristic elements of the Antarctic floral kingdom were found to be highly enriched with paleo-endemic plant genera and species and the entire landmass was found in a part of the erstwhile supercontinent, "Gondwana," and the modernization of the flora of this floristic kingdom was initiated during the Early Tertiary period. For further phytogeographical reviews in the micro level, this Antarctic floristic kingdom was further divided into four floristic regions and those four floristic regions were subsequently divided into sixteen floristic provinces at the regional level (Good, 1947; Bharucha, 1983; Takhtajan, 1986).
c) Meso-regional areas: Involved with one or two areas of floristic elements overlapping or adjacent to floristic provinces with large geographical areas, e.g., floristic region of the Sino-Japanese region comprised of identical floristic elements of Eastern China and the adjacent Japanese geopolitical territories until the unique Japanese floristic elements were evolved and the northward drifting and geological segregation of the Japanese land mass was completed in the Late Tertiary Period (Guowen, 1997).

Naturally, local and regional phytocorial studies involved floristic analysis of the micro-regional and meso-regional areas, whereas the analysis of biomes or intercontinental floristic studies involved with the analysis of floristic elements, an analysis of macro-regional areas. However, in the 80s, Bharucha (1983) presented a simple form of phytogeographical classification of the world in which the world was classified into six floristic realms:

1. Holarctic
2. Paleotropic
3. Neotropic
4. Capensis
5. Notogaeic and
6. Antarctic.

While it is an earnest effort to classify the world's phytogeographical scenario, the Indo-Malayan floristic identity is not clear from this classification. Pielou (1979); Muller (1980); Chapaman and Reiss (1995) proposed another classification, in which the world is divided under eight biogeographical regions –

(i) Nearctic
(ii) Palaearctic
(iii) Oceanic
(iv) Australasian
(v) Indo-Malayan
(vi) Afrotropic
(vii) Neotropic and
(viii) Antarctic

An outline of this revised classification of floristic regions of the world (Chapman and Reiss, 1995) is presented in Figure 22.

This classification is almost acceptable as the floristic identity of the respective regions have perfectly been depicted by the eight biographic regions. Though it seems troublesome, species-specific distributional records of the entire plant kingdom throughout the world are necessary for a perfect and lucid phytogeographical classification.

An effort towards a detailed classification of the world of flora under several floristic regions was initiated by Engler (1912, 1924) in the early 90s; he divided the flora of the world into 29 regions. A trend of critical analysis to classify the global floristic elements at the micro level were set to create more slots for the world of flora that were further visualized through the works of different phytogeographical analysts which are summarized hereunder:

- Engler (1912, 1924) divided the world of flora into 29 regions,
- Tolmatchev (1974) divided the world of flora into 30 regions,

- Good (1947, 1974) and Schmithusen (1961) divided the world of flora into 37 regions,
- Mattick (1964) divided the world flora into 43 regions.

This trend is further continued by Takhtajan (1969, 1974, 1986) and Udvardy (1975), who proposed another classification of the world of flora based on the distributional pattern of endemic floristic elements. Takhtajan (1974, 1986) classified the world of flora into six floristic kingdoms, subsequently subdivided under eight-sub kingdoms, followed by 25 floristic regions. As Takhtajan's (1986) approach towards classification of world of flora into 35 regions is chiefly based on diversity as well as concentration of endemic floristic elements, it is comprehensive and has greater merit to be taken into consideration than the others.

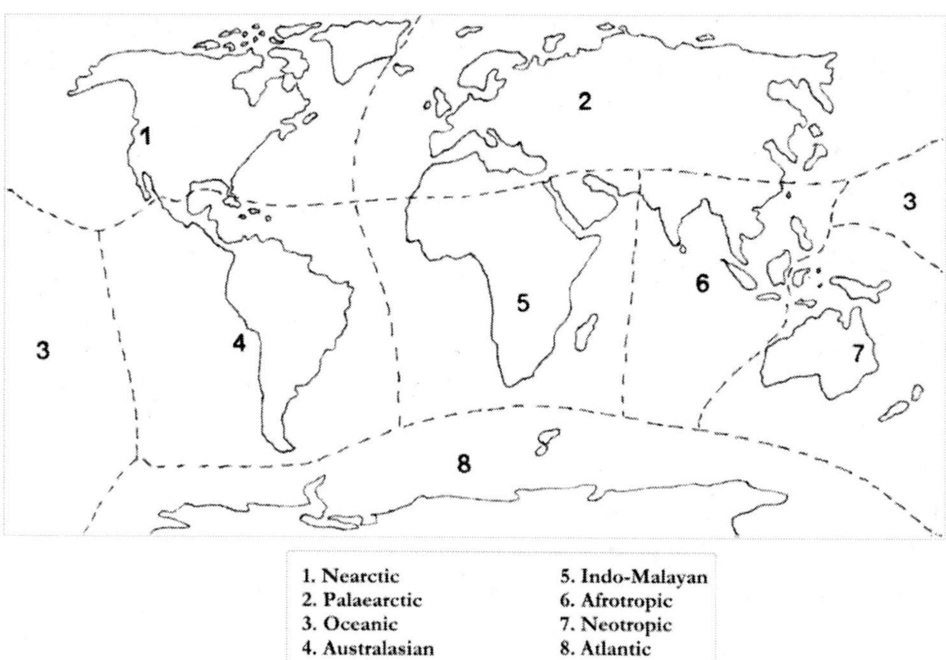

Figure 22. An outline of the floristic regions of the world (Chapman and Reiss, 1995).

Chapter 25

PHYTOENDEMISM OF FLOWERING PLANTS: CATEGORISATION, IN RELATION TO EVOLUTION OF PLANTS ON EARTH

Endemism in plants is a usual feature in different closed or confined habitats throughout the world. The term 'Endemic' is applied to any taxon (plant or animal) isolated by a barrier. The term is taken from the Greek Word – "En" meaning within and "Demos" meaning population. Conceptually, it has a long history of origin, that dates back to 1855 that starts with publication of DeCandolle's treatise. The span of endemism relatively depends upon the range of confinement of the taxonomic group in a particular matrix. Therefore, it could be narrated as a "rarity" or a "range restricted rarity" (Cowling et al., 1994). On the contrary, a taxonomic group which is widely distributed, is considered to be a pandemic taxon. The perception of endemism to evolutionary biologists, synecologists (who dealt with fossils and extant biota found that range sizes could not be considered as independent traits at the species level as the related species may have the same range size; the variance in size seems to have started at a sub-species level, but conservation biologists consider that range restricted rarity is an attribute that predisposes a taxon to extinction and proper understanding of the correlation between interactive abiotic and biotic factors (like plant-carbon relationships) to understand the process of extinction dynamics in the paleobotanical perspective and help undertake strategies that would be an execution of ideal management strategies to reduce the rate of extinction) and community ecologists who considered it is a form of rarity or range of restricted rarity (which depends on specific ecological traits like local population size, body size, reproductive fitness and dispersal fitness etc.) (Cowling and Samways, 1995; Cowling and Hilton-Taylor, 1997; Cowling, 2001) in their discreet observations. Motility of the animal species was found to be slightly advantageous over plants, as the6y could leave the habitat found unsuitable to survive or exceeding its carrying capacity (i.e., availability of the food with respect to the growth of population size to survive on food resources in a spatio-temporal perspective) threshold of that habitat but plant species are devoid of this advantage so the effect of the same selective pressures or changing factors, impacting the population dynamics of any habitat impact the plant species a little more than animals and this was noticed in the difference of the endemic rate of plants, generally found to be higher than the rate of endemism in animals. There are many factors acting upon a particular transducing taxon individually or synergistically to make that taxon endemic (Kundu, 2004), which may be categorized under the following heads:

1) *Physical barrier:* Isolated landmass, separated by seas; island; lofty mountains or forest patches separated by grasslands; land-locked natural lakes etc.
2) *Ecological barrier*: Climatic changes, atmospheric instability topographic alterations.
3) *Atmospheric interference:* Overexploitation, habitat modification, and deterioration etc.
4) *Genetic incompatibility:* Interbreeding, genetic errosion, genetic drifting, genetic intermixing created hassles to retain M. V. P. (Minimal Viable Population) level and finally affect reproductive biology.
5) *Biological factors:* Overgrazing by cattle, lack of pollen-specific vectors, pathogens, exotic weeds push aside a widely distributed taxon to a confined locality.
6) *Natural calamities:* Floods, droughts, landslides, glaciations, population etc.

The matrix occupied by endemic taxa should be regarded as an 'endemic matrix' or endemic area (DeCandolle, 1855). The concept of endemism of a particular taxon to a particular matrix is relative. The abundance or scarcity of endemic floristic elements in a particular matrix provide considerable information such as the:

a. Phytogeography of a particular region.
b. Suitability of the matrix to those endemic floristic elements.
c. Adaptive radiation of endemic floristic component to a particular matrix (viz. morphological adaptation, anatomical adaptation, cytological adaptation, ecological adaptation etc.).
d. Matrix of speciation and speciation pattern.
e. Matrix of extinction and reason behind it.
f. Trace out the evolutionary pathway from origin to extinction.
g. Methodology to undertake scientific exploitation of endemic taxa in a particular matrix.
h. Adopting conservation management strategy for a particular endemic taxon or taxa as a whole.

Considering all these relevant factors playing actively to render a taxon endemic, one can throw light on its origin and other details like:

1) whether the taxon is a living remnant of ancient, widely distributed taxon.
2) being newly developed from parental stock through speciation.
3) an intermediate state between origin to extinction.
4) its habitat.
5) its extent of ploidy.
6) adaptive radiation range.
7) possible fossil evidence.

An outline of classification of endemic plants has been presented in Table 4.

From the classical point of view, endemics are broadly classified into three groups, (Ahmedullah and Nayar, 1986).

A. PALEOENDEMICS: The term was coined by Chevalier and Guenot (1925). From time to time it has gone through different names, Engler (1882), Drude (1890) mentioned it as a "relic," whereas Diels (1909) called it a "connective," Herzog (1926) designated it as "relic or ancient endemism." The Paleoendemic remnants or relicta representative of ancient, widely disturbed taxa, characterized by the following characters:

a. Remnant of ancient widespread taxon might have fossil ancestors.
b. Arborescent habit.
c. Taxonomic isolation.
d. Low adaptive radiation.
e. Low level of ploidy.
f. High level of genetic erosion.

e.g., *Ginkgo biloba, Shorea tumbuggaia, Mangolia pterocarpa* etc.

B. NEOENDEMICS: The term was coined by Herzog (1926). Different works mentioned this group of endemics in different terms, Engler (1882) mentioned it as "autochthonous endemics"; Drude (1890) ascribed it as "secondary endemics"; Briquet (1908) noted it as "endemisme par novation"; and Diels (1909) mentioned it as "progressive endemics." The Neoendemics are characterized by:

a. Herbaceous habit.
b. Predominant vicariance.
c. High rate of adaptive radiation.
d. High level of ploidy, chromosomal aberrations, natural mutation etc.
e. Low rate of genetic erosion.

e.g., *Nilgirianthus beddomei, Meconopsis robusta, Arenaria ferruginea* etc.

Table 4. The categories of Phytoendemism of plants in the world

	TYPE	DEFINITION	SOURCE
1.	PALAEOENDEMICS	The Palaeoendemics are relict representative of widely distributed taxa with constancy in chromosome numbers.	Chevalier and Guenot (1925).
2.	NEOENDEMICS	The Neoendemics are newly evolved taxa of recent Origin from active genetic stalk, marked by polyploidy.	Herzog (1926).
3.	HOLOENDEMICS	The Holoendemics are intermediate state between Neoendemics and Palaeoendemics.	Richardson (1978).
4.	SCHIZOENDEMICS	The Schizoendemics are occurring in diverse ecological niches, having identical chromosome numbers.	Faverger and Contandrio-poulos. (1961).
5.	PATROENDEMICS	The Patroendemics are parent endemics, usually, diploids, give rise to polyploids.	Faverger and Contandriopoulos. (1961).
6.	APOENDEMICS	The Apoendemics are polyploids of hybrid origin, arising from crossing of taxa with identical chromosome number, distributed widely.	Faverger and Contandriopoulos. (1961).
7.	EDAPHIC ENDEMICS	The Edaphic endemics are result of specific physiographic requirements of a taxon met by a particular geologic formation.	Kruckeberg (1984).
8.	ISLAND-ENDEMICS	The Island-endemics are confined and existed together in an Island habitat having diverse taxonomic identity.	Gentry (1986).
9.	ANTHROPO-GENIC-ENDEMICS	The Anthropogenic-endemics are evolved due to human interference (e.g., Overexploitation, habitat transformation, habitat destruction etc.	Gentry (1986).
10.	PSEUDOENDEMICS	The Pseudoendemics due to inadequate knowledge of their geographical distribution.	Nelson et al. (1990).
11.[27]	NARROW RANGE ENDEMICS	The N.R.E.s are restricted in a particular political boundary (e.g., Country/State).	Kundu (2000)
12.[28]	BROAD RANGE ENDEMICS	The B.R.E.s are restricted in a region comprising of more than one state.	Kundu (2000)

[27] World Institute for Conservation & Environment, WICE (2004) have categorized the endemic taxa into three categories i. Site Endemics (just found on a particular mountain, like Mountain Celaque), ii. National Endemics (found in a particular country, like Honduras) and

[28] iii. Geographical Range Endemics (found in Himalayan region, covering several countries). Source: WWW.birdlist.org/nature_management/ecomapping/endemism.htm

C. HOLOENDEMICS: The term was coined by Richardson (1973), it includes intermediate states of endemic taxa between any state from origin to extinction. There are several intermediate phases, mentioned below:

a. Origin
b. Expansion
c. Stabilization
d. Diversification
e. Migration
f. Fragmentation
g. Contraction
h. Extinction

e.g., *Blepharistigma membranifolia, Corchorus pseudoolittorius, Asmischophacalus cucullata.*

Based on cytological characteristics of endemic plants, Favarger and Contandriopoulos (1961); Contandriopoulos (1962) classified endemics under four heads:

1. *PALEOENDEMICS:* Having little changes in chromosomal configuration.
2. *SCHIZOENDEMICS:* Occurring in diverse ecological niches but having identical chromosomal numbers.
3. *PATROENDEMICS:* Parent endemics having diploids, give rise to polyploids, mainly euploids.
4. *APOENDEMICS:* Polyploids of hybrid origin, arising from crossing of taxa with identical chromosome numbers, distributed widely. The relative percentage of diploids and polyploids helps ascertain the relative age of the floras (Ahmedullah and Nayar, 1986). According to Ahmedullah and Nayar (1986) the older flora have higher percentage of diploids whereas the younger flora has higher percentage of polyploids.

EDAPHIC ENDEMICS: Jenny (1941, 1980) stated that the soil formation process is highly influenced by the function of climate, organisms, topography, parent material, time and organisms. Mason (1946) and Kruceberg (1969) found that in a part of California is enriched with endemic floristic elements that are linked to important pedological factors like deficiencies in calcium and high concentrations of magnesium, chromium, and nickel and this kind of endemic plants in edaphically stressful soil habitats is also recognized as "Serpentine endemism" (Anacker, 2013). Tadros (1957) and Wicklow (1964) observed that spread of *Emmenanthe rosea* (which is endemic to serpentine soil) is inhibited due to increased microbial activities on non-serpentine soil. Kruckeberg (1984) finally coined a new term for these types of endemics "Edaphic endemics," where physiological requirements are matched to geologic formations.

Gentry (1986) has further proposed another classification of endemics, the four groups which he mentioned are:

a. ISLAND ENDEMISM.
b. RELIC ENDEMISM.
c. NEO ENDEMISM.
d. ANTHROPOGENIC ENDEMISM.

Very recently Daniel et al. (1995) appreciate Gentry's (1986) classification rather than Ahmedullah and Nayar's (1986); also considered "Pseudoendemism" a termed coined by Nelson et al. (1990).

Based on the existing distributional record of endemic taxa, Kundu (2000, 2004) has categorized all endemic species into two broad categories:

1. NARROW RANGE ENDEMICS: These are restricted in a particular country (within a particular political boundary) and are subjected to be protected under specific national legislative safeguards. All endemic taxa, belonging this category, have high habitat specificity, low ecological tolerance, low adaptive radiation and narrow range of distribution.

e.g., *Aponogeton satarensis, Nepenthes khasiana, Elaeocarpus blascoi* etc.

2. BROAD RANGE ENDEMICS: These are restricted in a particular geographical location comprising of a number of countries and where a coordinated effort is required to adopt a common conservation management program (bilateral or multilateral) in order to protect these resources. All endemic taxa belonging to this category have less habitat specificity, high ecological tolerance, high adaptive radiation and wide range of distribution.

e.g., *Denrocalamus parishii, Gironuiera lucida, Elatostema decipiens* etc.

Now it is a matter of consideration, whose classification is more appropriate. Apart from Ahmedullah and Nayar's (1986) classification, which has enough content to take in account but the classification of endemic categories by Faverger and Contandriopoulos (1961), Contandriopoulos (1962) classification is more tenable, in which, "Paleoendemics" is identical with "Paleoendemics" of Ahmedullah and Nayar's (1986); whereas "Schizoendemics" show its flexibility to be considered as "Holoendemics" and it may achieve the status of "Panendemic" or cosmopolitan in near future if high adaptive radiation exists. "Patroendemics" and "Apoendemics" may be taken under

"Neoendemics" of Ahmedullah and Nayar's (1986) system of classification of endemic plants. However, Genetry's (1986) proposition cannot be ignored either. Gentry's (1986) "Relict-endemics" and "Neoendemics" are identical with Ahmedullah and Nayar's (1986) "Paleoendemics" and "Neoendemics" respectively. But the status of "Island endemics," is not conspicuous much. Mac Arthur and Wilson (1967) in his "Theory of Island Biogeography" mentioned that Islands are ideal resorts for endemism; a fact is evidenced by presence of higher number of endemics than adjacent territories. Mac-Arthur and Wilson (1963) in their theory "An equilibrium theory on Insular Zoogeography" opined that all the species present on Island, come from mainland, some of them will be extinct and rest of them will colonize it.

Therefore, the number of species on an Island is equilibrium between numbers of extinction of number of colonization (Ward and Balausetein, 1994). This equilibrium may become unstable by any sort of interference anthropogenic, physical, biological or climatic.

According to Carlquist (1965), woody habit in islands is due to insular isolation rather than relict from. Naturally, island population, considering habit, does not give any impression regarding age factor. Hence, the status "Island-endemics" is not feasible as "Paleoendemics," "Neoendemics" and "Holoendemics," are existed altogether on an Island habitant - "Anthropogenic-endemics" are evolved due to habitat disturbance, overexploitation of a particular taxon or taxa as a whole in a population.

The contemporary phytogeographical reviews in the scale of endemism, on two different habitats like mainland habitat and island habitat shows that the rate of endemism of plants and vertebrate in island habitat are 9.5 and 8.1 respectively in the scale of 10, and it has been observed that habitat loss due to anthropogenic activities in the islands and mainlands wasfound almost identical and the extrapolation indicates that the rate would be considerably higher in islands than mainland in the year of 2100 (Kerr et al., 2009). Naturally, priority-based conservation management should be taken into consideration to conserve the highly impacted endemic floristic elements in low areas of distribution in islands in comparison to mainland's. The anthropogenic interference may shift an endemic status of a taxon, which is presented in Table 5.

Table 5. The shifting status of a taxon due to anthropogenic interference

From	To
a) Palaeoendemic	Epibiotic/Extinct Taxon
b) Holoendemic	Palaeoendemic
c) Neoendemic	Holoendemic
d) Panendemic	Neoendemic

Hence, any taxon whether Paleoendemic, Holoendemic or Neoendemic subjected to human interference, should be considered as "Anthropogenic-endemic" e.g., *Aconitum deinorrhizum*, *Vanda coerulea*, *Cycas beddomei* etc. "Pseudo-endemism" - the term which is recently used to designate a taxon "apparently endemic" due to inadequate knowledge of its distributional or geographical ranges. The term "Pseudo-endemism" (It is due to lack of taxonomic knowledge as well as distributional pattern) is used by Daniels et al., (1955). In the near future, it may be lifted from this transitional state to "Paleoendemics," "Holoendemics," "Neoendemics" or "Panendemic" after completion of the total investigation (phytogeographical, cytological, geological, etc.). e.g.,

Piper barberi (extended distribution)
Clematis bourdilloni (Type collection only)
Bulbophyllum rothschildianum (rediscovery after a long period, near about 100 years).

However, an effort of Kundu (1999) to categorize all endemic taxa into Narrow Range Endemics and Broad Range Endemics has been found very straightforward and which has a close resemblance with WICE (2004) classification of endemics. The "Site Endemics" and "National Endemics" could be considered as "Narrow Range Endemics" and "Geographical Range of Endemics" could be considered as Broad Range Endemics." Considering different categories of endemics, suggested and explained by various scientists, it seems that the following categories should be taken into considerations in the context of reviewing "Phytoendemism" of any geographical region.

a. Paleoendemics.
b. Neoendemics.
c. Holoendemics.
d. Anthropogenic-endemics.
e. Pseudoendemics.

In some points, it seems that the entire studies on phytoendemism is embodied with theoretical perspective. Phytoendemism of a country is not only studied and evaluated for academic interests (e.g., studies on paleoendemics helps for better understanding of evolutionary pattern, different changes through geological time scale etc.) but it has immense importance for adopting conservation measures of N. R. E. in national level or B. R. E. and in regional level. The developing countries (having plenty of natural resources) could not give-up its hope to utilize its natural resource in order to make its fiscal profile much stronger.

In such condition, proper management of that resource is required, which further pushed towards drafting a suitable model to start with. It validates such studies to help to trace out the center of endemism, center of speciation, especially for neo-endemics (So far its distributional pattern is completely known) of a particular family or a number of families as well as new taxonomic group, which is directly related to species diversity or biodiversity as a whole. Enrichment with endemic living resources should be considered as a sort of added mileage to improve the fiscal profile, in comparison to other countries deficient with endemic resources. As the economic aspects and commercial exploitation of endemic plant resources of a country play important roles in improvement of fiscal profile, the present studies have imponderable significance.

Endemism of plants in general and flowering plants, in particular, are partially interrelated to the occurrence of plate tectonics. The endemic plants of peninsular India, which are not extended further or have fossil ancestry, towards other parts of Gondwanaland to which Indian plate was attached once, may be regarded as 'paleoendemic', provided, anthropogenic factor does not play any key role. Those endemic plants should be regarded as 'holoendemics', which have covered wide distributional matrix or extended distribution e.g., endemic plants found in peninsular India as well as in Himalayan region.

On the other hand, when Indian plate merged with Laurasia and Himalayan mountain ranges were formed the corridor of floristic migration was initiated. As a result, Asian and European floristic elements have intruded through NorthWest corridor. Hence the endemic plants should be generally, treated as 'holoendemics', except those having fossil ancestry or supposed to be extinct in the near future. If the endemic plants of this region emerged through active speciation or emerged through anthropogenic inference they should be treated as 'neoendemics' as a number of such plants are found in the Eastern Himalayan region. The plant communities in the mountain tops are apparently identical to isolated plant populations of insular regions and the dynamics of phytoendemism, which was easily observed in the meso-regional areas (two adjacent matrix of phytoendemic species) has normally been influenced by the factors like the degree of isolation, extent of intrapopulation and interpopulation competition for space and food, habitat fragmentation and loss etc.; which determines the expansion of endemic plants in one habitat with enough abiotic resources and less competition and accelerates contraction of endemic plants in the adjacent habitat with more competition and scarce resources (Medail and Veralque, 1997).

From a theoretical viewpoint of Carlquist's (1965) "Island Biogeography," the isolated mountain and insular habitat both are physically alike so both of the endemic matrices are enriched with endemic floristic elements but Medail and Veralque, (1997) observed in their comparative studies on the endemic floristic elements of "Mesoregional enemic habitats"[29] that mountain region are a better habitat for the long-term settlement of endemic species rather than am insular habitat. The relationship of floristic richness and phytoendemism was seen as inversely proportional (Medail and Veralque, 1997) and it is convincible that

[29] The flora in the Mediterranean sea regions with endemic floristic elements of adjacent mountain regions.

floristic richness is directly related to the frequency and mode of diversification and the range-bound expansion of flowering plants in the Early Cretaceous Period ensured that the overall floristic richness all over terrestrial habitats of this Blue planet did not give enough opportunities to the endemic floristic elements to expand but humans controlled or micro-managed as well as manipulated the regime of Holocene or Anthopocene Epochs of the Tertiary Period and came up with new "sapatio-temporal" challenges wherethe ecolological equilibrium was tilted to one side and the evolution and existence of the "radiation of angiosperms" in the Cretaceous Period is on the survival threat of the "Sixth Extinction" (Kolbert, 2014) in Anthropocene.

Chapter 26

EVOLUTIONARY REVERSE SWING AND DOMESTICATION OF THE WILD FLOWERING SPECIES TO CULTIVATED CROPS THROUGH THE ANTHROPOCENE PERIOD: *PAEONIA* –A CASE STUDY

Apparently, the domestication of flowering plants in the Tertiary Period had no straightforward relation to the evolutionary progress of this specialized group of transducing organisms since their origin to attain the diversification phase. But, it is the truth that our blue planet in the Holocene and the Anthropocene went through few important phases like exponential growth of *Homo sapiens sapiens* over all other animal species to establish its supreme authority to perform the micromanagement and manipulation of all ecological functions to control over the population size of the other organisms. This was followed by giving-up the nomadic life to set up human settlement, called human civilization in different parts of the world; followed by domestication of wild animals to make those domestic animals and wilds plant species to convert these cultivated crops to utilize those as foods and successful execution of Green Revolution in order to achieve sustainability in crop production, as well as food production by means of application of cytogenetical, know how. Un-interrupted, exponential growth has tilted the equilibrium of biodiversity on this blue planet and in one point the yield of Green Revolution failed to cater the over-populated centers of earliest civilizations due to exceeding the carrying capacity and lead to the beginning of human migration, a shift from agricultural practices to industrial practices.

The intensive industrial practices in the end of the Eighteenth-Century led to make Industrial Revolution successful; apparently, Industrial Revolution have managed the population growth in preliminary phase but it could not cater to the excessive demands of food. The destruction of wild habitats has been started during formation of human settlements and conversion of wild habitat to cultivated lands and it has gone up during the period of Industrial Revolutions due to settlement of industrial settlements, formation of more human settlements and infrastructure developments led to severe shrinkage and fragmentation of natural habitats for wildlife (plants and animals). Thus, for green plants, particularly the angiosperms, those radiated during the Cretaceous Period lead to the reversal of evolutionary directions from the earlier mode of diversifications by the metaphoric firewall screenings like habitat shrinkage and inbreeding depression, genetic bottleneck due to drastic shrinkage of wild population size, lead to expediting genetic erosion. The backup support of co-evolution between plants and animals (as wild animals were in survival jeopardy since the regime of domestication, became acute issues during the era of Industrial Revolution) was found to be non-functional and it has not only stunted the mode of diversification and further expansion of the flowering plants but it also expedited the rate of extinction and leading to imbalance of biological diversity of life on earth. The apparent success of industrialization has left some permanent black holes like huge atmospheric pollution by emitting excessive carbon, that led to Greenhouse effects, discharging effluents in the ground level creating Ozone Hole in the stratospheric level to stuck the last hurdle to stop diversification and expansion of angiosperm populations. Hence, domestication is an important phase in the evolutionary history of flowering plants as it is the first step of the reverse swing of evolution as well as the paradigm shift from the mode of diversification to the mode of extinctions.

While maintenance of biodiversity is the ultimate goal to achieve sustainable development of human society, the relationship between the magnitude of agricultural and industrial growth and loss of biodiversity is directly proportional.

On the other hand, the loss of biodiversity is proportional to the rate of extinction of species. Thus, it's convincible that the rate of extinction is inversely proportional to the rate of diversification of flowering plants. At the same time the relationship of agricultural process (which started with the history of domestication of wild crop around 10,000 years ago) '(Gross and Olsen, 2010)' and conservation of biodiversity, as well as sustainability, is inversely proportional, which has been evidenced by the conversion of one-tenth of wild, terrestrial ecosystems to the habitat of cultivated crop replacing the wild crop progenitors by domesticated cultivars '(Arora and Nayar, 1983; Gross and Olsen, 2010)'.

The change of mode of adaptation and diversification process of any wild plant, which has been found later in the cultivation went through the history of domestication. Domestication is an integral part of human-dominated ecosystem management process but it is the primary impediment against the further continuation of "mode of diversification of angiosperm scot-free" as it has reversed the swing of evolutionary progress from the mode of diversification to the mode of extinction. It is necessary to understand from a spatiotemporal aspect, the evolutionary process of cultivated plant concerning its geographic origin; whether it is the outcome of single or multiple domestication and occurrence of "genetic bottleneck" in that process '(Gross and Olsen, 2010)'. While, origin of a cultivated plant of a particular geographic location from a single progenitor is a conventional idea (Yuan et al., 2013) which is known as "single domestication," the contemporary works on molecular biology support multiple independent geographic origins, e.g., Asian rice, Barley '(Yuan et al., 2013)' commonly known as "multiple domestication" and "independent domestication" of different species from different geographical locations. For example, two cultivated species of rice e.g., *Oryza Sativa* and *O. glaberrina* (Yuan et al., 2013); two cultivated species of cotton e.g., *Gossypium hirsutum* and *G. barbadense* (Westengen et al., 2005) went through the path of multiple domestications.

While a genetic bottleneck is a common feature in annual, seed plant but it is not very clear for perennial trees or herbs with vegetative propagules (Besnard et al., 2013; Yuan et al., 2013). The tree peonies which is worldwide popular for its ornamental and ethnomedicine potential, was cultivated in China (its center of origin) for 1600 years but its history of domestication remains unresolved (Yuan et al., 2013). *Paeonia sufruticosa*, "King of flowers" is one of the important tree species which form section *Moutan* Dc. with other nine tree species (Hong, 2010, 2011).

To trace out the history of domestication of tree peonies (the living fossil), Zhou et al., (2014), carried out an extensive phylogenetic investigations on the basis of sequences of ca. 14 fast evolved chloroplast regions and ca. 25 single-copy nuclear markers identified from RNA sequenced data of fairly large samples (ca. 441 accessions of ca. 37 populations) comprised of wild tree peonies, well-established cultivars of tree peonies supposedly was involved in a natural hybridization process and it has been found that tree peony cultivars originated as a result of random homoploid hybridization (the result of one and half millennia old anthropogenic practice to raise ornamental fertile hybrid peony in the household gardens by gathering wild germplasm of peonies from different parts of China). For the five wild tree peonies *P. cathayana, P. rockii, P. qiui, P. ostii* and *P. jishanensis* (Zhou et al., 2014) the process of hybridization in the backyard garden were chiefly man-made initiatives, whereas mostly allotetraploid species formation through natural hybridization has been evidenced in herbaceous peonies (Sang et al., 1995; Kundu, 2017). By using 14 nuclear microsatellite markers, Yuan et al., (2013) carried out an extensive investigation of history of domestication issues of tree Peonies in molecular level, on ca. 553 tree peonies, collected across China comprising of "Common Tree Species" or CTP, "Flare Tree Species" or FTP, wild species and subspecies of peonies: *Paeonia rockii* ssp. *rockii, P. rockii* ssp. *atava, P. jishanensis* and *P. decomposita*, the potential contributors to the domesticated tree peonies and noticed "independent domestications" of "CTP"s and "FTP"s from two distinct wild, allopatric species: *P. jishanensis* and *P. rockii* ssp. *atava*, respectively from different locations. It's quite interesting to found that genetic diversity in "CTP"s and "FTP"s is comparatively higher than four wild species and subspecies of peonies. This is mainly due to wild habitat destruction and human overexploitation result in smaller, isolated population, raising the inbreeding depression and ultimately leads to extinction '(Hong, 2010; Yuan et al., 2013)'. It has been noticed that selection of seedlings of cultivated tree peonies is in practice from the orchard under open pollination; Yuan et al., (2013) have opined that this particular selection breeding process offsets the loss of genetic diversity. The genetic bottleneck is a usual symptom of domestication process as allelic diversity in domesticated species is considered to be a form of replication of that found in their wild progenitors (Piry et al., 1999; Gross and Olsen, 2010).

The "genetic bottleneck" analysis shows considerable "heterozygosity excess" in wild taxa of *P. jishanensis, P. rockii ssp. rockii,P.rockii* ssp. *atava* in comparison to "CTP"s and "FTP"s (Yuan et al., 2013); instead of cultivated tree peonies, the genetic bottleneck has been noticed in some population of wild tree peonies. According to Austerlitz et al., (2000), long generation times of tree a peony (which is around 5 years) and overlapping generations could offset the domestication bottlenecks of peonies. The estimated "time of divergence" between "CTP"s and *P. jishanensis* has been almost double in comparison to "CTP"s and *P. rockii* ssp. *atava* which helps to infer recent domestication of "FTP"s in comparison to "CTP"s which happened earlier time and which has been historically affirmed: as "CTP" cultivars was developed in the Tang dynasty (627-850 AD) whereas "FTP" cultivars was developed in Ming to Qing dynasties (1398-1797 AD) (Yuan et al., 2013).

Most important trait in process of domestication is the number and size of flowers and number and colour of petals (Yuan et al., 2013); "CTP"s and "FTP"s have shown similar traits: (single vs. multiple flowers); (white petal vs. patchy white, pink, purple, red petals). In spite of independent domestications, "CTP"s and "FTP"s are two ideal group of peonies, ideal instances of parallel and convergent evolution from different locations and different species (Yuan et al., 2013). While process of domestication and conservation of biodiversity are two antagonistic processes (Zhou et al., 2014) in the backdrop of climatic changes, anthropogenic changes, over-exploitation of wild resources, the domestication of hybrid peonies from wild progenitors is a natural consolation in form of precious gene pool where a part of unique genomes conserved from wild endemic germplasm of peonies (Lopez-Puzol and Zhao, 2004) on the verge of extinction.

Chapter 27

PLANT EXTINCTIONS- IS IT PLUNGING IN THE DARKNESS OR EMERGING FROM THE SHADOWS OF THE REVERSE SWING OF EVOLUTION?

REVIEWING THE CONCEPT OF PLANT EXTINCTIONS SINCE THE LOWER ORDOVICIAN TO THE HOLOCENE PERIODS

According to Darwin (1859), "The Origin of Species by Means of Natural Selection" portrayed that a new life form emerges through divergence of character, struggle for existence when space and food becoming the limiting factors; ultimately, the competent one would survive (survival of the fittest) and other inhabitants would be pushed on the way to extinction. As the survivors are screened through the process of "natural selection" this process is considered to be the key factor of extinction. In this way, extinction is defined in the background of natural evolution and this is considered as a natural phenomenon.

From the ecological viewpoint, seral succession is a continuous process which has attributed dynamism to the ecological niche, which is gradually over-burdened in the due course of time and space; leads to disappearance of unfit ones from this ecological niche and welcome the new entrants.

Nayar (1987), has found that "the present-day changes in the environment and habitat are so unnatural and drastic that plant species could not get evolutionary time span for survival or adaptive radiation." Hence, the process of extinction should not only be considered as a natural process but conditional process depends upon the environmental limiting factors as well as anthropogenic interests towards a species.

In the policy statement of the formation of conservation areas under in-situ conservation management, Hooper (1971), has opined "....... that a minimal viable population of any given habitat is the smallest isolated population having 99 percent chance of extant for 1000 years despite the foreseeable effects of demographic, environmental and genetic stochasticity and natural catastrophes." Gomez Pompa et al., (1972) have visualized that reduced population have failed to sustain the single catastrophe or the reproductive biology of these extinction-pronged species are not sound, evidenced through the setting of flower in late, production of non-viable seeds etc.

The advancement of molecular botany since the end of the Nineties, helped to decipher the phylogenetic issues of different plant group. Yet absolute reliance on the contemporary data, available in the cladistics analysis process, comprised of DNA sequencing, computer algorithms, advanced graphical analysis etc. could not depict the entire phylogenetic profile of any group of living organisms, without any back-up of any paleobotanical evidence. The phylogenetic analysis (based on critical analysis of data available on the molecular biology of a group of plants) present a picture on the interrelationships of the basal group of plants, serving as ancestral stock, to the out-group and in-group plants. Particularly for a group of plants, like angiosperms, the evolutionary relationship of the relict type of basal angiosperms, the outcome of molecular botanical analysis, needs to be established to the paleobotanical evidences even the analytical data on geochemical analysis of those fossils to understand the mode of evolution of origin and diversifications of angiosperms (Coetzee and Muller, 1984; Crane et al., 2004). At the same time, the simulative back-tracking of the evolutionary foot-steps of a group of plants could help to reach at the ancestral phylogenetic root-stock transcending through the nodes of evolution. For any plant group, if the history of its origin, the phase of its expansion and the evidence of extinction (either paleobotanical or palynological) put together, that would help to focus on the tentative cause of disappearance or the extinction of that group of plants (Doyle and Donoghue, 1986). From contemporary phylogenetic aspects, it has been evidenced that the consequent disappearance of out-group plants (Graham and Iles, 2009) have more biogeographical as well as evolutionary impacts than extinction of core-group of plants however both issues have an irreversible impact of disappearance of the entire plant group in long-run.

From the ecological viewpoint, overlapped seral succession leads towards the formation of a new matrix, in which new entrants settled down and these types of overlapped succession have compelled the existing species to became the "extant species," before completion of its life-span. Any sort of limiting factor, either natural or anthropogenic is enhancing the overlapped-seral-succession. Hence the process of extinction is related to the process of evolution and which is not absolutely independent, as the limiting factors of environment, as well as human beings, have played the key role. The perspective of evolutionary biology, suggests the mode of diversification is inversely proportional to the mode of extinction hence the end of diversification of angiosperm and the exponential rate of extinction in Anthropocene should better be considered as "evolutionary reverse swing" of "Cretaceous radiation of angiosperms" rather it should be designated as "Anthropocene reverse swing of angiosperms." According to Myers (1980) "so long as human-kind is committed to playing God, on the grounds that we are determining the extinction of a large number of species, we might as well do it in a more rational, selective and considered manner, than by haphazardly condemning whole communities to oblivion."

THE ENHANCEMENT OF THE EXTINCTION OF FLOWERING PLANTS: SEVERAL ISSUES IN THE ANTHROPOCENE PERIOD

Miasek and Long (1978); Nayar, (1984); Ahmedulla and Nayar (1986); Guarino (1995) and Kundu (2004) have stated about the following factors, responsible for species extinction: -

I. *Environmental factors:* Each species has a tolerance limit for environmental factors, and lack of adaptive radiation beyond that tolerance limit enhanced extinction. The fossil evidence of *Aponogeton* (leaf impression of Tertiary Period) has been reported from (Aral Sea of C.I.S.) Northern Hemisphere, which indicates the shifting of its place of occurrence due to change of paleoclimatic conditions. It is now distributed over tropical and subtropical regions of Tropical Africa, South-east Asia, and Australia-Oceania region.

II. *Ecological substitute:* A species of a community is replaced from the matrix by competitive species or by a community (Prance and Ellias, 1977) having more vigor than the replaced one, as evidenced by seral succession.

III. *Biological factors:* Also play one of the most important roles in dwindling away of a species and their ultimate extinction.

IV. *Over-exploitation*: For medicinal, economic and other commercial potentialities many species are exploited from their natural habitat causing extinction. (e.g., *Paphiopedilium druryi* has been reported to be almost extinct in wild in Western Ghats of India (Ravi, 1994) due to over exploitation). Overgrazing is also another important biological factor. Plants of horticultural interests are also collected recklessly from their natural habitats placing them under threat to extinction (Sagreiya, 1978; Fitzerald, 1989).

V. *Weed invasion:* Species of *Mikania, Parthenium, Lantana, and Eichhornia* are many of the intruding exotic weeds, which jeopardized the very existence of the indigenous flora of the country.

VI. *Pathological causes:* Timber yielding plants of Pine forests are subjected to threats due to rising temperature when microbial attack increases and needles (leaves) decomposed slowly (Vats and Semwal, 1995).

VII. *Decrease in pollinator populations:* Habitat destruction, and increasing of usage of insecticide, limits the chances of pollination and retards viability of population, affects reproductive biology and finally pushed the species to be rare and finally extinction. e.g., pollen supply from male plant to female plant, of *Garcinia talbotii* (in Bhimashankar wildlife sanctuary, India) is affected, carried out by pollinators or vectors, due to forest fragmentation, which makes the plant vulnerable to survive.

VIII. *Genetic erosion:* According to Stebbins, (1938, 1942, 1980); Janaki Ammal, (1960); Sharma, (1979); Holubec (2006), "Genetic contamination" by hybridization with other species leads to changing of genetic configuration, which has direct relation to adaptation of a species in a particular habitat that may somehow lead to disappearing of the wild species.

IX. *Habitat destruction:* Natural habitat of wild plants are gradually fragmented and gradually destructed leading to affecting the stability of natural ecosystems by disrupting the food webs, tilting the equilibrium state of biodiversity and ultimate extinction of species (Kundu, 2004). There are several factors, responsible for habitat destruction.

X. *Natural catastrophe:* Earthquake, landslide, volcanic eruptions, glaciation, floods, drought, forest fire etc. leads to extinction of any species (Goodrich, 1987; Guarino, 1995) of a narrow range of distribution.

XI. *Man-made construction work:* Drainage works of humid regions, lower water table, dam-building results in flooding, road construction, mining, industrialization and urbanization, development of tourism etc. causes the destruction of natural habitat of wild plant and enhances extinction of species from its matrix (Bhat, 1985; Guarino, 1995).

XII. *Change in agriculture technique:* The change in agricultural techniques, increasing the usage of chemical fertilizers make the matrix fertile, instantly but barren ultimately "Slash and burn" technique is also responsible for habitat destruction (Jyrwa, 1992) and leads to disappearing of wild species.

XIII. *Environmental pollution:* Contamination of the soil, air pollution, water pollution, nuclear pollution which changes the physiology of wild plants, are also partially responsible for disappearing of species (Guarino, 1995).

Nowadays, theoretically, one unique species might be created through biotechnological manipulations (Sharma, 2002) but resurrection of an extinct species, diverged through million years effort of evolutionary progress, seems to be impossible hence development of conservation management is inevitable by offsetting the impediments on the way to the stunted mode diversification of flowering species by slowing down the high pace of extinction.

The Estimation of the Rate of Extinctions of Flowering Plants

In the late Eighties, the eminent phytogeographers, conservation-biologists and plant phylogeneticists have stressed on the simulative biogeographic analysis and extrapolation of population dynamics and forecasted on the rate of extinction 20-50% by the end of the Century i.e., by 2000 A.D. (Myers, 1979; Ehrlich and Ehrlich, 1981) whereas, the conservation biologists have keep their fingers crossed that the predictive results would never come true. Fortunately, the qualitative figure shows that the reality of Anthropocene did not exceed the predictive threshold limit but the contemporary projection of IUCN has not made us optimistic either as the current rate of extinction, influenced by anthropogenic activities were found to be 100-200 times faster than regular rate of extinction. The reality of the reverse swing of evolutionary history of flowering plants and other organisms, which might better be considered as "possibility of Sixth Extinction of flowering plants in Anthropocene," have pushed the conservation biologists and phylogeneticists to work out on an assessment model to quantify the rate of extinction.

Apparently, it is a difficult task to figure out or having a prediction of the negative surge or back-tracking of rate of diversification of an organism as well flowering plants. Seemingly, the relationship of plant richness, as well as diversification and rate of extinction, is to some extent inversely proportional, so from the contemporary rate of extinction of a group of flowering plants it is possible a qualitative assessment about the negative surge of the diversification of flowering plants and that might helped to undertake a broad-range conservation management (macro management) for offsetting the impediments stunted the tempo of diversification of flowering plants unless and until conservation biologists, molecular phylogeneticists, climatologists, autecologists could figure out specific cause-effect factor responsible for such negative surge of diversification for a particular species. Reid (1992) have coupled the data of habitat loss with biogeographic analysis of a population in relation to population dynamics in spatiotemporal background and adopted a model to determine the quantitative rate of extinction. The number of species recorded in an area increases with the size of the area that this increased leads to follow a predictable pattern, known as Arrhenius relationship:-

$$\text{Log } S = C + Z \text{ Log } A.$$

Whereas S = Number of species

A = Area
C = Constant
Z = Constant (values of Z = 0.15 – 0.40).

The most widely quoted generalization from this finding is that a tenfold reduction in an area (i.e., loss of 90% of habitat) will result in the loss of C. 30% (with Z= 0.15) to C. 60% (with z=0.40) of the species present. This is often reduced to the rule of thumb that a ten-fold decrease in area leads to a loss of half the species present. By using the rates of loss of forest covers of Latin America, Africa, Asia and according the difference of species diversity in three regions, Reid (1992) has estimated the rate of extinction, which is 1-5% per decade and his view is more tenable than earlier workers.

THE ANTHROPOGENIC IMPACT OF THE "SIXTH EXTINCTION" IN THE ANTHROPOCENE PERIOD OF THE QUATERNARY ERA PLANT LIFE

On the evolutionary trail, the flowering plants along with other lucky extant organisms that are still alive in the "Anthroprocene regime" (the unique geological regime, highly manipulated as well as ecologically micro-managed by the species *Homo sapiens sapiens)*, traversed through the fine screenings of five mega-extinctions. When the supremacy of *Homo sapiens sapiens* over other living organisms in this regime of Anthropocene is comparable to the "Act of God," it is a matter of debate how much time this Blue Plant would need to wait to witness the end of the journey of the unique, reticulate evolution of the "Green cover" on this blue planet as the expedited rate of anthropogenic extinction has already surpassed by 100 to 200 times more than the rate of biological extinction. Sometimes, self-destruction becomes a part of living, so the actions of *Homo sapiens sapiens* might cause the end of the evolutionary journey of "Green" and subsequently the end of the evolutionary journey of most living animals including the "God of Anthropocene." Such over-simplified, predictive, pessimistic, exaggerated assumptions of the end of the epic journey of biological evolution on this Blue plant is debatable, but the series of negative features ranges from opening the lid of the carbon sink for an indefinite period of time, stoically watching the Ozone Hole in the stratosphere, expediting the exponential growth of urban ecosystems (incomplete ecosystems) at the cost of destructing natural habitat, overexploitation of renewable natural resources (comprising of plants and animals) not only disrupting the equilibrium of biological diversity but making the co-evolutionary evolution (required for further diversification of any organism, benefited out of this mutual relationship) non-functional. Around 20,000, species belonging to 7 families played a key role in pollinating the wild and cultivated species of flowering plants to continue the progress of diversification of angiosperms on terrestrial ecosystems of the world (Michener, 2007). Based on the nature of foraging patterns the mutualistic relationship between bees and host plants can be compartmentalized into-

a) Monolecty: when the pollinators forage on a specific plant species for nectars. e.g., Female *Euglossa* orchid bees (e.g., *Euglossa sapphirina*) forage the nectars of Brazil nut tree (*Bertholletia excelsa*), exclusively (Bhattacharya and Chakraborty, 2017). Hence disappearance of monolectic pollinators could push a particular species to immediate extinction.
b) Polylecty: When the pollinator bees forage on a diverse group of plant species for nectars (e.g., Bumblebees, honey bees etc.). Due to the wideopen chance of collecting nectars from a diverse species of flowering plants, the chance of survival of polylectic pollinators is more likely than monolectic counterparts so the chance of pollination of the host plants by polylectic bees is more or the chance of depletion of these species due to lack of enough pollinators is less (Bhattacharya and Chakraborty, 2017).
c) Oligolecty: When the pollinator bees forage more than one flowering plants of phylogenetically related species, there is a potential threat of disappearance of a group of plants of closely related species in the potential threat of disappearance of oligolectic pollinators (e.g., *Lasioglossum nelumbonis* forage on a group of aquatic plant species related to Nymphaeaceae, Nelumbonaceae, and Cabombaceae etc.) (Bhattacharya and Chakraborty, 2017).

While, there is no global statistical record available on the plant-pollinator relationship, North American pollination biologists and synecologists observed recently that the pollination of 75% of the angiosperm populations in Canada and U.S.A. depend upon the pollinator bees (Moisset and Buchmann, 2011). In another study, related to quantitative appraisal of plant-pollinator relationship (ecosystem services) in terms of dollar value, contributed in North American economy, was found to be around $15.12 billion in 2009 (Calderone, 2012). So, if we don't pay attention the vast array of "vices of Anthropocene" like- uninterrupted air pollution, using of pesticides and insecticides, habitat destruction, fragmentation of forests, they could directly affect the coevolutionary equilibrium of the plant-pollinator equilibrium; stunting the evolution and diversification of flowering plants and jeopardizing the journey of biological evolution in long run.

The previously generated Carbon-di-Oxide is a product of respiration metaphoric slow-down of diversification of the dominant group of transducers, the flowering plants, and this kind of over-simplified narration of the end of evolutionary diversification of life is a valid issue. As far as key abiotic, ingredients of biological evolution on this blue planet is concerned, "Carbon" is the main ingredient and driving force behind the organic evolution on Earth, which has been released by living organisms in the process of respiration, allied metabolic processes and the mainly released by fuel burning (main source of carbon). The equilibrium level of atmospheric carbon, which has mainly been maintained by continuous process of absorbance by Green transducing plants (act like carbon sinks) to convert it to biochemical energy, and releasing oxygen in atmosphere, chiefly after its photosynthetic utilization, and the cardinal level of atmospheric carbon is found to be around 0.03 percent level. However, the alarming factors like-

(i) the anthropogenic involvements in promoting high pace of industrialization and urbanization,
(ii) over-exploitation of mineral resources, buried in the Earth and burning of excessive fossil fuels,
(iii) destruction of forests in the name of promoting urban sprawl and unsustainable infrastructure development, create the high degree of preliminary ecological imbalance, by exceeding the cardinal level of CO_2 in air.

This carbon dioxide gas and some others which are responsible for causing Ozone Hole in stratosphere, enabling to escape excess U. V. radiation on earth; enhanced U. V. radiation leads to mutagenesis of life on earth particularly the plants. The U. V. mutagenesis might be suitable for further survival in stressed condition on it may be lethal. Chlorofluorocarbon (CFC), Methane (CH4), Nitrous oxide (N2O), Ozone (O3) and some other trace gases with water vapor in the atmosphere have allowed solar radiation to pass through and reach the surface on the Earth, heated up by irradiated heat energies; the out-going heat waves or infrared waves are trapped by these, also called Green House Effect. The alarming situation created by current, increasing green-house-effect caused by increasing of green-house-gases in the atmosphere, leads to further increase of warmth to higher temperature. Hoffman and Wells (1987) have noted that CO2 content of atmosphere. Hoffman and wells (1987) have noted that CO_2 content of atmosphere has been increased 25 percent in course of evolution, elapsed since Industrial Revolution and further be doubted (100%) by the middle of the next century (2056 A. D.). By using the latest predictive techniques, the mean surface equilibrium temperature would rise by 1.5°C to 4.5°C on doubling of CO2 or equivalent of trace Greenhouse gases (Ambasht and Ambasht, 1992). According to the report of Radian corporation (1988) prepared for U.S. Environmental Protection Agency, the globally increasing of atmospheric temperature may have orchestrated little fluctuations in different regions. It might be more warming-up in higher latitudes during late autumn and winter than in tropics. In tropics the expected rise would be less than world average and in temperate regions more than this. The rise of temperature by 0.3° to 0.7°C during last century is easily corresponded the increase in CO2 content. The consequence of 1.5°C to 4.5°C rise from the present day average temperature may further lead to –

I. high rate of evaporation and transpiration in moist tropics.
II. the rainfall would also increase the temperate in the region.
III. the excessive melting of polar ice would lead to a much greater natural catastrophe due to further rise in ocean water level and submerging of coastal area.

In the past two decades, average annual temperatures have climbed down as much as 4^{0}C in Alaska, Siberia and parts of Canada. Sea ice has been found 40% thinner and covers 6% less area than in 1980 and it has been recorded further that global warming-induced sea-level rise has been recorded all-time high in last 2,800 years history of evolutionary history of the Earth (Kopp et al., 2016). In the span of temporal break-up, Kopp et al., (2016) has extrapolated the data on paleoclimatic reconstruction model and noticed that between the time span of 0 to 1900 AD, sea level rise has likely to be found in the range between 7-11cm. which has been decelerated 1/5th of millimeter in between the time span of year 1000 to 1400 AD, mainly due to substantial drop of global temperature.

Furthermore, Kopp et al., (2016) and other scientists involved in critical analysis of geological, climatic and atmospheric science of the Earth, observed that rise of the sea-level owing global warming in 20th Century is around 14cm; which is found to be extremely high, since 800BC and it is quite convincible that unsustainable roleplay and micromanagement of global ecosystems in Anthropocene by *Homo sapiens sapiens* just made it happen. This gradual warming is related to increase burning of fossil fuels, such as gasoline, coal that increased the volume of CO_2 and other Greenhouse gases in atmosphere. In order to cut-down the excessive emission level of atmospheric carbon, a global initiative has been taken by global leaders of developed and developing countries and finally the representatives from ca. 160 countries signed the 1997 Kyoto Protocol ratification, which set-up the guidelines for developed, industrial nations to reduce their greenhouse emissions (5.2% below 1990 level) between the years 2008 and 2012 (Anonymous, 2001a).

In order to arrive at some valid conclusion about the impact of alleviated level of Greenhouse gases, the United States EPA (Radian Corporation, 1988) have carried out critically analyze wide range of data on the past climatic changes related to atmospheric temperatures and change of forest vegetation pattern during the last 20,000 years and the efforts of CLIMAP simulation project-

1) Long-range Investigation Climate,
2) Mapping and
3) Prediction) and

COHMAP (Cooperative Holocene Mapping Project), General Circulation Models (GCM) are coupled with known dynamic vegetation models for 21sites in North America (Solomon et al., 1984). Boreal forests show declines in biomass as in last 50 years of increased global warming, several forest species would be dead, but biomass in less volume may then be recovered due to establishment of new hardy tree species; wheat and maize suffer from moisture stress due to increased evapotranspiration. Drought tolerant varieties will also be evolved. Naturally, warmer temperature may enhance biological nitrogen fixation but reduced moisture may retard the decomposition of organic matter, leads to change in fertility status of soil and may leads to imbalance of inorganic N, P. K. content.

As a result of melting of polar ice, coastal areas and seashore adjacent terrestrial lands would be submerged, leads to adaptive and evolutionary changes of vegetation pattern of seashore ecotone zone. The C4 and CAM vegetation probably evolved and diversified during the Cretaceous Period as a response to atmospheric changes. The Cretaceous atmosphere, was richer in Oxygen, could have favored the evolution of C4 systems because average global temperate were higher and more uniform than at present (Venkatachala, 1991). It seems that Green House Effect would bring-back the atmospheric condition of Cretaceous Period, which would be exploited by C4 plants, but increasing of CO2 level or deficiency in O2 level would not create favorable condition for C4 plants. Either the plants with CAM system predominantly succulents, xerophytes would survive and proliferate in that atmospheric condition or hopefully, another group of plants with new carbon-assimilating system would evolve to lead the continuation of plant life.

The evolution of green plants on this planet Earth might be matter of chance but is not a miraculous, isolated event of "appearance of life on the blue planet." It's not just merely a narration on the incidence of divergence of life on the Earth, once it has appeared in an ancient dark planet or passing through the hostile and transitional phases of continental plate movement, transmigration from water to land, went through the phase of metamorphosis like Thallophytes to embryonic Thallophytes to embryophytes to early spermatophytes to attained finally to the flowering plants with closed ovule in Cretaceous Period. The radiation of flowering plants has passed through the temporal node in the post-Cretaceous Period to enter into the phase of diversification to occupy all over terrestrial biome in Early Tertiary and continued its mode of diversification till attaining the period of Holocene, dominated and controlled by human beings, till passing through the of period of nomadic lives, era of domestications of animals and plants, settlement of civilizations, Agricultural Revolutions, Industrial Revolutions to reach to the age of information technology, Cyber Revolution and global warming. After traversing a long spatiotemporal evolutionary trail of divergence, metamorphosis, and diversification, the mode of diversification flowering plants and evolutionary progress of life as well or evolution of Green has been slowed down in Holocene or Anthropocene.

The evolutionary progress of life is not an isolated issue as the mode and frequency of organic evolution on the blue planet and metamorphosis of dark proto-earth should always be considered as an integral part of co-evolutionary metamorphosis of planetary earth from primitive, dark form to modern blue form along with the evolution of all living organisms, the living spectrum of biosphere: virus, bacteria, plants, animals and humankind to maintain the equilibrium state of biosphere, once evolved, in an well orchestic form of organic-inorganic macro-evolution on the Earth in the Universe. The shift of paradigms from "age of science" to the "era of philosophy" in the bank of a metaphoric "Blue Civilization" has come across a long way passing through the temporal nodes of "Green Revolution," "Industrial Revolution" to reach in the doorstep of "Cyber Revolution" of Anthropocene. Is there anyone who did not love his or her mother, most likely none;

perhaps people mistakenly said "Blue" is color of sorrow as they did not see their "Mother Earth" properly in the midst of thousands of thousand shades of clouds of "Anthropocene'. It took a long-time millions of years to diverge with a nice mosaic of greens in the vast tapestry of "Blue," where blue sky nicely merged with blue planet in the blue horizon of infinity.

Finally, it's up to the discretionary choice of *Homo sapiens sapiens* to fix the tilted equilibrium of biological diversity in Anthropocene as there are not enough open-ended options to us either accepting the dominance and diversification of flowering plants, which could give sustainability to the legacy of 'organic evolution," by giving up self-pride of supremacy or the end of epical journey of unique organic evolution through the abominable consequences of self-destruction. After traversing the spatio-temporal matrix of around 4 billion years of topsy-turvy roller-coaster rides, interrupted by five mega-extinctions, the enigmatic quest of maiden evolution of green cellular entity of life is a valid and earnest introspect to establish self-identity on the "Mother Earth" of in the labyrinth of roaring laughter of lightning clouds in the paleoclimatic environment; in the deep frosted suspended crystal of water vapors and dry ice in the cosmogenic clouds; in the dark fumes of melting rocks from monstrous volcanoes; in the cumulative, scorching heat of hundred Saharan deserts, frequently inundated by solar winds; in the midst of thousands of thousand blooming flowers in the galactic space, triggered by nuclear explosion.

Unless and until the irrational acts of human beings in Anthropocene will be controlled, environmental conservation, sustainability and long-term survival opportunity, which could better be addressed as "continuation of journey of biological evolution" on this blue Planet would remain as a matter of jeopardy of "Sixth Extinction." All our perceptions, explanations, ideas and opinion about continuity of progress of biological evolution are merely predictive and heuristic rather than deterministic and conclusive and the progress of diversification of flowering plants should better be recognized as integral part of biological evolution rather than nurturing it in the compartment of seclusion. In the temporal journey through Anthropocene to the infinity, we all found ourselves in the "Noah's Ark" and time would determine whether it is a "Sinking Ark" or it would take us to a new horizon.

REFERENCES

Adam, H., Jouannic, S., Morcillo, F., Verdeil, J. L., Duval, Y. & Tregrear, J. W. (2007). Determination of flower structure in *Elaeis guineensis*: do palms use the same homeotic genes as other species? *Ann. Bot., 99*(6), 1-12.

Adams, J. (2009). *Species richness: Patterns in the diversity of life*. Chichester: Springer-Praxis Publishing.

Adanson, M. (1763). *Familles des Plantes*. [*Families of Plants*.] In C. Vincent (Ed.) Paris: Partie I.

Ahmedullah, M. & Nayar, M. P. (1986). *Endemic plants of Indian region (Peninsular India)* Vol. 1. Calcutta: Botanical Survey of India.

Albert, V. A., Oppenheimer, D. G. & Lindqvist, C. (2002). Pleiotropy, redundancy and the evolution of flowers. *Trends in pl. Sci*., *7*(7), 297-301.

Anderson, E. (1934). Origin of angiosperms. *Nature, 133*, 462.

Ambasht, R. S. & Ambasht, P. K. (1992). *Environment and Pollution: An ecological approach.* (2nd ed.). Varanasi: Student's Friends and Co.

Anacker, B. L. (2013). The nature of serpentine endemism. *Am. J. Bot*., *101*(2), 219-224.

Anderson, J. M., Anderson, H. M. & Cleal, C. J. (2007). *Brief history of Gymnosperms: classification, biodiversity, phytogeography, and ecology*. (Strelitzia 20). Pretoria: South African National Biodiversity Institute.

Andrews, H. N. (1961*). Studies in paleobotany*. New York, London: John Wiley.

Anonymous. (2001). Big meltdown in Arctic. *Enviro news, 7*, 6.

APG I (Angiosperm Phylogeny Group). (1998). An ordinal classification for the families of flowering plants. *Ann. Missouri Bot. Gard*., *85*(4), 531-553.

APG II (Angiosperm Phylogeny Group). (2003). Classification of Flowering Plants. *Bot. J. Linn. Soc., 141*(4), 399-436.

APG III (Angiosperm Phylogeny Group). (2009). An update of Angiosperm Phylogeny Group classification for the orders and families of flowering plants: APG III. *Bot. J. Linn. Soc., 161*, 105- 121.

APG IV. (Angiosperm Phylogeny Group). (2016). An update of the Angiosperm Phylogeny Group classification for the orders and families of flowering plants: APG IV. *Bot. J. Linn. Soc. 181*, 1-20.

Arber, A. (1950). *The natural philosophy of plant ferns.* New York: Cambridge University Press.

Arber, E. A. N. & Parkin, J. (1907). On the origin of angiosperms. *Bot. J. Linn. Soc.* 38: 28-80.

Archibald, J. M. (2015). Endosymbiosis and eukaryotic cell evolution. *Curr. Biol., 25*(19), R911-R921.

Arlt, T. (1907). *Die Entwicklung der Kontinente und iher Leewelt.* [*The development of the continents and their Leewelt.*] Leipzig.

Arnold, C. A (1947). *An introduction to paleobotany*. New York: McGraw Hill.

Arnold, C. A. (1948). Classification of Gymnosperms from the point of view of paleobotany. *Bot. Gaz., 100,* 2-12.

Arnold, C. A. (1953). Origin and relationships of the Cycads. *Phytomorphology, 3*(1-2), 51-65.

Arora, R. K. & Nayar, E. R. (1983) 1985. Wild relatives, and related species of crop plants in distribution. *Bull. Bot. Surv. India., 25*, 35-45.

Arrhenius, S. (1908). *Worlds in the making: The evolution of universe*. New York: Harper and Row.

Asama, K. (1960). Evolution of the leaf forms through the ages explained by successive retardation and neoteny. *Sci. Reports Tohoku Univ. 2nd Ser. (Geology)*., *4*, 252-280.

Asama, K. (1982). Evolution and phylogeny of vascular plants based on the principles of growth retardation: part 5. Origin of angiosperms inferred from the evolution of leaf forms. *Bull. Nat. Sci. Mus. Tokyo, Ser. C. (Geology), 8*, 43-58.

Ash, S. R. (1976). Occurrence of the controversial plant fossil *Sanmiguelia* in the upper Triassic of Texas. *Journ. Paleont, 50*(5), 799-804.

Ash, S. R. (1986). Fossil plants and the Triassic-Jurassic boundary, In K. Padian, (Ed.) *The beginning of the age of dinosaurs* (pp.21-30). Cambridge: Cambridge University Press.

Ash, S. R. (1997). Evidence of arthropod-plant interactions in the Upper Triassic of the Southwestern United States. *Lethala, 29*, 43-58.

Asphaug, E., Agnor, C. B. & Williams, Q. (2006). Hit-and-Run planetary collisions. *Nature, 439*, 155-160.

Austerlitz, F., Mariette, S., Machon, N., Gouyon, P. H., & Godelle, B. (2000). Effects of colonization processes on genetic diversity: Differences between annual plants and tree species. *Genetics, 154*, 1309-1324.

Axelius, B. (1991). Areas of distribution and areas of endemism. *Cladistics, 7*, 197-199.

Axelrod, D. I. (1952). A theory of angiosperm evolution. *Evolution, 6*(1), 29-60.

Axelrod, D. I. (1961). How old are the angiosperms? *Am. J. Sci., 259*(6), 447-459.

Axelrod, D. I. (1970). Mesozoic paleogeography and early angiosperm history. *Bot. Rev., 36*, 277-319.

Ayre, B. G. & Turgeon, R. (2004). Graft transmission of a floral stimulant derived from CONSTANS. *Plant Physiol., 135*, 2271-2278.

Bahadur, K. (1954). Photosynthesis of amino acids of paraformaldehyde and potassium nitrate. *Nature, 173*, 1141.

Bahadur, K. (1965). About a few experiments on preparation of 'Jeewanu' from thermal peptides. *Agra Univ. J. Res., 14*(1), 41-48.

Bakker, R. (1978). Dinosaur feeding behaviour and the origin of flowering plants *Nature, 274* (5672), 661-663.

Banks, H. P. (1970). *Evolution of plants of the past.* (Fundamentals of Botany Series) Belmont, CA: Wadsworth Publishing Company.

Banks, H. P. (1975). The oldest vascular land plants: A note of caution. *Rev. Paleobot. & Palynology, 20*(1-2), 13-25.

Barghoorn, E. S. (1971). The oldest fossils. *Scientific Am., 224*, 30-54.

Barghoorn, E. S. & Schopf, J. W. (1965). Microorganisms from the Late Precambrian of Central Australia. *Science, 150*, 337-339.

Barghoorn, E. S. & Schopf, J. W. (1966). Microorganisms three billion years old from the Precambrian of South Africa. *Science, 152*, 758-763.

Barkman, T. J., Chenery, G., Mcneal, J. R., Lyons-Weller, J., Ellisens, W. J., Moore, G. M., Wolfe, A. D.& Depamphilis, C. W. (2000). Independent and combined analyses of sequences from all three genomic compartments converge on the root of flowering plant phylogeny. *Proc. Natl. Acad. Sci. USA, 97*(24), 13166-13171.

Barrett, P. & Willis, K. (2001). Did dinosaurs invent flowers? Dinosaur–angiosperm coevolution revisited. *Biol. Rev. Cambr. Phil. Soc., 76* (3), 411-447.

Barrett, S. C. H. & Forno, I. W. (1982). Style Morph distribution in new world populations of *Eichhornia crassipes* (Mart.) Solms-Laubach (Water Hyacinth). *Aquatic Botany, 13*, 299-306.

Bateman, R. A., Hilton, J. & Rudall, P. J. (2006). Morphological and molecular phylogenetic context of angiosperms: contrasting the "top-down" and "bottom-up" approaches used to infer the likely characteristics of the first flowers. *Journ. Exp. Bot., 57*(13), 3471-3503.

Baum, D. A. & Hileman, L. C. (2006). A developmental genetic model for the origin of the flower. In C. Ainsworth (Ed.) *Flowering and its Manipulation.* (Vol. 20), Sheffield: Blackwell.

Beard, J. S. (1967). Some vegetation types of tropical Australia in relation to those of Africa and America. *Journ. Ecol., 55*(2), 271-290.

Beck, C. B. (1960). Connection between Archaeopteris and Callixylon. *Science, 131*, 1524-1525.

Becker, B. & Marin, B. (2009). Streptophyte algae and the origin of embryophytes. *Ann. Bot., 103*(7), 999-1004.

Becker, A. & Theiben, G. (2003). The major clades of MADS-box genes and their role in the development and evolution of flowering plants. *Mol. Phylogen. & Evol., 29*, 464-489.

Beerling, D. J. & Berner, R. A. (2002). Biogeochemical constraints on the Triassic-Jurassic boundary carbon cycle event. *Glob. Biogeochem. Cycles., 16* (3), 10-1 – 10-13.

Beerling, D. J. & Berner, R. A. (2004). Feedbacks and the coevolution of plants and the atmospheric CO_2. *Proc. Natl. Acad. Sci. USA, 102*, 1302-1305.

Bell, C. D., Soltis, D. E. & Soltis, P. S. (2005). The age of angiosperms: a molecular timescale without a clock. *Evol., 59*, 1245-1258.

Bell, C. D., Soltis, D. E. & Soltis, P. S. (2010). The age and diversification of angiosperms revisited. *Am. J. Bot., 97*(8), 1296-1303.

Bell, E. A. & Evans, C. S. (1978). Biochemical evidence of a former link between Australia and Mascarene islands. *Nature, 273*, 295-296.

Bello, M. A., Hawkins, J. A. & Rudall, P. J. (2010). Floral ontogeny in Polygalaceae and its bearing on the homologies of keeled flower Fabalaes. *Intl. J. Pl. Sci., 171*, 482-498.

Bennett, K. D. (1990). Milankovitch cycles and their effects on species in ecological And Evolutionary Time. *Paleobotany, 16*, 11-21.

Benson, M. (1904). *Telangium scotia*, a new species of *Telangium* (Calymmatotheca) showing structure. *Ann. Bot., 18*, 161-177.

Benson, M. (1914). *Sphaerostoma ovale* (*Conostoma ovale et intermedium* Williamson), a lower carboniferous ovule from Pettycur, Fifeshire, Scotland. *Trans. Roy. Soc. Edinburgh, 50*, 1-15.

Bessey, C. E. (1915). The phylogenetic taxonomy of flowering plants. *Ann. Missouri Bot. Gard., 2*(1-2), 109-164.

Bentham, G. & Hooker, J. D. (1862-1883). *Genera Plantarum* (pp. 1-240), Vol.1. London: L. Reeve and Co.

Benton, M. J. & Twichett, R. J. (2003). How to kill (almost) all life: the end-Permian extinction event. *Trends in Ecol. & Evol., 18*(7), 358-365.

Beraldi-Campesi, H. (2013). Early life on land and the first terrestrial ecosystem. *Ecol. Processes, 2*, 1.https://doi.org/10.1186/2192-1709-2-1.

Berner, R. A. (1999). Atmospheric oxygen over Phanerozoic time. *Proc. Natl. Acad. Sci. USA, 96*(20), 10955-10957.

Berner, R. A. & Kothavala, Z. (2001). Geocarb III: A revised model of atmospheric CO_2 over Phanerozoic time. *Amer. J. Sci., 301*(2), 182-204.

Bernhardt, H. S. (2012). The RNA world hypothesis: The worst theory of the early evolution of life (except for all the others). *Biol. Direct., 7,* 23. doi: 10.1186/1745-6150-7-23.

Besnard, G., Khadari, B., Navascues, M., Fernandez-Mazuecos, M., El-Bakkali, A., Arrigo, N., Baali-Cherif, D., Brunini-Bronzini De Caraffa, V., Santoni, S., Vargas, P. & Savolainen, V. (2013).The complex history of olive tree: From late Quaternary diversification of Mediterranean lineages to primary domestication in the Northern Levant. *Proc. R. Soc. B., 280*(1756), 2012-2833.

Bessey, C. E. (1907). A synopsis of plant phyla. *Univ. Nebr. Stud.*, 7(4), 275-373.

Bharucha, F. R. (1983). *A textbook of the plant Geography of India*. Bombay: Oxford University Press.

Bhat, K. G. (1985). Eco-development of Western Ghats – Role of wood processing. *Indian. Journ. Forestry, 8*, 102-108.

Bhattacharyya, M. & Chakraborty, S. K. (2017). Citizen science can benefit native bees. *Everyman's Sci., 51*(6), 384-387.

Bierhost, D. (1971). *Morphology of vascular plants*. London: Macmillan.

Binggeli, P. (2003). Ponteridaceae, *Eichhornia crassipes*, water hyacinth, Jacinthe d'eau, telezanalika, tsikafokafona. In S. M. Goodman, J. P. Benstead (Eds.) *The natural history of Madagascar* (pp. 476-478). Chicago: University of Chicago Press.

Blackwell, W. H. (2003). Two theories of origin of the land-plant sporophyte: Which is left standing? *Bot. Rev.*, *69*(2), 125-148.

Blench, R. M. (2007). The intertwined history of the silk-cotton and baobab. In R. T. J. Carpenters (Ed.). *Fields of change; progress in African ethnobotany* (pp.1-20). Groningen: Barkhuis and Groningen University.

Bold, H. C. (1948). The occurrence of chlorophyll in the sporophyte of *Riciocarpus natans. Am. J. Bot.*, *35*, 440-443.

Bold, H. C. & Wynne, M. J. (1978). *Introduction to the algae: structure and reproduction.* New Jersey: Prentice Hall.

Bonamo, P. M. (1975). The Progymnospermopsida: Building a concept. *Abst. Papers xii. Int. Bot. Congr. Leningrad.* P. 92.

Bonneton, F., Zelus, D., Iwema, T., Robinson-Rechavi, M. & Lauder, V. (2003). Rapid divergence of the ecdysone receptor in Diptera and Lepidoptera suggests coevolution between ECR and USP-PXR. *Mol. Biol. & Evol., 20*(4), 541-553.

Borg, J., Mcdade, L. A. & Schonenberger, J. (2008). Molecular phylogenetics and morphological evolution of Thunbergioideae (Acanthaceae). *Taxon, 57,* 811-822.

Bose, K. & Ganguly, J. (1995). Experimental and theoretical studies of the stabilities of talc, antigorite, and phase A at high pressures with applications to subduction processes. *Earth and Planetary Science Letters*, *136*(3-4), 109-121.

Bower, F. O. (1889). The comparative examination of the meristems of ferns, as a phylogenetic study. *Ann. Bot.*, *3*, 305-392.

Bower, F. O. (1890). On antithetic as distinct from homologous alternation of generations in plants. *Ann. Bot., 4,* 347–370.

Bower, F. O. (1908). *The origin of a land flora*. London: Macmillan.

Bower, F. O. (1935). *Primitive land plants, also known as the archegoniate*. London: Macmillan.

Braun, A. (1864). Uebersicht des naturlichen Systems nach der Anordnung desselben. In. P. ASCHERSON (Ed.). *Flora Brandenburg, 1*(2) Apr. 1864, 22-67(Einleitung).

Brenner, G. J. (1996). Evidence for the earliest stage of angiosperm pollen evolution: a paleoquatorial section from Israel. In. D. W. Taylor & L. J. Hickey (Eds.) *Flowering plant origin, evolution & phylogeny* (pp. 91-115). New York. Chapman & Hall.

Briggs, D. & Walters, S. M. (2016). *Plant variation and evolution* (4TH ed.). Cambridge: Cambridge University Press.

Briquet, J. (1908) Les reimmigrations postglaciaires des flores en Suisse. *Act. Soc. Helv. Scii. Nat.,* Vol. *1*.

Brongniart, A. (1828-1837). *A histoire des vegetaux fossiles, ou, Researches botaniques et geologiques sur les vegetaux renfermes dans les diverses couches du globe[A history of fossil plants, or, Botanical and geological researches on plants contained in the various layers of the globe]*. In Chez G. Dufour & Ed. D'Ocagne (Eds.). A Paris et a Amsterdam.

Brown, G. L & Terry, R. G. (1992). Petal appendages in Bromeliaceae. *Am. J. Bot., 79,* 1051-1071.

Brown, D. & Jenvey, K. (2012, April 25). *NASA scientists find history of asteroids impacts in Earth rocks.* https://www.nasa.gov/home/hqnews/2012/apr/HQ12-135_Asteroid_Imapcts_Earth_Rocks.html.

Browne, M. P. (1913). A colonial contribution to our knowledge of the genus *Lycopodium*. *New Phytol., 12*, 222-225.

Brown, R. & Linnean Society (London). (1827). *Character and description of Kingia; a new genus of the plants found on the southwest coast of New Holand with observations on the structure of its unimpregnated ovulum, and on the female flower of Cycadaceae and Coniferae*. (Reprinted from a narrative of a survey of the intertropical and Western coasts of Australia performed between the years 1818 and 1822 by Captain Philip P. Ling). Vol.2 London.

Brugman, W. A. (1983). *Permian-Triassic palynology*. Utrecht: Laboratory of Paleobotany and Palynology. State University Utrecht.

Burger, W. C. (1981). Heresy revived: the monocot theory of angiosperm origin. *Evol. Theory, 5*, 189-225.

Burleigh, J. G. & Mathews, S. (2004). Phylogenetic signal in nucleotide data from seed plants: implications for resolving the seed plant tree of life. *Am. J. Bot., 91*(10), 1599-1613.

Busov, V. B., Brunner, A. M. & Strauss, S. H. (2008). Genes for control of plant stature and form. *New Phytol. 177*, 589-607.

Byng, J. W. (2014). *The flowering plants handbook: A practical guide to families and genera of the world*. Hertford: Plant gateway.

Cain, M. L., Milligan, B. G. & Strand, A. E. (2000). Long-distance seed dispersal in plant populations. *Am. J. Bot., 87*(9), 1217-1227.

Calderone, N. W. (2012). Insect pollinated crops, Insect pollinators, and US agriculture: Trend analysis of aggregate data for the period 1992-2009. *PLoS ONE, 7*(5), e37235.

Cameron, A. G. W. (1978). Physics of the primitive Solar Nebula and of giant gaseous protoplanets. In T. Gehrels (Ed.) *Protostar and Planets* (pp. 453-487). Tuscon: University of Arizona Press.

Campbell, D. H. (1891). On the relationships of the Archegoniatae. *Bot. Gazet.*, Xvi, 323.

Campbell, D. H. (1903). Antithetic *versus* homologous alternation. *Amer. Nat.*, *37*, 153 - 169.

Campbell, D. H. (1905). *The structure and development of mosses and ferns*. New York: MacMillan Company.

Campbell, D. H. (1928). The phylogeny of angiosperms. *Bull. Tor. Bot. Club, 55,* 433-478.

Campbell, D. H. (1940). *The evolution of land plants (Embryophyta)*. California: Stanford University Press.

Carlquist, S. (1965). *"Island Life": A natural history of islands of the World.* New York: Natural history Press.

Cassen, P. (2001). Nebular thermal evolution and the properties of primitive planetary materials. *Meteorites and Planetary Science, 36*, 671-700.

Cavers, F. (1910). The inter-relationships of the bryophytes. I. Sphaerocarpales. *New Phytol.*, *9,* 81-92.

Celakovsky, L. (1874). Ueber die verschiedenen Formen und die Bedeutung des Generationswechsels der Pflanzen. *Sitzungsberichte der königlich böhmischen Gesellschaft der Wissenschaften in Prag 1874[About the different forms and the meaning of the generation change of the plants. Meeting reports of the Royal Bohemian Society of Sciences in Prague 1874]*, 21-61.

Chamberlain, C. J. (1935). *Gymnosperms, structure, and function*. Chicago: University of Chicago Press.

Chapman, J. L. & Reiss, M. J. (1995). *Ecology: Principles and applications.* Cambridge: Cambridge Univ. Press.

Chapman, V. J. (1962). *The Algae*. London: Palgrave Macmillan.

Chapman, V. J. & Chapman, D. J. (1973, 2nd Edn.). *The Algae*. London: Macmillan.

Chase, M. W., Soltis, D. E., Olmstead, R. G., Morgan, D., Les, D. H., Mishler, B. D., Duvall, M. R., Price, R. A., Hills, H. G., Qiu, Y. L., Kron, K. A., Rettig, J. H., Conti, E., Palmer, J. D., Manhart, J. R., Sytsma, K. J., Michaels, H. J., Kress, W. J., Karol, K. G., Clark, W. D., Hedren, M., Gaut, B. S., Jansen, R. K., Kim, K. J., Wimpee, C. F., Smith, J. F., Furnier, G. R., Strauss, S. H., Xiang, Q. Y., Plunkett, G. M., Soltis, P. S., Swensen, S. M., Williams, S. E., Gadek, P. A., Quinn, C. J., Eguiarte, L. E., Golenberg, E., Learn, G. H., Graham, S. W., Barett, S. C. H., Dayanandan, S. & Albert, V. A.

(1993). Phylogenetics of seed plants: an analysis of nucleotide sequences from the plastid gene rbcL. *Ann. Missouri Bot. Gard.*, *80*, 528-580.

Chaw, S. M., Chang, C. C., Chen, H. L. & Li, W. H. (2004). Dating the monocot-dicot divergence and the origin of core eudicots using whole chloroplast genomes. *Journ. Mol. Evol.*, *58*(4), 424-441.

Chehab, E. W., Eich, E. & Braam, J. (2009). Thigmomorphogenesis: a complex plant response to mechano-stimulation. *J. Exp. Bot., 60*, 43-56.

Chen, I. A. & Walde, P. (2010). From self-assembled vesicles to protocells. *Cold Spring Harb. Perspect. Biol.* *2*(7), a002170.

Chevalier, A. & Guenot, L. (1925). *Biogeographie, chep x.* Pueplement des espaces vides[Stacking empty spaces], Paris.

Chailakhyan, M. K. (1985). Hormonal regulation of flowering plants. In M. Bopp (Ed.). *Plant Growth Substances (*pp. 303-307). (Proceedings of the 12th International Conference on Plant Growth Substances Held at Heidelberg, Germany. Aug. 26-31, 1985). Heidelberg: IPGSA and the University of Heidelberg.

Christenhusz, M. J. M., Vorontsova, M. S., Fay, M. F. & Chase, M. W. (2015). Results from an online survey of family delimitation in angiosperms and ferns: recommendations to the Angiosperm Phylogeny Group for thorny problems in plant classification. *Bot. J. Linn. Soc., 178*, 501-528.

Christenhusz, M. J. M. & Byng, J. W. (2016). The number of known plant species in the world and its annual increase. *Phytotaxa, 261*(3), 201-217.

Christensen, G. (1954). Some considerations on the phylogeny of Bryophyta. *Bot. Tidskr.*, *51*, 53-58.

Chirstopherson, R. W. (2011). *Geosystems: An introduction to physical geography* (8th ed.). New Jersey: Prentice Hall.

Church, A. H. (1919). Thallasiophyta and the subaerial transmigration. *Oxford Botanical Memoirs, 3*, 1-95.

Chrysler, M. A. (1945). The shoot of *Botrychium* interpreted as a series of dichotomies. *Bull Torrey Bot. Club, 72*, 491-505.

Clay, B. L., Kugrens, P. & Lee, R. E. (1999). A revised classification of Cryptophyta. *Bot. J. Linn. Soc.*, *131* (2), 131-151.

Cleal, C. J. & Thomas, B. A. (2009). *An introduction to plant fossils*. Cambridge: Cambridge University Press.

Coen, E., Rolland-Lagan, A. G., Matthews, M., Bangham, A. & Prusinkeiwicz, P. (2004). The genetics of geometry. *Proc. Natl. Acad. Sci. USA, 101*(14), 4728-4735.

Coetzee, J. A. & Muller, J. (1984). The phytogeographic significance of some extinct Gondwana pollen types from the Tertiary of the southwestern Cape (South Africa). *Ann. Missouri Bot. Gard.*, *71*(4), 1088-1099.

Contandriopoulos, M. J. (1962). Essai de classification des endemiques corse [Classification test of Corsican endemic]. *Bull. De la Soc. De France, 109*(2), 449-459.

Copeland, E. B. (1947). *Genera Filicum*. Waltham, MA: Chronica Botanica Company.

Cornet, B. (1977). Preliminary investigation of two Late Triassic conifers from York County, Pennsylvania. In R. C. Romans (Ed.), *Geobot.,* (New York, Plenum vol.), 165-172.

Cornet, B. (1989a). Late Triassic angiosperm-like pollen from the Richmond rift basin of Virginia, USA. *Paleontogr. Abt., B213*, 37-87.

Cornet, B. (1989b). The reproductive morphology and biology of *Sanmiguelia lewisii* and its bearing on angiosperm evolution in the late Triassic. *Evol. Trends in pl., 3*, 25-51.

Cornet, B. (1993). Applications and limitations of palynology in age, climatic, and paleoenvironmental analysis of Triassic sequences in North America. *New Mexico Mus. of Nat. Hist. & Sci. Bull., 3*, 75-93.

Cornet, B. (2002). *When did angiosperms first evolve? (Part I)* (Updated on 30th April 2016). Retrieved from https://www.researchgate.net/publication/301726157_Part_I_When_Did_Angiosperms_First_Evolve. (DOI: 10./RG.2.1.4674.1366).

Cornet, B. & Habib, D. (1992). Angiosperm-like pollen from the ammonite-dated Oxfordian (Upper Jurassic) of France. *Rev. Paleobot. & Palynol., 71*, 269-294.

Cornet, B. & Traverse, A. (1975). Palynological contributions to the chronology and stratigraphy of the Hartford basin in Connecticut and Massachusetts. *Geosci. & Man, 11*, 1-33.

Cornet, B. & Waanders, G. (2006). Palynomorphs indicate Hettangian (Early Jurassic) age for middle Whitemore Point Member of the Moenave Formation, Utah, and Arizona. *New Mexico Mus. of Nat. Hist. & Sci. Bull., 37*, 390-406.

Cowling, R. M. & Hilton-Taylor, C. (1997). Phytogeography, flora, and endemism. In R. M. Cowling, D. M. Richardson & S. M. Pierce (Eds.). *Vegetation of Southern Africa*. Cambridge: Cambridge University Press.

Cowling, R. M. & Samways, M. J. (1995). Predicting global patterns of endemic plant species richness. *Biodiversity Letters, 2*(5), 127-131.

Cowling, R. M., Witkowski, E. T. F., Milenwiski., A. V. & Newbey, K. R. (1994). Taxonomic, edaphic and biological aspects of narrow plant endemism on matched sites in Mediterranean South Africa and Australia. *Journ. Biography, 21*, 651-664.

Cowling, S. A. (2001). Plant-carbon balance, evolutionary innovation, and extinction of land plants. *Global Change Biology, 7*(3), 231-239.

Crandall-Stotler, B. (1980). Morphogenetic designs and theory of bryophyte origins and divergence. *Bioscience,* 30, 580-585.

Crandall-Stotler, B. & Stotler, R. E. (2000). Morphology and classification of Marcahntiophyta. In A. J. Shaw & B. Goffinet (Eds.) *Bryophyte Biology* (pp. 21-70). Cambridge: Cambridge University Press.

Crane, P. R. (1985). Phylogenetic analysis of seed plants and the origin of angiosperms. *Ann. Missouri Bot. Gard.*, 72, 716-793.

Crane, P. R. (2013). *Ginkgo: The tree that time forgot.* New Haven, London: Yale University Press.

Crane, P. R., Friis, E. M. & Pedersen, K. R. (1995). The origin and early diversification of angiosperms. *Nature, 374*, 27-33.

Crane, P. R., Herendeen, P. & Friis E. M. (2004). Fossils and plant phylogeny. *Am. J. Bot., 91*(10), 1683-1699.

Crepet, W. L. (2000). Progress in understanding angiosperm history, success, and relationships: Darwin's abominably "perplexing phenomenon." *Proc. Natl. Acad. Sci. USA, 97*(24), 12939-12941.

Cronquist, A. (1968). *The evolution and classification of flowering plants.* Boston: Houghton Mifflin.

Cronquist, A., Takhtajan, A. & Zimmermann, W. (1966). On the higher taxa of Embryophyta. *Taxon, 1966*, 129-134.

Croizat, L. (1952). *Manual of Phytogeography*. Uitgeverij, The Hague: W. Junk.

Dahlgren, R. (1938). General aspects of angiosperm evolution and macrosystematics. *Nord. J. Bot., 3*, 119-149.

Dahlgren, R. M. & Clifford, H. T. (1982). *The monocotyledons: A comparative study.* New York, Academic Press.

Darrah, W. C. (1939). *Principles of Paleobotany*. Leiden: Chronica Botanica Company.

Darwin, C. (1859). *On the Origin of Species by Means of Natural Selection, or the Preservation of Favoured Races in the Struggle for Life.* London: John Murray.

Dawson, A. (2016). *Extinction: A radical history*. New York, O R Books.

Dawson, J. W. (1859). On fossil plants from the Devonian rocks of Canada. *Q. J. Geo. Soc. Lond.,* 15, 477-488.

DeBodt, S., Marie, S. & Van Der Peer, Y. (2005). Genome duplication and the origin of angiosperms. *Trends. in Ecol. Evol., 20*(11), 591-597.

DeCandole, A. P. (1855). *Geographie Botanique Raisonnee [Reasoned Botany Geography]*. Vols. *I* & *II*. Paris: Masson.

Delevoryas, T. (1962). *Morphology and evolution of fossil plants.* New York: Holt, Reinhart, and Winston.

Delsuc, F., Brinkmann, H. & Philippe, H. (2005). Phylogenomics and the reconstruction of the tree of life. *Nature Reviews Genetics, 6*(5), 361-375.

Diels, L. (1909). *Geographie der pflanzen [Geography of the plants]*. Berlin.

Diels, L. (1936). A. Engler's *Syllabus der Pflanzenfamilien: eine Übersicht über das ganze Pflanzensystem mit besonderer Berücksichtigung der Medizinal- und Nutzpflanzen nebst einer Übersicht über die Florenreiche und Florengebiete der Erde zum Gebrauch bei Vorselungen und Studien überspezielle und medizinisch-pharmazeutische Botanik [Syllabus of the plant families: an overview of the whole plant system with special consideration of the medicinal and useful plants together with an overview of the flora and floral habitats of the earth for use in prescriptions and studies of special and medical-pharmaceutical botany]*, (posthumous publication, 11th ed.). Berlin: Gebrüder Borntraeger Verlag.

Dilcher, D. (1974). Approaches to the identification of angiosperm leaf remains. *Bot. Rev.*, *40*, 1-157.

Dilcher, D. (1995). In P. C. Hock & A. G. Stephenson (Eds.). *Experimental and molecular approaches to plant biosystematics* (pp. 187-198). St. Louis: Missouri Botanical Garden.

Dilcher, D. (2000). Towards a new synthesis: Major evolutionary trends in the angiosperm fossil record. *Curr. Sci.*, *97*(13), 7030-7036.

Donoghue, M. J., Ree, R. H. & Baum, D. A. (1998). Phylogeny and the evolution of flower symmetry of Asteridae. *Trends in Pl. Sci., 3*, 311-317.

Donoghue, M. J. & Doyle, J. A. (2000). Seed plant phylogeny: demise of the anthophyte hypothesis? *Curr. Biol.*, *10*(3), R106-R109.

Doyle, J. A. (1978). Origin of angiosperms. *Ann. Rev. Ecol. & Syst., 9*, 365-392.

Doyle, J. A. (1998). Palynology of vascular plants. *Annual. Rev. Ecol. Syst.*, *29*, 567-599.

Doyle, J. A. (2001). Significance of molecular phylogenetic analyses for paleobotanical investigations on the origin of angiosperms. *Paleobot.*, *50*, 167-188.

Doyle, J. A. (2006). Seed ferns and the origin of angiosperms. *Journ. Torrey Bot. Soc.*, *133*(1), 169-209.

Doyle, J. A. (2008). Integrating molecular phylogenetic and paleobotanical evidence on origin of the flower. *Intl. Journ. Pl. Sci.*, *169*, 816-843.

Doyle, J. A. (2010). Function and evolution of saccate pollens. *New Phytol.*, *188*(1), 6-9.

Doyle, J. A. (2012). Molecular and fossil evidence on the origin of angiosperms. *Annu. Rev. Earth Planet. Sci., 40*, 301-326.

Doyle, J. A. & Donoghue, M. J. (1986). Seed plants phylogeny and the origin of angiosperm: an experimental cladistic approach. *Bot. Rev.*, *52*(4), 321-431.

Doyle, J. A. & Endress, P. K. (2000). Morphological phylogenetic analysis of basal angiosperms: comparison and combination with molecular data. *Intl. Journ. Pl. Sci.*, *161*(6), S121-s153.

Drude, O. (1890). *Handbuch der pflanzengeographie [Handbook of plant geography]*. Stuttgart: J. Engelhorn.

Dunay, R. E. & Fischer, M. J. (1974). Late Triassic palynofloras of North America and their European correlatives. *Rev. Paleobot. & Palynol., 17*, 179-186.

DuToit, A. L. (1937). *Our wandering continents- An hypothesis of continental drifting.* Edinburgh: Oliver & Boyd.

Eames, A. J. (1936). *Morphology of vascular plants: Lower Groups (Psilophytales to Filicales).* New York: McGraw-Hill Book Company.

Edwards. D. S. (1986). *Agalophyton major*, a non-vascular land plant from the Devonian Rhynie chart. *Bot. J. Linnean Soc.*, *93*(2), 173-204.

Ehrlich, P. R. & Ehrlich, A. H. (1981). *Extinction: the causes and consequences of the disappearance of species*. New York: Random House.

Eichler, A. W. (1883) *Syllabus der Vorlesungen über Specielle und Medicinisch Pharmaceutische Botanik [Syllabus of the lectures on Special and Medicinal Pharmacic Botany]* (3rd ed.). Berlin: Borntraeger.

Eldridge, N. & Gould, S. J. (1972). Punctuated equilibria an alternative to phyletic gradualism. In T. J. M. Schopf (Ed.). *Models in Paleobiology* (pp82-115). San Fransisco: Freeman Cooper.

Elias, M. K. (1942). *Tertiary Prairie grasses and other herbs from the high plains*. Geol. Soc. Amer. Special Paper no. 41. New York: The Geological Society of America, USA.

Elias, M. K. & Moore, R. C. (1965). *Introduction to Paleoecology*. (Scripta Technica, Trans.). In H. R. Federovich (Ed.). New York: American Elsevier.

Endress. P. K. (1994). *Diversity and evolutionary biology of tropical flowers.* Cambridge: Cambridge University Press.

Endress, P. K. (1995). *Diversity and evolutionary biology of tropical flowers.* Cambridge: Cambridge University Press.

Endress, P. K. (2001). Origins of flower morphology. *J. Exp. Zool. (Mol. Dev. Evol.), 291*(2), 105-115.

Endress, P. K. (2010). Synorganization without fusion in the flowers of *Geranium robertianum* (Geraniaceae) and it's not so trivial obdiplostemony *Ann. Bot.*, *106*, 687-695.

Endress, P. K. (2011). Evolutionary diversification of the flowers in angiosperms. *Am. J. Bot.*, *98*(3), 370-396.

Engel, M. H., Macko, S. A. & Silfer, J. A. (1990). Carbon isotope composition of individual amino acids in the Murchison meteorite. *Nature*, *348*, 47-49.

Engler, A. (1879), *Versuch einer Entwicklungsgeschichte der Pflanzenwelt[Attempt of a developmental history of the plant world]* (Vols. 1-2). Leipzig.

Engler, A. (1882). *Versuch einer Entwick Lungsgeschichte der Pflangenwelt [Attempt of a developmental history of the world of the pharynx]* Leigpzig.

Engler, A. (1912). *Syllabus der Pflanzenfamilien [Syllabus of plant families]* (7 Aufl.). Berlin.

Engler, A. (1924). Ubersicht uber die Florenreiche and Florengebiete der Erde [Overview of the floral kingdoms and flora of the earth]. In A. Engler & E. Gilg. (Eds.) *Syllabus der Pflazenfamilier [Syllabus of plant families]* (9-10 Aufl.). Berlin.

Engler, A. (1926). Gymnospermae (Geographical distribution). In A. Engler & K. Prantl (Eds.) *Die Naturlichen Pflanzenfamilien [The natural plant families]* (2nd ed.; Vol.13). Leipzig: Verlag von Wilhelm Engelmann.

Ernst, W. G., Sleep, N. H. & Tsujimori, T. (2016). Plate tectonic evolution of the Earth: bottom-up and top-down mantle circulation. *Canadian Journal Earth Science, 53*(11), 1103-1120.

Erwin, D. H. (2006). *Extinction: How life on earth nearly ended 250 Million years ago*. Princeton: Princeton University Press.

Evans, A. W. (1939). The classification of Hepaticae. *Bot. Rev., 5*, 49-96.

Ezer, D. & Cameron, A. G. W. (1965). A study of solar evolution. *Canadian Journal Physics., 43*(8), 1497-1517.

Favarger, C. & Contandriopoulos, T. (1961). Essai sur l'endemisme. *Bull. Soc. Bot. Suisse, 71,* 384-408.

Felix, C. J. & Burbridge, P. P. (1977). A new Riccisporites from the Triassic of Arctic Canada. *Paleont., 20*(3), 581-587.

Field, T. S., Arens, N. C., Doyle, J. A. M Dawson, T. E. & Donoghue, M. J. (2004). Dark and disturbed: a new image of early angiosperm ecology. *Paleobiol., 30*(1), 82-107.

Field, T. S. & Arens, N. C. (2005). Form, function, and environments of the early angiosperms: Merging extant phylogeny and ecophysiology with fossils. *New Phytol., 166*(2), 383-408.

Fishbein, M. (2001). Evolutionary innovation and diversification in the flowers of Asclepiadaceae. *Ann. Missouri Bot. Gard., 88*, 603-623.

Fischer, M. J. (1972). The Triassic palynofloral succession in England. *Geosci. & Man, 4,* 101-109.

Fischer, O. (1866). On the disintegration of a chalk cliff. *Geol. Mag., 3,* 354-356.

Fitzgerald, S. (1989). *International Wildlife Trade: whose business is it?* Washington DC: WWF.

Florin, R. (1933). Studien uber die Cycadales des Mesozoikums nebst Erorterungen uber die Spaltoffnungsapparate der Bennettitales [Studies on the Cycadales of the Mesozoic with discussions on the Bennettitales fission opening apparatus]. *K. Svenska Vetensk Akad. Handl.*, (Ser. 3), *12*(5), 1-134.

Florin, R. (1936). Pollen grains of Cordaitales. *Svensk. Bot. tidskr., 30*, 624-651.

Florin, R. (1948). On the morphology and relationships of the Taxaceae. *Bot. Gaz., 110*, 31-39.

Florin, R. (1949). The morphology of *Trichopitys heteromorpha* Saporta, a seed plant of Paleozoic Age, and the Evolution of the female flowers in the Ginkgoinae. *Acta Horti Bergiani., 15*, 80-109.

Florin, R. (1951). Evolution in Cordaites and Conifers. *Acta Horti Bergiani., 15*, 285-388.

Florin, R. (1958). Notes on the systematics of the Podocarpaceae. *Acta Horti Bergiani., 17*, 403-411.

Florin, R. (1963). The distribution of conifer and taxad genera in time and space. *Acta Horti. Bergiani., 20*, 121-312.

Floyd, S. K. & Bowman, J. L. (2007). The ancestral developmental tool kit of land plants. *Intl. Journ. Pl. Sci., 168*(1), 1-35.

Fox, S. W. & Foster, J. F. (1957). *Introduction to protein chemistry*. New York: John Wiley and Sons.

Fowell, S. J., Cornet, B. & Olsen, P. E. (1994). Geologically rapid Late Triassic extinctions: Palynological evidence from the Newark Supergroup. In G. D. Klein (Ed.). *Pangea: Paleoclimate, Tectonics, and Sedimentation during accretion, zenith, and breakup of a supercontinent.* Boulder, Colorado: Geological Society of America Special Paper no. 288.

Föyn, B. (1929). Untersuchungen über die Sexualität und Entwicklung von Algen. IV. Vorläufige Mitteilung über die Sexualität und den Generationswechsel von *Cladophora* und *Ulva*. *Berichte der Deutschen Botanischen Gesellschaft* [Studies on the sexuality and development of algae. IV. Provisional Communication on the Sexuality and Generational Change of *Cladophora* and *Ulva.*] *Reports of the German Botanical Society*, *47*, 495-506.

Fricke, H. C. & Pearson, D. A. (2008). Stable isotope evidence for changes in dietary niche partitioning among hadrosaurian and ceratopsian dinosaurs of the Hell Crick Formation, North Dakota. *Paleobot. 34*(4), 534-552.

Friedman, W. E. (1992). Evidence of a pre-angiosperm origin of endosperm. *Science, 225*, 336-339.

Friedman, W. E. (2006). Embryological evidence for developmental liability during early angiosperm evolution. *Nature, 441*, 337-340.

Friedman, W. E. (2008). Hydatellaceae are water lilies with gymnospermous tendencies. *Nature, 453*, 94-97.

Friis, E. M., Pederson, K. R. & Crane, P. R. (2000). Reproductive structure and organization of basal angiosperms from the Early Cretaceous (Barremian or Aptian) of western Portugal. *Intl. Journ. Pl. Sci., 161*(6), S169-S182.

Friis, E. M., Pedersen, K. R. & Crane, P. R. (2001). Fossil evidence of water lilies (Nymphaeales) in the early Cretaceous. *Nature, 410*, 357-360.

Friis, E. M., Pedersen, K. R. & Crane, P. R. (2006). Cretaceous angiosperm flowers: innovation and evolution in plant reproduction. *Paleogeogr. Paleoclimat. and Paleoecol., 232*, 251-293.

Friis, E. M., Pedersen, K. R. & Crane, P. R. (2011). *Early flowers and angiosperm.* Cambridge: Cambridge University Press.

Fritsch, F. E. (1916). The algal ancestry of the higher plants. *New Phytologist*, *15*, 233-250.

Fritsch, F. E. (1935). *The structure and reproduction of algae* (repr. ed. 2016; Vol.1.). Delhi: Gyan Books Pvt. Ltd.

Fritsch, F. E. (1945). *The structure and reproduction of algae* (repr. ed.2016; Vol. 2.). Delhi: Gyan Books Pvt. Ltd.

Frohlich, M. W. (2002). The mostly male theory of flower origins: summary and update regarding the Jurassic pteridosperm *Pteroma*. In Q. C. B. Cronk, R. M. Bateman & J. A. Hawkins (Eds.) *Developmental genetics and Plant evolution* (pp. 85-108). London: Taylor and Francis.

Frohlich, M. W. (2003). An evolutionary scenario for the origin of flowers. *Nature Rev. Genetics, 4*, 559-566.

Frohlich, M. W. & Parker, D. S. (2000). The mostly male theory of flower evolutionary origins: from genes to fossils. *Syst. Bot., 25*, 155-170.

Gaidukov, N. (1903). Weitere Untersuchungen uber den Einfluss farbigen Lichtes auf die Farbung der Oscillerien. [Further studies on the influence of colored light on the coloring of the Oscillerien.] *Ber Dt. Bot. Ges., 21*, 484-493.

Garrison, W. M., Morrison, D. C., Hamilton, J. G., Benson, A. A. & Calvin, M. (1951). Reduction of CO_2 in aqueous solutions by ionizing radiations. *Science, 114*, 416.

Gastaldo, R. A. Adendroff, R., Bamford, M., Labandeira, C. C., Neveling, J. & Sims, H. (2005). Taphonomic trends of macrofloral assemblages across the Permian- Triassic boundary, Karoo basin, South Africa. *Palaios, 20*(5), 479-497.

Gellion, G. & Mcginnis, W. (1998). Shaping body plans in development and evolution by modulation of Hox expression patterns. *BioEssays, 20*(2), 116-125.

Gensheng, Y., Honglang, X. & Wanquan, T. (2001). Mitigating the effects of disastrous sand-dust storms: A Chinese perspective. (Y. Youlin, Trans.). In Y. Youlin, V. Squires & L. Qi (Eds.). *Global Alarm: Dust and stand-storms from the world's dryland* (pp. 283-317). Bangkok: UNCCD Report, Thailand.

Gentry, A. H. (1986). Endemism in tropical vs temperate plant communities In M. SOULS (Ed.) *Conservation biology: The science of scarcity and diversity* (pp. 153-181). Sunderland, Massachusetts: Sinauer Associates.

Gleason, H. A. & Cronquist, A. (1965). *Natural geography of plants*. New York: Columbia University Press.

Glover, B. (2007). *Understanding flowers and flowering- An integrated approach.* Oxford: Oxford University Press.

Goldschimdt, R. (1940). *The material basis of evolution*. New Haven: Yale University Press, New Haven.

Goebel, K. (1930). *Organographie der Pflanzen [Organography of the plants]*. (3rd ed.). Jena: Gustav Fischer Verlag.

Goldring, W. (1924). The upper Devonian forest of seed fern in eastern New York. *New York State Mus. Bull., 251*, 49-72.

Gomez, B., Daviero-Gomez, V., Coiffard, C., Martin-Closas, C. & Dilche, D. L. (2015). *Montsechia,* an ancient aquatic angiosperm. *Proc. Natl. Acad. Sci. USA, 112*(35), 10985-10988.

Gomez-Pompa, A., Vazauezyanes, G., Guevara, A. (1972). The tropical rain forest: A renewable resource. *Science, 117*, 762-765.

Good, R. (1947). *The geography of the flowering plants*. London: Longman Green.

Goodrich, W. J. (1987). *Monitoring Genetic Erosion: Detection and assessment.* (Unpublished Consultancy Document Report, 86/99). Rome: AGPG: IBPGR.

Gordon-Smith, C. (2012, October 28). *Origin of Life: Twentieth Century Landmarks-* The Oparin- Haldane hypothesis. Retrieved from http://www. Rim soup. Info/Origin_Landmarks_ Oparin_Haldane.html.

Goremykin, V. V., Hansman, S. & Martin, W. F. (1997). Evolutionary analysis of 58 proteins encoded in six completely sequenced chloroplast genomes: revised molecular estimates of two seed plant divergence times. *Pl. Syst. & Evol.*, *206*, 337-351.

Gorniak, M., Paun, O. & Chase, M. W. (2010). Phylogenetic relationships within Orchidaceae based on a low-copy nuclear coding gene, Xdh: Congruence with organellar and nuclear ribosomal DNA results. *Mol. Phylogen. & Evol.*, *56*, 784-795.

Goswami, S. & Das, M. (2016). Extremophiles- A clue to origin of life and biology of other planets. *Everyman's Sci., 51*(1): 17-25.

Gould, S. B., Waller, R. R. & Mcfadden, G. I. (2008). Plastid evolution. *Ann Rev. Plant Biol.*, *59*, 491-517.

Graham, L. E., Arancibia-Avila, P., Taylor, W. A., Strother, P. K. & Cook, M. E. (2012). Aeroterrestrial Coleochaete (Streptophyta, Coleochaetales) models early plant adaptation to land. *Am. J. Bot., 99*(1): 130-144.

Graham, S. W. & Iles, W. J. D (2009). Different gymnosperm outgroups have (mostly) congruent signal regarding the root of flowering plants phylogeny. *Am. J. Bot., 96*, 216-227.

Grand'eury, F. C. (1877). Mémoires sur la flore Carbonifère du Département de la Loire et du centre de la France, étudiée aux trois points de vue botanique, stratigraphique et géognostique [Memories on the Carboniferous flora of the Loire and Central France, studied from three botanical, stratigraphic and geognostic points of view]. *Mém. Acad. Sci. Inst. Fr.*, *24,* 1-624.

Grant, V. (1971). *Plant Speciation*. New York: Columbia University Press.

Grauvogel-Stamm, L. & Ash, S. R. (2005). Recovery of the Triassic land flora from the end-Permian life crisis. *Comptes Rendus Palevol.*, *4*(6-7): 593-608.

Green, B. R.(2011). Chloroplast genomes of photosynthetic eukaryotes. *The Pl. Journ.*, *66*(1), 34-44.

Grimaldi, D. A. (1999). The co-radiations of pollinating insects and angiosperms in the Cretaceous. *Ann. Missouri Bot. Gard.*, 86, 373-406.

Groombridge, B. (Ed.) (1992). *Global Biodiversity- Status of Earth's living resources.* A report compiled by the "World Conservation Monitoring Centre." New York: Hall & Chapman.

Gross, B. L., & Olsen, K. M. (2010). Genetic perspectives on crop domestication. *Trends in Pl. Sci., 15*, 528-537.

Guarino, L. (1995). Assessing the threat of genetic erosion. In L. Guarino, V. R. Rao & R. Reid (Eds.). *Collecting plant genetic diversity: Technical guideline* (pp. 67-74). UK: Cab International.

Guowen, X. (1997). On phytogeographical affinities of the forest floras between east China and Japan. *Chinese Geogr. Sci., 7*(3), 236-242.

Guppy, H. B. (1917). *Plants, Seeds, and Currents in West Indies and Azores* (pp. 448-450). London: Williams and Norgate.

Hagner-Holler, S. Pick, C., Girgenrath, S., Marden, J. H. & Burmester, T. (2007). Diversity of stonefly hexamerins and implication for the evolution of insect storage proteins. *Insect Biochem. & Mol. Biol., 37*(10), 1064-1074.

Halle, G. (1935). The position and arrangement of the spore-producing members of the Paleozoic pteridosperms. *Proceed. 6th Intr. Bot. Congr.*, 2, 230.

Halliday, A. N. (2004). Mixing, volatile loss and compositional change during impact-driven accretion of Earth. *Nature, 427*, 505-509.

Hansen, A. (1920). *Die pflanzendecke der Erde*. [*The plant cover of the earth*.] Leipzig.

Harper, J. T. & Saunders, J. W. (2001). Molecular systematics of the Florideophyceae (Rhodophyta) using nuclear large and small subunit rDNA sequence data. *Journ. Phycol., 37*, 1073-1082.

Harper, J. T. & Saunders, J. W. (2002). A reclassification of the Acrochaetiales based on molecular and morphological data, and establishment of the Colaconematales ord. nov. Florideophyceae (Rhodophyta). *European Journ. Phycol., 37*, 463-476.

Harvey, W. (1651). *Excercitationes de generat one animalium: quibus accedunt de partu, de membranis ac humonibus uteri; & de conception*. [The generation of animals. Added to these are some essential facts concerning the birth; of the membranes and humors of the uterus; and the Conception.] London, United Kingdom.

Hassler, M. (2017, July 26). World ferns: Checklist of Ferns and Lycophytes of the World. In Y. Rosokov, L. Abucay, T. Orrell, D. Nicolson, N. Bailly, P. M. Kirk, T. Bourgoin, R. E. Dewalt, W. Decock, A. De Wever, Van E. Nieukerken, J. Zarucchi & L. Penev (Eds.) *Species 2000 & ITIS Catalogue of Life.* Retrieved from http://worldplants.webarchiv.kit.edu/ferns/.

Hayakawa, T., Fox, S. W. & Windsor, C. R. (1967). Co-polymerization of the Leuchs anhydrides of the eighteen amino acids common to protein. *Arch Biochem. Biophys., 118*, 265-272.

Hayashi, C. (1966). Evolution of protostars. Annual Rev. *Astronomy & Astrophysics. 4*, 171-192.

Heywood, V. H. (1978). *Flowering plants of the world*. Oxford: Oxford University Press.

Hickey, L. J. (1973). Classification of the architecture of Dicotyledonous leaves. *Am. J. Bot., 60*(1), 17-33.

Hickey, L. J. & Doyle, J. A. (1977). Early Cretaceous fossil evidence for angiosperm evolution. *Bot. Rev., 43*, 3-104.

Hill, K. W., Borsch, T., Muller, K., Soltis, D. E., Solits, P. S., Savolainen, V., Chase, M. W., Powell, M. P., Alice, L. A., Evans, R., Sauquet, H., Neinhuis, C., Slotta, T. A. B., Rohwer, J. G., Campbell, C. S. & Chatrou, L. W. (2003). Angiosperm phylogeny based on <011> matK sequence information. *Am. J. Bot.*, *90*(12), 1758-1776.

Hochuli, P. A. & Feist-Burkhardt, S. (2004). A boreal early cradle of angiosperms? Angiosperm-like pollen from the Middle Triassic of the Barents Sea (Norway). *Journ. Micropaleontol.*, *23*, 97-104.

Hochuli, P. A. & Feist-Burkhardt, S. (2013). Angiosperm-like pollen and Afropolis from the Middle Triassic (Anisian) of the Germanic Basin (Northern Switzerland). *Front Plant Sci.*, *4*, paper no. 344. doi:10.3389/fpls. 2013.00344.

Hoffman, J. S. & Wells, J. B. (1987). Changes in Greenhouse gases. In W. E. Shands & J. S. Hoffman (Eds.) *The conservation foundation* (pp. 19-42). Washington DC: US Government Printing Office.

Holloway, J. E. (1910) 1909. A comparative study of the anatomy of six New Zealand species of *Lycopodium*. *Trans. N. Z. Inst.*, *42*, 356-370.

Holttum, R. E. (1949). Classification of Ferns. *Biol. Rev.*, *24*, 267-296.

Holttum, R. E. & Sen, U. (1961). Morphology and classifications of the tree fern. *Phytomorphol.*, *11*, 406-420.

Holubec, V. (2006). Genetic erosion and extinction threat to old world cottons, *Gossypium* L.: a case study. In B. V. Ford-Lloyd, S. R. Dias & E. Bettencourt (Eds.) *Genetic erosion and pollution assessment methodologies* (pp.65-78). (Proceedings of PGR forum workshop 5, Terceira Island, Autonomous regions of the Azores, Portugal, 8-11 September 2004.). Rome: Biodiversity International.

Hong, D. Y. (2010). *Peonies of the world: Taxonomy and Phytogeography*. St. Louis: Royal Botanic Garden, Kew, and Missouri Botanic Garden.

Hong, D. Y. (2011). *Peonies of the world: Polymorphism and diversity*. St. Louis: Royal Botanic Garden, Kew, and Missouri Botanic Garden.

Hooper, M. D. (1971). The size and surroundings of nature reserves. In E. Duffey & A. S. Watt (Eds.) *The Scientific Management of Animal and plant committees in Conservation* (pp. 555-561). Oxford: Blackwell Scientific Publication.

Hopping, C. A. (1957). II.- On a specimen of *Trimerphyton robustious* Dawson, from the lower Devonian of Canada. *Proc. Roy. Soc. Edinburgh (Sec.B), 66*(1), 10-28.

Howe, M. A. (1898). The Anthocerotaceae of North America. *Bull. Torr. Bot. Club, 25*, 1-24; 321-326.

Hummel, J., Gee, C., Sudekum, K., Sander, P., Nogge, G. & Clauss, M. (2008). In vitro digestibility of fern and gymnosperm foliage: implications for sauropod feeding ecology and diet selection. *Proceed. Roy. Soc. B: Biol. Sci., 275* (1638), 1015-1021

Hutchinson, J. (1969). *Evolution and phylogeny of flowering plants. Dicotyledons; Facts and Theory*. London: Academic Press.

Huxley, T. H. (1870). Biogenesis and Abiogenesis. In (1968) *Collected Essays of Thomas H. Huxley* (Vol. viii) *Discourses biological and geological* (p-256). New York: Greenwood Press.

International Commission of Stratigraphy (2014). *International Chronostratigraphic Chart. Retrieved* on May 30, 2014. http://www.stratigraphy.org/index.php/ics-chart-timescale.

Igantov, M. S. (1990). Upper Permian mosses from the Russian platform. *Paleontographica., 217b*, 147-189.

Irish, V. F. (2006). Duplication, diversification and comparative genetics of angiosperm MADS-BOX genes. In D. E. Soltis, J. H. Lebens-Mack, & P. S. Soltis (Eds.) *Advances in Botanical Research., developmental genetics of the angiosperm flower* (Vol.44, pp. 129-161). Amsterdam: Elsevier.

Janaki Ammal, E. K. (1960). The effect of the Himalayan uplift on the genetic composition of the flora of Asia. *Journ. Ind. Bot. Soc., XXXIX* (3), 327-334.

Janzen, D. H. (1984). Dispersal of small seeds by big herbivores: foliage is the fruit. *Amer. Nat., 123*, 338-353.

Jardine, P. (2011). Patterns in paleontology: The Paleocene-Eocene Thermal Maximum. *Paleontology, 1*(5), 1-7.

Ji, Q., Li, H., Bowe, L. M., Liu, Y. & Taylor, D. W. (2004). Early Cretaceous *Archaefructus eoflora* Spec. Nov. with bisexual flowers from Beipiao, Western Liaoning, China. *Acta Geologica Sinica, 78*(4), 883-896.

Johnson, D. S. (1933). The curvature, symmetry, and homologies of the sporocarps of *Marsilea* and *Pilularia*. *Am. J. Bot., 25*, 141-156.

Joice, G. F. (1989). RNA evolution and the origins of life. *Nature, 338*, 217-224.

Jordan, R. & Green, J. C. (1994). A checklist of the extant haptophyte of the world. *J. Mar. Bios. Asso. U.K., 74*, 149-174.

Judd, W. S., Sanders, R. W. & Donoghue, M. J. (1994). Angiosperm family pairs: preliminary phylogenetic analysis. *Harvard. Pap. Bot., 5*, 1-51.

Julien, M. H. (2001). Biological control of water hyacinth with arthropods: a review to 2000. In M. H. Julien, M. P. Hill, T. D. Center & D. Jianqianging (Eds.). *Biological and integrated control of water hyacinth, Eichhornia crassipes* (pp. 8-20). (Proceedings of the 2nd meeting of the global working group for the biological and integrated control of water hyacinth, Beijing, China, 9-12 Oct. 2000). Beijing.

Jyrwa, L. R. (1992). Improvement of Jhum. *Arunachal Forest News, 10,* 26-28.

Kant, I. (1755). *Universal natural history and theory of heaven* (May 1, 2009) (Allgemeine Naturgeschichte und Theorie Himmels, Trans, Ian Johnston). Konigsberg: Richer Resources Publications.

Kashyap, S. R. (1919). Relationships of the Liverworts especially in the light of some recently discovered Himalayan forms. *Proceedings Asiat. Soc. Bengal* (New Series)., *15*, 152-166.

Kato, M. (1990). Taxonomic studies of pteridophytes of Ambon and Seram (Moluccas) collected by Indonesian-Japanese botanical expeditions III. Eusporangiate and some lower leptosporangiate families. *Acta Phytotax Geobot., 40*, 67-82.

Keeling, P. J. (2010). The endosymbiotic origin, diversification, and fate of plastids. *Philos. Trans. R. Soc. Lond. B Biol. Sci., 365*(1541), 729-748.

Kellman, M. C. (1975). *Plant Geography.* London: Methuen & Co. Ltd.

Kenrick, P. & Crane, P. R. (1997a). The origin and early diversification of land plants. *Nature, 389*, 33-39.

Kenrick, P. & Crane, P. R. (1997b). *The origin and early diversification of land plants: A cladistics study.* Washington DC: Smithsonian Press.

Kent, D. V. & Olsen, P. E. (2000). Implications of astronomical climate cycles to the chronology of the Triassic. *Zbl. Geol. Palaont.*, *1*, 1463-1473.

Kerr, G., Kreft, H., Lee, T. M., Jetz, W., Ibisch, P. L., Nowciko, C., Mutke, J. & Barthlott, W. (2009). A global assessment of endemism and species richness across island and mainland regions. *Proc. Natl. Acad. Sci. USA, 106*(23), 9322-9327.

Kidston, R. & Lang, W. H. (1917). On Old Red Sandstone Plants showing Structure, from the Rhynie Chert Bed, Aberdeenshire Part 1. *Rhynia Gwynne-Vaughani* Kidston and Lang. *Trans. Roy. Soc. Edinb.*, *51*, 761-784.

Kidston, R. & Lang, W. H. (1921). On Old Red Sandstone Plants showing Structure, from the Rhynie Chert Bed, Aberdeenshire Part 4. Restorations of vascular cryptogams, and discussions of their bearing on the general morphology of the Pteridophyta and the origin of the organization of land plants. *Trans. Roy. Soc. Edinb.*, *52*, 831-854.

Kim, D. H. & Sung, S. (2013). Coordination of vernalization response through a VIN3 and FLC gene family regulatory network in *Arabidopsis*. *Plant Cell, 25*(2), 454-469.

King, C. (1877). Catastrophism and Evolution. *The Amer. Nat., 11*(8), 449-470.

King, K. E., Moritz, T. & Harberd, N. P. (2001). Gibberellins are not required for normal stem growth in *Arabidopsis thaliana* in the absence of GAI and RGA. *Genetics, 159*(2), 767-776.

Kirkland, J. I., Lockley, M. & Milner, A. R. (2002). The St. George Dinosaur Tracksite. *Survey Notes, Utah Geol. Surv.*, *34*(3), 4-5.

Knoll, A. H., Bambach, R. K., Payne, J. L. Pruss, S. & Fischer, W. W. (2007). Paleophysiology and end-Permian mass extinction. *Earth Planet Sci. Lett.*, *256*, 295-313.

Kolbert, E. (March 2011). Enter the Anthropocene-Age of Man. *National Geographic*. http://ngm.nationalgeographic.com/2011/03/age-of-man/kolbert-text.

Kolbert, E. (2014). *The Sixth Extinction: An unnatural history*. London: Bloomsbury Publishing Plc.

Kopp, R. E., Kemp, A. C., Bittermann, K. Horton, B. P., Donnelly, J. P. Gehrels, W. R., Hay, C. C., Mitrovica, J. X., Morrow, E. D. &. Rahmstorf, S. (2016). Temperature-driven global sea level variability in the common era. *Proc. Natl. Acad. Sci. USA, 113*(11), E1434-E1441.

Korovin, E. P. (1934). Types of migration in the plant world. (In Russian) *Bull. Cent. As. State Univ., Ser. 8, No. 16.*

Kramer, E. M. (2009). New model systems for the study of developmental evolution of plants. *Current Topics in Developmental Biology, 86*, 67-105.

Krassilov, V. (1975). Dirhopalostachyaceae a new family of proangiosperms and its bearing on the problem of angiosperm ancestry. *Paleontographica, 153*B: 100-110.

Krassilov, V. (1996). Origin of angiosperm characters. *Paleobot.*, *45*, 400-406.

Krassilov, V. & Golovneva, L. B. (2004). A minute mid-Cretaceous flower from Siberia and implications for the problem of basal angiosperms. *Geodiversitas, 26*(1), 5-15.

Krassilov, V. & Schuster, R. M. (1984). Paleozoic and Mesozoic fossils. In R. M. Schuster (Ed.) *Manual of Bryology* (Vol. 2; pp.1172-1193). Nichinan: The Hattori Botanical Laboratory.

Krassilov, V., Tekleva, M., Meyer-Melikyan, N. & Rasnitsyn, A. P. (2003). New pollen morphotype from gut compression of a Cretaceous insect, and its bearing on palynomorphological evolution and paleoecology. *Cretaceous Res.*, *24*, 149-156.

Krausel, R. (1949). Koniferen und adere Gymnospermen aus der Trias von Lunz, Neider-Osterreich. Untersuchungen zur mesozoischen Florengeschichte des alpinen und suddeutschen Raumes III. *Paleontographica, 89*, 35-82.

Krausel, R. & Weyland, H. (1923). Beitrage zur Kenntnis der Devonflora I. *Senckenbergiana, 5*, 172-181.

Krausel, R. & Weyland, H. (1926). Beitrage zur Kenntnis der Devonflora II. *Abhandl. Senck. Naturf. Ges., 40*, 127-133.

Krausel, R. & Weyland, H. (1929). Beitrage zur Kenntnis der Devonflora III. *Ebenda, 4*, 319-325.

Kruckeberg, A. R. (1984). *California serpentines: Flora, vegetation, geology, soils, and management problems* (Vol.78). Berkley: University of California Press.

Kundu, S. R. (2000). A synopsis of Tiliaceae in Indian subcontinent: Its distribution and endemism. *Geobios, 27,* 49-56.

Kundu, S. R. (2001). Phytogeographical provinces (phytocorias) of India: An account. *Geobios, 29*, 69-88.

Kundu, S. R. (2004). *Endemic and threatened plants of India (Series; Thalamiflorae) with reference to economic, horticultural and biological potentialities* (Vol. 1 & 2). (Ph.D. Dissertation), Kolkata: University of Calcutta.

Kundu, S. R. (2017). *Peony: A metaphoric Journey through Time*. Beau Bassin: Lambert Academic Publishing.

Kunze, H. (2005). Morphology and evolution of the Corolla and corona of Apocynaceae s.l. *Botanische Jahrb. Systm., 126*, 347-383.

Kuprianova, L. A. (1979). On the possibility of the development of tricolpate pollen from monosulcate. *Grana, 18*, 1-4.

Kwiatkowska, D. (2006). Flower primordium formation at the Arabidopsis shoot apex: quantitative analysis of surface geometry and growth. *Journ. Exp. Bot.*, *57*(3), 571-580.

Labandeira, C. C. (1998). How old is the flower and the fly? *Science, 280*, 57-59.

Labandeira, C. C. (2002). The history of associations between plants and insects. In C. M. Herrera & O. Pellmyr (Eds.). *Plant-animal interactions: an evolutionary approach* (pp. 26-74). Oxford: Blackwell.

Labandeira. C. C. (2006). The four phases of plant-arthropod associations in deep time. *Geologica Acta, 4*(4), 409-438.

Labandeira, C. C., Dilcher, D. L., Davis, D. R. & Wagner, D. L. (1994). Ninety-seven million years of angiosperm-insect association: paleobiological insights into the meaning of coevolution *Proc. Natl. Acad. Sci., 25*, 12278-12282.

Lam, H. J. (1948). Classification and the new morphology. *Acta Biotheoretica, 8*(4), 107-154.

Lam, H. J. (1952). Classification of vascular plants. *Ann. Bio.*, *28*, 57-88.

Lam, H. J. (1959). Taxonomy- General principles and angiosperms, in vistas in Botany: A volume in the honor of the bicentenary of the Royal Botanic Garden, Kew. In W. B. TURRILL (Ed.) *International series of monographs on pure and applied biology* (Vol.2; pp. 3-75). Pergamon Press (Botany Division): London.

Lang, W. H. (1937). On the plant-remains from the Downtonian of England and Wales. *Philo. Transact. Roy. Soc. B*, *227*(544), 245–291,

Laskar, J., Joutel, F. & Robutel, P. (1993). Stabilization of the Earth's obliquity by the moon. *Nature, 361*, 615-617.

Leclercq, S. & Bonamo, P. M. (1971). A study of the fructification of *Milleria* (*Protopteridium*) *thomsonii* Lang from middle Devonian of Belgium. *Taxon, 22*, 435-437.

LeCorre, V., Machon, N., Petit, R. J., & Kremer, A. (1997). Colonization with long-distance seed dispersal of maternally inherited genes in forest trees: a simulation study. *Genetical Res.*, *69*, 117-125.

Lee, R. E. (1989). *Phycology* (2nd ed.). Cambridge: Cambridge University Press.

Lee, R. E. (1999). *Phycology* Cambridge: Cambridge University Press.

Lee, R. E. (2008). *Phycology* (4th ed.). Cambridge: Cambridge University Press.

Leebens-Mack, J., Raubeson, L. A., Cui, L., Kuehl, J. V., Fourcade, M. H., Chumley, T. W. M Boore, J. L., Jansenm R. K. & Depamphilis, C. W. (2005). Identifying the basal angiosperm node in chloroplast genome phylogenies: sampling one's way out of the Felsenstein zone. *Mol. Biol. & Evol.*, *22*(10), 1948-1963.

Leliaert, F., Verbruggen, H. & Zechman, F. W. (2011). Into the deep: New discoveries at the base of green plant phylogeny. *BioEssays*, *33*, 683-692.

Lemaitre, G. (1931). The evolution of the Universe: Discussion. *Nature, 128*(3234), 699-701.

Leroy, J. F. (1983). The origin of angiosperms: an unrecognized ancestral dicotyledons, *Hedyosmum* (Chloranthales), with a strobiloid flower is living today. *Taxon, 32*, 169-175.

Leslie, A. (2011). Predation and protection in the macroevolutionary history of conifer cones. *Proceed. Roy. Soc. B: Biol. Sci., 278*, 3003-3008.

Lewis, H. (1962). Catastrophic selection as a factor in speciation. *Evolution, 16*(3), 257-271.

Lewis, H. & Raven, P. H. (1958). Rapid evolution in *Clarkia*. *Evolution, 12*, 319-336.

Lewis, H. & Roberts, M. R. (1956). The origin of *Clarkia lingulata*. *Evolution, 10,* 126-138.

Lewis, L. A. & Mccourt, R. M. (2004). Green algae and the origin of land plants. *Amer. J. Bot., 91*, 1535-1556.

Li, X. G. & Yao, Z. Q. (1983). Fructifications of gigantopterids from South China. *Paleontographica Abt. B., 185*(1-3), 11-26.

Lignier, O. (1903). Equisetales et Sphenophyllales, leur origine filicineenne commune. [Equisetales and Sphenophyllus, their common filicinean origin.] *Bull. Soc. Linn. Norm.*, *7*(5), 93-137.

Linnaeus, C. (1753). *Species Plantarum*. Stockholm: Laurentius Salvius.

Lissauer, J. J., Barnes, J. W. & Chambers, J. E. (2012). Obliquity variations of a moonless Earth. *Icras*, *217*(1), 77-87.

Litwin, R. J., Traverse, A. & Ash, S. R. (1991). Preliminary palynological zonation of the Chinle Formation, southwestern U.S.A. and its correlation to the Newark Supergroup (Eastern U.S.A.). *Rev. Paleobot. & Palnol., 68,* 269-287.

Liu, J. W. & Wang, X. (2016). A perfect flower from the Jurassic of China. *Historical Biology, 28*(5), 707-719.

Livermore, P. W., Hollerbach, R. & Finlay, C. C. (2017). An accelerating high- latitude jet in Earth's core. *Nat. Geosci.*, *10*, 62-68.

Lopez-Puzol, J. & Zhao, A. M. (2004). China: A rich flora needed of urgent conservation. *Oris, 19*, 49-89.

Lotsy, J. P. (1909). Vorträge über botanische Stammesgeschichte (Band II)., *Cormophyta zoidogamia*. Jena: Gustav Fischer Verlag.

Looy, C. V., Brugman, W. A., Dilcher, D. & Visscher, H. (1999). The delayed resurgence of equatorial forests after the Permian- Triassic ecologic crisis. *Proc. Natl. Acad. Sci. USA,* 96(24), 13857-13862.

Looy, C. V., Twichett, R. J. Dilcher, D., Van Konijnenburg-Van Cittert, J. H. A. & Visscher, H. (2001). Life in the end-Permian dead zone. *Proc. Natl. Acad. Sci. USA,* 96(24), 13857-13862.

Lowe, P. (1998). *The boab tree*. Port Melbourne: Lothian.

Lundblad, B. (1954). Contributions to the geological history of the Hepaticae. Fossil Marchantiales from the Rhaetic-Liassic coal mines of Skromberga (Prov. Of Scania), Sweeden. *Svensk. Botanisk Tidskrift., 48*, 381-417.

Lunine, J. I. (2006, October 29). Physical conditions of the early Earth. *Philosophical Trans. Royal Soc. Series-B*. Retrieved from DOI: 10.1098/rstb.2006.1900.

Lupia, R., Lidgard, S. & Crane, P. R. (1999). Comparing palynological abundance and diversity: implications for biotic replacement during the Cretaceous angiosperm radiation. *Paleobiol., 25*, 305-340.

Mac Arthur, R. H. & Wilson, E. O. (1963). An equilibrium theory of insular zoogeography. *Evolution, 17,* 353 – 387.

Mac Arthur, R. H. & Wilson, E. O. (1967). *The theory of island biogeography*. Princeton: Princeton University Press.

Magallon, S. & Castillao, A. (2009). Angiosperm diversification through time. *Am. J. Bot.*, *96*(1), 349-365.

Magallon, S., Crane, P. R. & Herendeen, P. S. (1999). Phylogenetic pattern, diversity, and diversification of eudicots. *Ann. Missouri Bot. Gard.*, *86*, 297-372.

Magallon, S., Gomez-Acevedo, S. & Sanchez-Reyes, L. L. (2015). A metaclaibrated time-tree documents the early rise of flowering plant phylogenetic diversity. *New Phytol.*, *207*(2), 437-453.

Magallon, S. & Sanderson, M. J. (2002). Relationships among seed plants inferred from highly conserved genes: sorting conflicting phylogenetic signals among ancient lineages. *Am. J. Bot.*, *89*(12), 1991-2006.

Mahler, H. R. & Raff, R. A. (1972). The evolutionary origin of mitochondrion: A non-symbiotic model. In G. H. Bourne & J. F. Danielli (Eds.). *International Review of Cytology* (Vol. 43, pp 2-112). New York: Academic Press.

Mamay, S. H. (1976). *Paleozoic origin of the Cycads*. (Geological Survey Professional Paper; No. 934., P 66). Washington DC: United States Geological Survey, US Government Printing Office.

Marazzi, B. & Endress, P. K. (2008). Patterns and development of floral asymmetry in *Senna* (Leguminosae, Cassinae). *Am. J. Bot.*, *95*, 22-40.

Marguilis, L. (1970). *Origin of Eukaryotic Cells. Evidence and Research Implications for a Theory of the Origin and Evolution of Microbial, Plant, and Animal Cells on the Precambrian Earth*. London: Yale University Press.

Marguilis, L. & Schwartz, K. V. (1998). *Five Kingdoms. An illustrated guide to the phyla of life on earth* (3rd Ed.). New York: W. H. Freeman & Company.

Markgraf, F. (1930). Monographie der Gattung *Gnetum.* [Monograph of the genus *Gnetum*.] *Bull. Jardin Bot. Buitenz.*, *10*, 407-511.

Martin, W., Lydiate, D., Brinkmann, H., Forkmann, G., Saedler, H. & Cerff, R. (1993). Molecular phylogenies in Angiosperm evolution. *Mol. Biol. Evol.*, *10*(1), 140-162.

Mathews, S. & Donoghue, M. J. (1999). The root of angiosperm phylogeny inferred from duplicate phytochrome genes. *Science, 286*, 947-950.

Mathews, S. & Kramer, E. M. (2012). Plants: a re-examination based on insights from developmental genetics. *New Phytol., 194*(4), 910-923.

Mattick, F. (1964). Übersicht ber die Florenreiche and Florengebiete der Erde. [Overview of the floral kingdoms and flora of the earth.] In A. Engler (Ed.) *Syllabus der pflanzenfamilien* (Bd.2. 12 Aufl. pp. 626 – 630). Berlin: Borntraeger.

Mayer, E. (1954). Change of genetic environment and evolution. In. J. Huxley, A. C. Hardy & E. B. Ford (Eds.) *Evolution as a process* (pp 157-180). London: Allen & Unwin.

McElwain, J. C., Beerling, D. J. & Woodward, F. I. (1999). Fossil plants and global warming at the Triassic-Jurassic boundary. *Science, 285*, 1386-1390.

McElwain, J. C., Popa, M. E., Hesselbo, S. P., Haworth, M. & Surlyk, F. (2007). Macroecological responses of terrestrial vegetation to climatic and atmospheric changes across the Triassic/Jurassic boundary in east Greenland. *Paleobiol., 33*(4), 547-573.

McGhee, G. R. (1996). *The late Devonian mass extinction*. New York: Columbia University Press.

McLaughlin, S. (1994). Floristic plant geography: the classification of floristic areas and floristic elements. *Progress in Physical Geography, 18*(2), 185-208.

McLaughlin, S., Carpenter, R. J., Jordan, G. J. & Hill, R. S. (2008). Seed ferns survived the end-Cretaceous mass extinction in Tasmania. *Am. J. Bot., 95*(4), 465-471.

Medail, F. & Verlaque, R. (1997). Ecological characteristics and rarity of endemic plants from southeast France and Corsica: Implications for biodiversity conservation. *Biol. Conserv., 80*(3), 269-281.

Mehra, P. N. (1969). Evolutionary trends in Hepaticae. Panchanan Maheswari Memorial Volume. *Phytomorphology, 19*, 203-218.

Melchior, H. & Werdermann, E. A. (1954). In E. A. Engler (Ed.) *Syllabus der Pflanzenfamilien*. [*Syllabus of plant families*.] Berlin: Borntraeger.

Melville, R. (1969). Leaf venation patterns and the origin of angiosperms. *Nature, 224*, 121-125.

Meyen, S. V. (1988). Origin of the angiosperm gynoecium by gamoheterotrophy: transition of characters from one sex to another. *Zhurn. Obshchei Biologii, 47*, 291-309.

Miasek, M. A. & Long, C. R. (1978). *Endangered plant species of the world and their endangered habits: A compilation of the literature*. Bronx: New York Botanic Garden.

Michener, C. D. (2007). *The bees of the world* (2nd ed.). Baltimore, Maryland: John Hopkins University Press. Pp. 992.

Miller, J. M. (2008). *Origin of Angiosperms* (Updated 1st May 2017). Retrieved from http://www.gigantopteroid.org/html/angiosperm. htm.

Miller, S. (1953). A production of amino acid under possible primitive Earth conditions. *Science, 117*(3046), 528-529.

Milner, A. R. C., Lockley, M., Kirkland, J., Bybee, P. & Mickelson, D. (2004). St. George tracksite, southern Utah: remarkable early Jurassic (Hettangian) record of dinosaurs walking, swimming, and sitting provides a detailed view of paleoecosystem along the shores of lake Dixie. *Journ. Vertebrt. Paleontol., 24*, 94A.

Mishler, B. D., Lewis, L. A., Buchheim, M. A., Renzaglia, K. S., Garbary, D. J., Delwiche, C. E., Zechman, F. W., Kantz, T. S. & Chapman, R. L. (1994). Phylogenetic relationships of green algae and bryophyte. *Ann. Missouri Bot. Gard., 81*, 451-483.

Moisset, B. & Buchmann, S. (2011). *Bee basics: An introduction to our native bees.* Washington DC: A USDA Forest Service and Pollinator Partnership Publication.

Mojzsis, S. J., Harrison, T. M. & Pidgeon, R. T. (2001). Oxygen-isotope evidence from ancient zircons for liquid water at the Earth's surface 4,300 myr ago. *Nature, 409*, 178-181.

Moldowan, J. M. & Jacobsen, S. R. (2003). Chemical signals for early evolution of major taxa: biosignatures and taxon-specific biomarkers. In W. G. ERNST (Ed.). *Frontiers in Geochemistry, organic, and ore deposit geochemistry* (Vol.2; pp. 19-26). Columbia: Bellwether Publishing Ltd.

Moore, M. J., Bell, C. D., Soltis, P. S. & Soltis, D. E. (2007). Using plastid genome-scale data to resolve enigmatic relationships among basal angiosperms. *Proc. Natl. Acad. Sci. USA, 104*(49), 19363-19368.

Mukherji, H. & Ganguly, A. K. (1964) *Plant Groups (6th ed., Vol.1).* Calcutta: New Central Book Agency.

Muller, J. E. (1968). An introduction to hydraulic and topographic sinuosity index. *Ann. Assoc. Amer. Geogr., 58*(2), 371 – 385.

Muller, P. (1980). *Biogeography*. New York, Harper & Row.

Myers, N. (1979). *The Sinking Ark: a new look at the problem of disappearing species.* Oxford: Pergamon Press.

Myers, N. (1980). The problem of disappearing species: what can be done? *Ambio, IX* (5), 229 – 235.

Nakai, T. (1938). Indigenous species of conifers and taxads of Korea and Manchuria and their distribution. I. *Tyosen Sanrin Kayho, 158*, 1-29.

Nandi, W., Chase, M. W. & Endress, P. K. (1998). A combined cladistic analysis of angiosperms using rbcL and non-molecular data sets. *Ann. Missouri Bot. Gard., 85*, 137-212.

Nauholmykh, S. V. (2006). Foliar seed-bearing organs of Paleozoic ginkgophytes and the early evolution of the Ginkgoales. *Paleontological Journal, 41*(8), 815-859.

Nayar, M. P. (1984). Extinction of species and concept of rarity of plants. *J Econ. Tax. Bot., 5*(1), 1-6.

Nayar, M. P. (1987). *In situ* conservation of wild flora resources. *Bull. Bot. Surv. India., 29*, 319-333.

Nostrog, K. J. & Fawcett, P. K. S. (1989). Insect-cycad symbiosis and its relation to the pollination of *Zamia furfuracea* (Zamiaceae) by *Rhopalotria mollis* (Cuclinidae). *Am. J. Bot*. 76(9): 1380-1394.

Nelson, B. W. Carlos, A. C. A., daSilva, M. F. Kawasaki, M. L. (1990). Endemism Centres: refugia and botanical collection intensity in Brazillian Amazonia. *Nature, 345*, 714-716.

Neuberg, M. F. (1960). Mousses feuillees des couches Permiennes de l'Angeride. [Leaf moss Permian layers of the Angel.] *Trav. Institute geologique Acad. Sci. U.S.S.R., 107*, 321-324.

Niklas, K. J. (1997). *The evolutionary biology of plants*. Chicago: University of Chicago Press.

Nilsson, M. A., Churakov, G., Sommer, M., Tran, N. V., Zemann, A., Brosisus, J., Schmitz, J. (2010) Tracking Marsupial Evolution Using Archaic Genomic Retroposon Insertions. *PLOS Biology*, *8*(7), e1000436. https://doi.org/10.1371/journal.pbio.1000436.

Nilsson, T. (1958). Uber das Vorkommen eines mesozoischen Sapropelgesteins in Schonen [On the occurrence of a Mesozoic sapropelic rock in Schonen]. *Publ. Inst. Miner. Paleont. Quat. Geol. Lund., 53*, 1-111.

Nishida, M. (1957). Studies on the systematic position and constitution of Pteridophyta, 10. A further investigation on the vascular dichotomy in the phyllomophore of Ophioglossales, with special reference to phylogeny. *J. Coll. Arts Sci., Chiba Univ., 2*, 179-211.

Nozu, Y. (1950). On the so-called petiole of *Botrychium* (Preliminary Report). *Bot. Mag. Tokyo, 63*, 4-11.

Nuraliev, M. S., Oskolski, A. A., Sokoloff, D. D. & Remizowa, M. V. (2010). Flowers of Araliaceae: Structural diversity, developmental and evolutionary aspects. *Plant Diversity & Evol., 128*, 247-268.

Nutman, A. P.; Bennett, V. C.; Friend, C. R. L.; Kranendonk; Artin, J. V. & Chivas, A. R. (2016). Rapid emergence of life shown by discovery of 3.700 million-years-old microbial structures. *Nature, 537*, 535-538.

Oliver, F. W. & Scott, D. H. (1903). On *Lagenostoma lomaxi*, the seed of *Lyginodendron. Proceedings Roy. Soc*., *71*, 477-481.

Ollier, C. D. (1981). *Tectonics and Landforms*. London: Longmans.

Olsen, P. E., Kent, D. V., Cornet, B., Witte, W. K. & Schlische, R. W. (1996). High-resolution stratigraphy of the Newark rift basin (early Mesozoic, Eastern North America). *GSA Bull., 108*, 40-77.

Olsen, P. E. & Kent, D. V. (2000). High-resolution early Mesozoic Pangean climatic transect in lacustrine environments. *Zbl. Geol. Palaont., 1*, 1475-1495.

Oparin, A. I. & Morgulis, S. (1938). *Origin of Life* (pp.1-270). New York, The MacMillan Company.

Oreskes, N. (2008). *Plate Tectonics: An insider's history of the modern theory of the earth*. Emeryville CA: Avalon Publishing Group, Inc.

Oxburgh, E. R. (1979). Plate Tectonics. In I. G. Gass & P. J. Smith (Eds.) *Understanding the Earth*. Sussex: The Artemis Press, Ltd.

Pacini, E. (2008). Pollination Biology: Orchids pollen dispersal units and reproductive structures. In. T. Kull, J. Arditti & S. M. Wong (Eds.). *Orchid biology: Reviews and perspectives* (Vol. 10, pp. 187-218). [unknown location]: Springer.

Palfy, J., Demeny, A. Hass, J., Hetenyi, M., Orchard, M. & Vito, I. (2001). Carbon isotope anomaly and other geochemical changes at the Triassic-Jurassic boundary from a marine section in Hungary. *Geology, 29*, 1047-1050.

Papenfuss, G. F. (1946). Proposed names for the phyla of algae. *Bull. Torrey Bot. Club, 73*(3), 217-218.

Papenfuss, G. F. (1955). Classification of the algae. In *A Century of progress in Natural Sciences* 1853-1953 (pp.115-224). San Francisco: California Academy of Science.

Parui, S. M. & Mondal, A. K. (2016) 2017. Molecules of flowering. *Everyman's Sci., 51*(5), 306-310.

Payens, J. P. D. W. (1967). A monograph of the genus *Barringtonia* (Lecythidaceae). *Blumea, 15*(2), 157-263.

Pearson, L. C. (1995). *The diversity and evolution of plants*. New York, CRC Press.

Peck, W. H., Valley, J. W., Wilde, S. A. & Graham, C. A. (2001). Oxygen isotope ratios and rare earth elements in 3.3 to 4.4 Ga zircons: Ion microprobe evidence for high $\delta^{18}O$ continental crust of the Archaean. *Geochimica et Cosmochimica Acta., 65*, 4215-4229.

Petit, C. (2010). Continental hearts: Ancient expanses called cratonspose a geological puzzle. *Science News, 178*(13), 22-26.

Phillips, T. L., Andrews, H. N. & Gensel, P. G. (1972). *Two heterosporous species of Archaeopteris from the upper Devonian of West Virginia.* Paleontographica, *139*B, 47-71.

Pichi-Sermolli, R. E. G. (1958). The higher taxa of Pteridophyta and their classification. In O. Hedberg (Ed.) *Systematics of today* (Vol. 6, pp. 70-90). Uppsala: Uppsala Universitets Arsskrift.

Pielou, E. C. (1979). *Biogeography*, New York, Wiley.

Pilger, R. & Melchior, H. (1954). XVI. Abteilung: Gymnospermae Nackstamer (Archispermae). In. A. Engler (Ed.) *Syllabus de Pflanzenfamilien*. Band 1. *Allgemeiner Teil Bakterien bis Gymnospermen*. In H. Melchior & E. Werdermann (Chapter Eds.) Allgemeiner Teil Bakterien bis Gymnospermen (pp. 312-344). Berlin- Nikolasse, Gebruder Brontraeger.

Pimm, S. L. & Joppa, L. N. (2015). How many plant species are there, why are they, and what rate are they going extinct? *Ann. Missouri Bot. Gard., 100*, 170-176.

Piry, S., Luikart, G.,& Cornuet, J. M.(1999). “BOTTLENECK”: a computer program for detecting recent reductions in the effective population size using allele frequency data. *Journ. of Heredity*, *90*, 502-503.

Ponnasperuma, C., Sagan, C. & Mariner, R. (1963). Ultraviolet synthesis of Adenosine Triphosphate under possible primitive Earth conditions. *SAO Special Report, 128*(1963): 1-18. [Archive maintained by NASA Astrophysics Data System].

PPG-I (2016). A Community-derived classification for extant lycophytes and ferns. *Journ. Syst. Evol.*, *54*(6), 563-603.

Prance, G. T. & Ellias, T. E. (Eds.) (1977). *Extinction is forever. Threatened and Endangered species of plants in the American and their significance in ecosystems today and in the future.* (Proceedings of a symposium held at the New York Botanical Garden, May 11-13, 1976, in commemoration of the bicentennial of the USA). Bronx, NY: The New York Botanical Garden.

Preisig, H. R. (1995). A modern concept of chrysophyte classification. In C. D. Sanderson; J. R. Smol & J. Kristiansen (Eds.). *Chrysophyte Algae: ecology, phylogeny, and development*. Cambridge: Cambridge University Press, Cambridge.

Prenner, G. (2004). Floral development of *Polygala myrtifolia* (Polygalaceae) and its similarities with Leguminosae. *Plant Syst. & Evol., 249*, 67-76.

Prescott, G. W. (1968). *The Algae: A Review*. Boston: Houghton Mifflin.

Preston, J. C. & Hileman, L. C. (2009). Developmental genetics for floral symmetry evolution. *Trends in pl. Sci., 14*, 147-154.

Preyer, W. (1880). *Die Hypothesen uber den Ursprung des Lebens [The hypotheses about the origin of life]*. Berlin, Germany.

Priftes, D. & Tirrell, M. (2012). Phase behaviour and complex coacervation of aqueous polypeptide solutions. *Soft Matter*, *8*, 9396-9405.

Pringshiem, N. (1876). *Uber vegetative Sprossung der Moosfruchte. [Vorlaufige Mittheilung]. Monatsberichte der koniglich preussischen [About vegetative budding of the moss fruit. [Provisional communication]. Monthly reports of the royal Prussian]* (pp.42-430). Berlin: Akademie der Wissenschaften zu Berlin.

Pritzel, E. (1900). Lycopodiaceae. In: A. Engler & K. Prantl (Eds.) *Die Nat. Pflanzenfamilein* (Vol. 4(3), pp. 563-606). Leipzig: W. Engelman.

Proskauer, J. (1957). Studies on Anthocerotales. V. *Phytomorphology*, *7*, 113-135.

Qiu, Y. L., Cho, Y. Cox, J. C. & Palmer, J. D. (1998). The gain of three mitochondrial introns identifies liverworts as the earliest land plants. *Nature*, *394*, 671-674.

Qiu, Y. L., Jungho, L., Bernasconi-Quadroni, F., Soltis, D. E., Zanis, M., Zimmer, E., Ziduan, C., Savolainen, V. & Chase, M. W. (1999). The earliest angiosperms: analyses of five genes from three genomes. *Intl. Journ. Pl. Sci., 161*(6), S3-S27.

Qiu, Y. L., Li, L., Hendry, T. A., Li, R., Taylor, D. W., Issa, M. J., Ronen, A. J., Vekaria, M. L. & White, A. M. (2006). Reconstructing the basal angiosperm phylogeny: evaluating information content of mitochondrial genes. *Taxon, 55*(4), 837-856.

Radian Corporation (1988). *Emissions and cost estimates for globally significant anthropogenic combustion sources NO_2, N_2O, CH_4, CO and CO_2.* (Prepared for United States Environmental Protection Agency). Research Triangle Park, North Carolina: National Service Center for Environmental Publications (NSCEP).

Ramirez, S. R., Gravendeel, B., Singer, R. B., Marshall, C. R. & Pierce, N. E. (2007). Dating the origin of the Orchidaceae from fossil orchid with its pollinator. *Nature, 448*(7157), 1042-1045.

Randhawa, M. S., Singh, J., Dey, A. K., Visnu Mittre. (1969). *Evolution of Life*. New Delhi: Centre for Scientific & Industrial Research (C.S.I.R.).

Rangan, H., Bell, K. L., Baum, D. A., Flower, R., Mcconvell, P., Saunders, T., Spronck, S., Kull, C. A. & Murphy, D. J. (2015). New genetic and linguistic analyses show ancient human influence on baobab evolution and distribution in Australia. *PLOS ONE, 10*(4), e0127582.

Rashid, A. (1999). *An introduction to Pteridophyta* (2nd ed.). New Delhi: Vikas Publishing House.

Raven, P. H. (1979). Plate tectonics and southern hemisphere biogeography. In K. LARSEN, & L. B. HOLM-NIELSEN (Eds.) *Tropical Botany*. London: Academic Press.

Raven, P. H. & Axelrod, D. I. (1974). Angiosperm biogeography and past continental movements. *Ann. Missouri Bot., Gard*., *61*(3), 539-673.

Ravi, N. (1994). Threatened plants and their conservation. *Zoo's Print.,* 9, 2-3.

Raymond, B. H. & Ward, P. D. (2005). Hypoxia, global warming, and terrestrial Late Permian extinctions. *Science, 308*, 398-401.

Rees, P. M. (2002). Land-plant diversity and the end-Permian mass extinction. *Geology, 30*(9), 827-830.

Reid, W. V. (1992). How many species will there be? In J. C. Whitmore & J. A. Sayer (Eds.) *Tropical Deforestation and Species Extinction* (pp. 55-73). London: Chapman Hall.

Reimers, H. (1954). XV. Abteilung: Pteridophyta. Farnpflanzen [Department: Pteridophyta. Ferns], In H. Melchior & E. Werdermann (Eds.) *A. Engler's Syllabus der Pflanzenfamilien* (pp. 269-311), Berlin: Borntraege.

Renault, B. & Zeiller, R. (1885). Sur des Mousses de l'epoque houillere. [On mosses from the coal era.] *C. R. Acad. Sci., 100*, 660-662.

Renvoise, S. A. (1979). The origin of Indian ocean island floras. In D. Bramwell (Ed.) *Plants and islands* (pp. 107-129). London: Academic Press.

Retallack, G. J. (2009). Greenhouse crisis of the past 300 million years. *Geol. Soc. Am. Bull., 121*, 1441-1455.

Retallack, G. J. & Dilcher, D. (1981). A coastal hypothesis for the dispersal and rise to dominance of flowering plants. In K. J. Niklas (Ed.) *Paleobotany, Paleoecology, and Evolution*. Praeger, New York. Pp. 27-77.

Retallack, G. J., Greaver, T., Smith, R. M. H. & Sheldon, N. D. (2006). Middle-Late Permian extinction on land. *Geol. Soc. Am. Bull., 118*, 1398-1411.

Retallack, G. J. & Jahren, A. H. (2008). Methane release from igneous intrusion of coal during Late Permian extinction events. *J. Geol., 116*, 1-20.

Retallack, G. J., Smith, R. M. H. & Ward, P. D. (2003). Vertebrate extinction across the Permian –Triassic boundary in the Karoo basin of South Africa. *Geol Soc. Amer. Bull., 115*, 1133-1152.

Reveal, J. L. (2012). An outline of a classification scheme for extant flowering plants. *Phytoneuron, 37*, 1-221.

Reyre, Y. (1973). Palynologie du Mesozoque Saharien. *Memoires du Museum National d'Historie Naturelle [*Palynology of the Saharan Mesozoic. *Memories of the National Museum of Natural History]*, (Series C)., *27*, 284.

Richardson, I. B. K. (1978). Endemic Taxa and the Taxonomist In H. Street (Ed.) *Essays in plant Taxonomy* (pp. 245-262). London: Academic Press.

Richter, H. (1865). Zur Darwinschen Lehre [To the Darwinian doctrine]. *Schmidts Jahreb. Ges. Med., 126*, 243-249.

Ridley, H. N. (1923). *Dispersal of plants, throughout the world.* Kent.: L. Reeve & Co. Ltd.

Riedl, R. (1978). *Order in living organisms: A systems analysis of evolution.* New York: Wiley.

Rieppel, O. (1988). *Fundamentals of comparative biology*. Boston: Birkhauser Verlag.

Rogers, B. T., Peterson, M. D. & Kaufman, T. C. (1997). Evolution of the insect body plan as revealed by the sex combs reduced expression pattern. *Development, 124*, 149-157.

Rodin, R. J. (1958). Leaf anatomy of *Welwitschia*. II. *Am. J. Bot., 45*, 96-103.

Roth, J. L. & Dilcher, D. (1979). Investigations of angiosperms from the Eocene of North America: stipulate leaves of the Rubiaceae including a probable polyploid population. *Am. J. Bot., 66*, 1194-1207.

Rothmaler, W. (1951). Die Abtielungen und Klassen der Pflanzen [The abboties and classes of plants], *Feddes Repert., 54*(2-3), 256-266.

Rothwell, G. W., Mapes, G. & Mapes, R. H. (1996). Anatomically preserved Vojnovskyalean seed plants in Upper Pennsylvanian (Stephanian) marine shales of North America. *J. Palynol., 70*(6), 1067-1079.

Rothwell, G. W. & Lev-Yadun, S. (2005). Evidence of polar auxin flow in 375 million-year-old fossil wood. *Am. J. Bot., 92*, 903-906.

Round, F. E. (1973). *The biology of the Algae* (2nd ed.). London: Edward Arnold Publisher.

Rudall, P. J. & Bateman, R. M. (2002). Roles of synorganisation, zygomorphy, and heterotrophy in floral evolution: The gynostemium and labellum of orchids and other lilioid monocots. *Biol. Rev. Cambridge Phil. Soc., 77*, 403-441.

Rudall, P. J. & Bateman, R. M. (2004). Evolution of zygomorphy in monocot flowers: iterative patterns and developmental constraints. *New Phytol., 162*, 25-44.

Rudall, P. J., Sokoloff, D. D., Remizowa, M. V., Conran, J. G., Davis, J. I., Macfarlane, T. D. & Steenson, D. W. (2007). Morphology of Hydatellaceae, an anomalous aquatic family recently recognized as an early-divergent angiosperms. *Ann. Bot.*, *101*, 941-956.

Saarela, J. M., Rai, H. S., Doyle, J. A., Endress, P. K., Mathews, S., Marchant, A. D., Briggs, B. G. & Graham, S. W. (2007). Hydatellaceae identified as a new branch near the base of angiosperm phylogenetic tree. *Nature, 446*, 312-315.

Sagreiya, K. P. (1978). Our dwindling forests and national welfare. *Indian. Journ. Forestry, 1*(4), 281-286.

Sahni, B. (1920). On certain Archaic features in the seed of *Taxus baccata*, with remarks on the antiquity of the Taxineae. *Ann. Bot., 34*(1), 117-133.

Sahni, B. (1932). A petrified *Williamsonia* (*W.sewardiana Sp. Nov.*) from Rajmahal hills, India. *Paleontologica Indica*. (New Series), *20*(3), 1-19.

Sahni, B. (1948). The Pentoxylae: A new group of Jurassic Gymnosperms from the Rajmahal hills of India. *Bot. Gaz.*, *110*, 47-80.

Sahni, B. (1991). Revolutions in the plant world (Presidential address delivered by B. Sahni at annual meeting of the National Academy of Sciences, 5 Much 1938; (repr. ed.). *Curr. Sci. 61*(9-10), 594 – 600.

Sajo, M. G., deMello-Silva, R. & Rudall, P. J. (2010). Homologies of floral structures in velloziaceae with particular reference to corona. *Intl. J. Pl. Sci., 171*, 595-606.

Sambamurty, A. V. S. S. (2005). *A textbook of Algae*. New Delhi: I. K. International Publishing House Pvt. Ltd.

Sampson, F. B. (1993). Pollen morphology of the Amborellaceae and Hortoniaceae (Hortonioideae: Monimiaceae). *Grana, 32*, 154-162.

Sanderson, M. J. & Doyle, J. A. (2001). Sources of error and confidence intervals in estimating the age of angiosperms from rbcL and 18S rDNA data. *Am. J. Bot., 88*, 1499-1516.

Sanderson, M. J. Thorne, J. L., Wikstrom, N. & Bremer, K. G. (2004). Molecular evidence on plant divergence times. *Am. J. Bot., 91*, 1656-1665.

Sang, T., Crawford, D. J. & Stuessy, T. F. (1995). Documentation of reticulate evolution of peonies (*Paeonia*) using internal transcribed spacer sequences of nuclear ribosomal DNA implications for biogeography and concerted evolution. *Proc. Natl. Acad. Sci. USA.*, *92*, 6813-6817.

Sargent, R. D. (2004). Floral symmetry affects speciation rates in angiosperms. *Procd. Roy. Soc. Lond., B, Biol. Sci., 271*, 603-608.

Sauer, J. D. (1988). *Plant Migration: The dynamics of geographic patterning in seed plant species.* Berkley: University of California Press.

Saxena, S. K. (1966). Evolution of zircons in sedimentary and metamorphic rocks. *Sedimentology*, *6*, 1-33.

Scharf, C., Virgo, N., Cleaves, H. James Ii, Aono, M., Aubert-Kato, N., Aydinoglu, A., Barahona, A., Barge, L. M., Benner, S. A., Biehl, M., Brasser, R., Butch, C. J., Chandru, K., Cronin, L., Danielache, S., Fischer, J., Hernlund, J., Hut, P., Ikegami, T., Kimura, J., Kobayashi, K., Mariscal, C., Mcglynn, S., Menard, B., Packard, N., Pascal, R., Pereto, J., Rajamani, S., Sinapayen, L., Smith, E., Switzer, C., Takai, K., Tian, F., Ueno, Y., Voytek, M., Witkowski, O. & Yabuta H.(2015). A Strategy for origins of life research. *Astrobiology*, *15*(12), 1031-1042.

Scheckler, S. E. (1973). *New data on Aneurophytales and Cladoxylales from upper Devonian strata of New York State*. (Ph.D. dissertation, submitted in 1944) Ithaca, New York: Cornell University.

Scheckler, S. E. (1976). Ontogeny of Progymnosperms I. Shoots of upper Devonian Aneurophytales. *Canad. J. Bot.*, *54*, 202-219.

Scheckler, S. E. (1999). Progymnosperms. In R. SINGER (Ed.). *Encyclopedia of paleontology* (M-Z), (Vol. 2, pp.992-995). Chicago: Fitzroy Dearborn Publ.

Scheuring, B. W. (1974). 'Protosaccate' Strukturen, ein weit verbreitetes Pollenmarkmal zur fruhen und mittleren *Gymnospermen zeit.* ['Protosaccate' structures, a widely used pollen mark for early and middle gymnosperms time *Gymnospermen zeit.*] *Geologisch-Palontologische Mitteilungen Innsbruck, 4*(2), 1-30.

Schimithusen, J. (1961). *Allgemeine Vegetations-Geographie*. [General vegetation geography.] Berlin: Walter de Gruyter & Co.

Schimper, K. (1860). *Synopsis Muscorum Europaeorum*. Stuttgart: Schweizerbart.

Schmitt, C. & Turgeon, S. L. (2011). Protein/Polysaccharide complexes and coacervates in food systems. Adv. *Colloid Interface Sci., 167*(1-2), 63-70.

Schneider, H. E., Schueltpelz, E., Pryer, K. M., Cranfill, R. & Lupia, R. (2004). Ferns diversified in the shadows of angiosperms. *Nature, 428*, 553-557.

Schopf, J. W. (2001). *Cradle of life: The discovery of Earth's earliest fossils* (repr. Ed.). Princeton, New Jersey: Princeton University Press.

Schopf, J. W. (2012). The fossil record of cyanobacteria. In B. A. Whitton & M. Potts (Eds.). *Ecology of Cyanobacteria-II* (pp. 15-36). Dordrecht: Springer Science-Business Media.

Schouw, J. F. (1823). *Grunzage einer allgeminen pflanzengeographic*. [Basic features of a general plant geography.] Berlin: Reimer.

Schuster, R. M. (1966). *The Hepaticae and Anthocerotae of North America* (Vol.1). New York: Columbia University Press.

Scotese, C. R. (2005). *Paleomap project*. Retrieved from http://www. scotese.com 2005 [Updated July 27, 2008].

Scotland, R. W. & Wortley, A. H. (2003). How many species of seed plants are there? *Taxon, 52*(1), 101-104.

Scott, A. C. & Glasspool, I. J. (2006). The diversifications of Paleozoic fire systems and fluctuations of atmospheric oxygen concentration. *Proc. Natl. Acad. Sci. USA, 103*, 10861-10865.

Scott, D. H. (1895). Nathanael Pringsheim. *Nature, 51,* 399-402.

Scott, D. H. (1908). Alternation of generations in plants. *Nature, 79*, 1-4.

Scott, D. H. (1911). *The Evolution of plants* (repr.ed. 2017). Delhi, Relnk Books.

Scott, E. R. D. & Krot, A. N. (2001). Oxygen Isotopic compositions and origins of calcium-aluminum-rich inclusions and chondrules. *Meteoritics and Planetary Science, 36*, 1307-1319.

Senter, P. (2011, May-June). The defeat of flood geology by flood geology: The ironic demonstration that there is no trace of the Genesis Flood in geological record. *Reports NCSE.com*. 31.3,1.1.

Sergeev, V. N., Sharma, M. & Shukla, Y. (2012). Proterozoic fossil bacteria. *The Paleobot., 61,* 189-358.

Shapiro, R. (1987). *Origin: A skeptic's guide to the creation of life on Earth*. New York: Bantam Books.

Sharma, A. K. (1979) 1980. Chromosomes and Adaptation of Species. *Bull. Bot. Soc. Bengal, 33*, 1-4.

Sharma, C. (1993). Palynostratigraphy of lake deposits of Himalaya and Paleoclimate in Quaternary Period. *Curr. Sci., 64*(11-12), 930-932.

Sharma, M. (2002). Biodiversity conservation and socio-economic development: role and relevance of biotechnology. In. S. K. Nandi, L. M. S. Palni, & A. Kumar, (Eds.). *Role of plant tissue culture in biodiversity conservation and economic development* (pp. 1-9). [Himavikas Occasional publication no.15.] Almora: G. B. Pant Institute of Himalayan environments and development.

Sharma, O. P. (2006). *Pteridophyta*. New Delhi: Macmillan Publishers India Ltd.

Sheikh, M. T. & Kapgate, D. K. (1982). First petrified fossil *Riccia* l. from Indian beds. *Curr. Sci., 51*, 241-242.

Shiu, P. K. T., Raju, N. B., Zickler, N. B. & Metzenberg, R. L. (2002). Meiotic silencing by unpaired DNA. *Cell, 107*, 905-916.

Singh, S. (1998). *Geomorphology.* Allahabad: Prayag Pustak Bhawan.

Sinha, N. (1999). Leaf development in angiosperms. *Ann. Rev. Pl. Physio. & Mol. Biol., 50*, 419-446.

Sleep, N. H. & Zahnle, K. (2001). Carbon dioxide cycling and implications for climate on ancient Earth. *Journ. Geophysical Res., 106*, 1373-1399.

Smith, A. R., Pryer, K. M., Schuettpelz, E., Korall, P. Schneider, H. & Wolf, P. G. (2006). A classification for extant ferns. *Taxon*, *55*, 705-731.

Smith, D. S. & Goldberg, E. (2015). Tempo and mode of flower color evolution. *Am J. Bot., 1*02(7): 1014-1025.

Smith, G. M. (1955). *Cryptogamic Botany* (Vol-2). New York: Mc Graw Hill.

Smith, S. A., Beaulieu, J. M. & Donoghue, M. J. (2010). An uncorrelated relaxed-clock analysis suggests an earlier origin for flowering plants. *Proc. Natl. Acad. Sci. USA, 107*, 5897-5902.

Smoot, E. L. & Taylor, R. N. (1986). Structurally preserved fossil plants from Antarctica: II. A Permian moss from the Transantarctic mountains. *Am. J. Bot.*, *73*, 1683-1691.

Solomon, A. M., Tharp, M. L., West, D. C., Taylor, G. E., Webb, J. M. & Trimble, J. L. (1984). *Response of unmanaged forests to Co_2 induced climate change: available information, Initial tests, and data requirements.* (TR009). US Department of Energy (Carbon dioxide research division).

Soltis, D. E., Soltis, P. S., Chase, M. W., Mort, M. E., Albach, D. C., Zanis, M., Savolinen, V., Hahn, W. H., Hoot, S. B., Fay, M. F., Axtell, M., Swensen, S. M., Prince, L. M., Kress, W. J., Nixon, K. C. & Farris, J. S. (2000). Angiosperm phylogeny inferred from 18S rDNA, rbcL, and atpB sequences. *Bot. J. Linn. Soc., 133*, 381-461.

Soltis, D. E., Albert, V. A., Leebens-Mack, J., Bell, C. D., Paterson, A. H., Zheng, C., Sankoff, D., Depamphilis, C. W., Karr Wall, P. & Soltis, P. S. (2009). Polyploidy and angiosperm diversification. *Am. J. Bot.*, *96*(1), 336-348.

Soltis, P. S., Soltis, D. E. & Edwards, C. E. (2005). *Angiosperms: Tree of life web project.* Retrieved from http://tolweb.org/tree?group=angiosperms.

Sonnerat, M. (1782). *Voyage ans Indus Orientalis et la chine fait par order due Roi [Travel years Indus Orientalis and China made by order of King]* Depuis 1774 jasquen 1781, Paris.

Spallanzani, L. (1765). *Dissertazioni due. la. Saggio di Osservazioni microsscophie concernenti il Sistema della generazionede signoridi Needham e Buffon. 2a [Dissertations two. there. Wise of microsscophie Observations concerning the System of the Generation of Misses Needham and Buffon. 2a.]. De Lapidibus ab aqua resilientibus dissertation [The precious water recoil dissertation.]*. Modena, Eredi di Bart. Soliani.

Sporne, K. R. (1959). On the phylogenetic classification of plants. *Am. J. Bot.*, *46*(5), 385-394.

Sporne, K. R. (1962). *The morphology of Pteridophytes*. Hutchinson University Library. London, U.K. Pp. 189.

Sporne, K. R. (1991). *The morphology of Gymnosperms* (repr. ed.). Delhi: B. I. Publications Pvt. Ltd.

Stanley, S. M. (1999). *Earth System History* (pp.297-302). New York: W. H. Freeman and Co.

Stebbins, G. L. (1938). Cytological characteristics associated with the different growth habits in the dicotyledons. *Am. J. Bot., 25*, 189-198.

Stebbins, G. L. (1942). The genetic approach to problems of rare and endemic species. *Madrona, 6*, 241-258.

Stebbins, G. L. (1950). *Variation and evolution in plants*. New York: Columbia University Press.

Stebbins, G. L. (1974). *Flowering plants: Evolution above the species level.* Cambridge: Harvard University Press.

Stebbins, G. L. (1980). Rarity of plant species: a synthetic viewpoint *Rhodora, 82*(829), 77-86.

Stebbins, G. L. (1984). Polyploidy and distribution of the arctic-alpine flora: new evidence and a new approach. *Bot. Helvetica, 94*, 1-13.

Stein, C. A. & Stein, S. A. (1992). A model for the global variation in oceanic depth and heat flow with lithospheric age. *Nature, 359*, 123-129.

Stein, W. E. (1998). Developmental logic: establishing a relationship between developmental process and phylogenetic patterns in primitive vascular plants. *Review of Paleobot. & Palynology, 102*, 15-42.

Stern, D. P. & Peredo, M. (2001). *The Sun and the solar wind-NASA*. Retrieved from https://www-spof.gsfc.nasa.gov/Education/Isun.html.

Stevens, P. F. (1980). Evolutionary polarity of character states. *Ann. Rev. Ecol. Sys., 11*, 333-358.

Stewart, W. N. & Rothwell, G. W. (1993). *Paleobotany and the evolution of plants* (2nd ed.). Cambridge: Cambridge University Press.

Stopes, M. C. (1905). On the double nature of the cycadean integument. *Ann. Bot., 19*, 561-566.

Strasburger, E. (1894). The periodic reduction of the number of the chromosomes in the life-history of living organisms. *Ann. Bot., 8*, 281-316.

Stuessy, T. F. (2004). A transitional– combinational theory for the origin of angiosperms. *Taxon, 53*(1), 3-16.

Sun, G. Zheng, S., Dilcher, D., Wang, Y. & Mei, S. (2001). *Early angiosperms and their associated plants from western Liaoning, China*. Shanghai: Scientific and Technological Education Publishing House.

Sun, G., Dilcher, D., Wang, H. & Chen, Z. (2011). A eudicot from the Early Cretaceous of China. *Nature, 471*, 625-628.

Suzuki, D. & Grandy, W. (2004). *Tree: A life story*. Vancouver: Greystone Books.

Szostak, J. W. (2012). The eightfold path to non-enzymatic RNA replication. *Journ. Syst. Chem.,* **3,** 2 https://doi.org/10.1186/1759-2208-3-2.

Takeda, H. (1913). Some points in the anatomy of the leaf of *Welwitschia mirabilis*. *Ann. Bot.*, *27*, 347-357.

Takhtajan, A. L. (1953). Phylogenetic principles of higher plants. *Bot. Rev., 19*(1), 1-45.

Takhtajan, A. L. (1958). *Origin of angiospermous plants* (Trans. Russian edition of 1954). Washington DC: American Institute of Biological Science.

Takhtajan, A. L. (1969). *Flowering plants: Origin and dispersal*. Edinburgh: Oliver & Boyd.

Takhtajan, A. L. (1974). Floristic division of the world (Vol. 1, pp. 117-153). In A. A. Fedorov (Ed.) *Plant Life*. (in Russian). Moscow: Prosweshenie.

Takhtajan, A. L. (in association with Corvello, T. J. & Cronquist, A. (1986). *Floristic regions of the world*. Berkley: University of California Press.

Takhtajan, A. L. (1991). *Evolutionary trends in flowering plants*. New York: Columbia University Press.

Taylor, D. W., Li, H., Dahl, J., Fago, F. J., Zinniker, D. & Moldowan, J. M. (2006). Biogeochemical evidence for the presence of the angiosperm molecular fossil oleanane in Paleozoic and Mesozoic non-angiospermous fossils. *Paleobiol., 32*(2), 179-190.

Taylor, G. (1940). Antarctica. In *Regionale Geologie der Erde.I,* (Sect. 8). Leipzig: Akademische Verlagsgesellschaft.

Taylor, T. N., Kerp, H. & Hass, H. (2005). Life history biology of early land plants: deciphering the gametophyte phase. *Procd. Nat. Acad. Sci. USA, 102*, 5892-5897.

Taylor, T. N., Taylor, E. L. & Krings, M. (2009). *Paleobotany: The biology and evolution of fossil plants.* Amsterdam: Academic Press.

Theiben, G. & Becker, A. (2004). Gymnosperm orthologues of class B floral homeotic genes and their impact on understanding flower origin. *Critic. Rev. Pl. Sci*., *23*(2), 129-148.

Theiben, G. & Meizer, R. (2007). Molecular mechanisms underlying origin and diversification of the angiosperm flower. *Ann. Bot*., *100*(3), 1-17.

Thomas, H. H. & Bancroft, N. (1913). On the cuticles of some recent and fossil Cycadean fronds. *Trans. Linn. Soc. Lond.* (2nd Ser.), *8*, 155-204.

Thomson, R. B. (1927). Evolution of seed habit in plants. *Trans. Roy. Soc. Can.* (Ser.3), *21*, 229-272.

Thomson, R. B. (1934). Heterothally and seed habit versus heterospory. *New Phytologist, 33*(1), 41-44.

Thummel, C. S. & Chory, J. (2002). Steroid signaling in plants and insects- Common themes, different pathways. *Genes & Development, 16*(24), 3113-3129.

Tiffney, B. H. (2004). The role of vertebrate herbivory in the evolution of land plants. *Paleobot. 41*, 87-97.

Tobie, G., Lunine, J. I. & Sotin, C. (2006). Episodic outgassing as the origin of atmospheric methane on Titan. *Nature*, *440*(7080), 61-64.

Tolmatchev, A. I. (1974). *Introduction to geography of plants*. Leningrad: University Press.

Thien, L., Benhardt, P., Gibbs, G. W., Pellmyr, O., Bergstorm, G., Groth, I. & Mcpherson, G. (1985). The pollination of *Zygogynum* (Winteraceae) by a moth, *Sabatinca* (Micropterigidae): an ancient association? *Science,* 227, 540-543.

Thien, L., Bernhardt, P., Devall, M. S., Chen, Z. D., Luo, Y. B., Fan, J. H. Yuan, L. C. & Williams, J. H. (2009). Pollination biology of basal angiosperms (ANITA grade). *Am. J. Bot*., *96*(1), 166-182.

Thomas, H. (1932). The old morphology and new. *Proc. Linn. Soc. London, 145*, 17-32.

Thorne, J. L. & Kishino, H. (2002). Divergence time and evolutionary rate estimation with multilocus data. *Syst. Biol., 51*, 689-702.

Townrow, J. A. (1959). Two Triassic bryophytes from South Africa. *J. S. Afr. Bot., 25*, 1-22.

Traverse, A. (1988). Plant evolution dances to a different beat. *Hist. Biol., 1*, 277-301.

Treub, M. (1890). Etudes sur les Lycopodiacees. [Studies on Lycopodiaceae.] *Ann. Jard. Bot. Buitenzorg, 8*, 1-37.

Truman, J. W. & Riddiford, L. M. (2002). 34. Insect developmental hormones and their mechanism of action. In D. W. Plaff, A. P. Arnold, A. M. Etgen, S. E. Fahrbach & R. T. Rubin (Eds.). *Hormones, Brain, and Behaviour* (Vol.2.). Boston: Academic Press.

Turrill, W. B. (1964). Plant taxonomy, phytogeography and plant ecology. In W. B. Turrill (Ed.) *Vistas in Botany* (Vol.4, pp. 187-224). London: Pergamon Press.

Tutin, T. G., Heywood, V. H., Burges, N. A., Moore, D. M., Valentine, D. H., Walters, S. M. & D. A. Webb (1968). *Flora Europea* (Vol.2). Cambridge: Cambridge University Press.

Udvardy, M. D. F. (1975). A classification of the Biogeographical Provinces of the world. *IUCN occasional paper,* No. *18*., p.49. Gland: IUCN.

Vakhrameev, V. A. (1970). Range and Paleoecology of Mesozoic Conifers, the Cheriolepidaceae. *J. Paleont*., *1*, 12-25.

Valdiya, K. S. (1999). Rising Himalayas: Advent and intensification of monsoon. *Curr. Sci., 76,* 514-517.

Valley, J. W., Peck, W. H., King, E. M. & Wilde, S. A. 2002. A cool early Earth. *Geology*., *30*(4), 351-354.

Van Konijnenburg-Van Cittert, J. H. A. (1975). *Angiopteris blackii* Van Cittert *nom. nov*. Rev. *Paleobot. & Palynology, 20*(3), 215.

Van Kranendonk, M. J., Philipot, P., Lepot, K., Bodorkos, S. & Pirajno, F. (2008). Geological setting of Earth's oldest fossils in the c.3.5 Ga Dresser Formation, Pilbara Craton, Western Australia. *Precambr. Res., 167*, 93-124.

Vats, S. K. & Semwal, R. L. (1966). Global Warming and forest fires in tropical Himalayas. *Curr. Sci*., *71*, 5-6.

Venkatachala, B. S. (1991). Paleobotany in India – the post-Sahni era in retrospect. *Curr. Sci., 61*, 586-593.

Venkatachala, B. S. & Goczan, F. (1964). The spore-pollen flora of Hungarian " Kossen facies." *Acta Geol. Sci. Hungary, 8*, 203-228.

Vigran, J. O. & Thusu, B. (1975). Illustrations and distribution of Jurassic palynomorphs of Norway. *Roy. Norw. Coun. Sci. Ind. Res. (Cont. Shelf Div. Publ.*), *65*, p56.

Vishnu-Mittre (1953). A male flower of the Pentoxyleae with remarks on the structure of the female cones of the group. *Paleobot*., *2*, 75-84.

Von Goethe, J. W. (1790). *Versuch die Metamorphose der Pflanzen.* [Metamorphosis of the plants.] Gotha: Carl Wilhelm Ettinger.

Von Herzog, T. (1926). *Geographie der Moose.* [*Geography of mosses.*] Jena: Gustav Fischer Verlag.

Walton, J. (1953). The evolution of ovule in pteridosperms. *Adv. Sci. London, 10*, 223-230.

Walton, J. (1957). On *Protopitys* Goppert: with a description of a fertile specimen *Protopitys scotica* Sp. Nov. from the Calciferous Sandstone series of Dunbartonshire. *Trans. Roy Soc. Edinburgh, 63*, 333-340.

Wang, X. (2010). *The dawn angiosperms: Uncovering the origin of flowering plants.* Switzerland: Springer Nature.

Wang, X., Duan, S., Geng, B., Cui, J. & Yang, Y. (2007). Schmeissneria: A missing link to angiosperms? *BMC Evol. Biol., 7*, 14-26.

Wang, Z. Q. (2000). Vegetation declination on the eve of P-T event in north China and plant survival strategies. *Chinese Sci. Bull., 43*(11), 889-897.

Ward, D. & Blaustein, L. (1994) The overriding influence of flash floods on species-area curves in ephemeral Negev Desert Pools a consideration of the value of island biogeography theory. *Journ. Biogeogr., 21*, 595-603.

Ward, P. D. (2006). *Out of thin air: Dinosaurs, birds and Earth's ancient atmosphere.* Washington DC.: Joseph Henry Press.

Wardlaw, C. W. (1955). *Embryogenesis in plants.* New York: John Wiley and Sons.

Wearn, J. A., Chase, M. W., Mabberley, D. J., Couch, C. (2013). Utilizing a phylogenetic classification for systematic arrangements in botanic gardens and herbaria. *Bot. J. Linnean. Soc.,* 172, 127-141.

Weber, R. (1985). Las plantas fosiles de la formacion Santa Clara (Triassico Tardio, Sonora, Mexico): Estabo actual de las investigaciones. In R. Weber (Ed.) *III Congresso Latinoamericano de Paleontologia. Mexico. Simposio Sobre Floras del Triassico Tarido, su Fitogeografia y Paleoecologia, Memoria [The fossil plants of the Santa Clara formation (Triassico Tardio, Sonora, Mexico): Current research status. In R. Weber (Ed.) III Latin American Congress of Paleontology. Mexico. Symposium on Floras of the Triassico Tarido, its Phytogeography and Paleoecology, Memory]* (P I-V, pp 67-81). Mexico DF: Inst. Geologia U.N.A.M.

Wegner, A. (1912). Die Enstehung der Kontinente- Dr. A Petermanns Mitteilungen aus Justus Perthes. *Geographischer Anstalt [*The emergence of the continents - Dr. A Petermann's messages from Justus Perthes. *Geographical institution].*, *63*, 185-195, 235-256, 305-309.

Wegner, A.(1924). *The origin of the continents and oceans.* (J. G. A. Skerl, Trans.). New York: E. P. Dutton & Co.

Westengen, O., Huaman, Z. & Heum, M. (2005). Genetic diversity and geographic pattern in early South American cotton domestication. *Theoretical & Applied Genetics, 110*, 392-402.

Wettstein, R. V. (1924) (1923-1924). *Handbusch der Systematichen Botanik [Handbusch of the systematic botany]* (3rd ed., Vol. I & II., p577). Vienna: Deuticke.

Wheeler, E. A. & Lehman, T. M. (2000). Late Cretaceous woody dicots from Aguja and Javelina formations, Big Bend National Parks, Texas, U.S.A. *Intl. Asso. Wood Anatom. Journ.*, *21*, 83-120.

Whiteside, J. H., Olsen, P. E., Sambrotto, R. N. & Cornet, B. (2004). Milankovitch modulation of d13 Corn and fish communities in the tropical great lakes of the Triassic-Jurassic Pangean rift system. *Geophysical Research Abstracts*, V.6, 07738.

White, M. E. (1986). *The greening of Gondwana: the 400 million year story of Australia's plants*. Frenchs Forest: Reed books Party limited, N.S.W. Pp 256.

Wickens, G. E. & Lowe, P. (2008). *The Baobabs: Pachycauls of Africa, Madagascar, and Australia.* Dordrecht: Springer Science & Business Media.

Wikstrom, N., Savolainen, V. & Chase, M. W. (2001). Evolution of the angiosperms: calibrating the family tree. *Proc. Roy. Soc. Lond. Ser. B. Biol. Sci., 268*(1482), 2211-2220.

Williamson, W. C. (1870). Contributions towards the history of *Zamia gigas* L and H. *Trans linn. Soc.*, *26*, 663- 674.

Williamson, W. C. (1887). On the organization of fossil plants of the coal measures. Part xiii: *Heterangium tiliaeoides* (Williamson) and *Kaloxlyon hookeri*. *Philosophical Transact. Roy. Soc. Lond.*, *178*B: 289-304.

Willis, J. C. (1923). The origin of species by large, rather than by gradual, change and by Guppys Method of differentiation. *Ann. Bot., 37,* 605-628.

Willis, J. C. (1918). Age and Area hypothesis. *Science* (New Series), *47*(1226), 626-628.

Wilmshurst, J. M., Hunt, T. L., Lipo, C. P. & Anderson, A. J. (2011). High-precision radiocarbon dating shows recent and rapid initial human colonization of East Polynesia. *Proc. Acad. Sci. USA, 108*(5): 1815-1820.

Wilson, E. O. (1992). *The Diversity of Life*. Cambridge: Belknap Press of Harvard University Press.

Wilson, T. J. (1965) A new class of faults and their bearing on continental drift. *Nature, 207,* 343-347.

Wolfe, K. H., Gouy, M., Yang, Y. W., Sharp, P. M. & Li, W. H. (1989). Date of the monocot-dicot divergence estimated from chloroplast DNA sequence data. *Proc. Natl. Acad. Sci. USA, 86*, 6201-6205.

Wooldridge, S. W. & Morgan, R. S. (1960). *An outline of geomorphology*. Cape Town: Longmans Green & Co.

Wulff, E. V. (1943). *An introduction to historical plant geography*. (E. Brissenden Trans.) Waltham, Massachusetts: Chronica Botanica co.

Xie Yan, Zheyu, L., Greg, W. P. & Dianmo, L. (2001). Invasive species in China: An overview. *Biodiversity & Conserv., 10*(8), 1317-1341.

Yang, Z. R. & Lin, Q. (2005). Comparative morphology of the leaf epidermis in *Schisandra* (Schisandraceae). *Bot. J. Linnean. Soc.*, *148*, 39-56.

Yoon, H. S., Hackett, J. D., Ciniglia, C., Pinto, G. & Bhattacharya, D. (2004). A molecular timeline for the origin of photosynthetic eukaryotes. *Mol. Biol. Evol.*, *21*(5), 809-818.

Yuan, J. H., Cornille, A., Giraud, T. Cheng, Y. F. & Hu, H. Y. (2013). Independent domestications of cultivated tree peonies from different peony species. *Molecular Ecology, 23*(1), 82-95.

Zalasiewicz, J., Williams, M., Smith, A., Barry, T. L., Coe, A. L., Bown, P. R., Brenchley, P., Cantrill, D., Gale, A., Gibbard, P., Gregory, F. J., Hounslow, M. W., Kerr, A. C., Pearson, P., Knox, R., Powell, J., Waters, C., Marshall, J., Oates, M., Rawson, P. & Stone, P. (2016). Are we now living in the Anthropocene? *GSA Today, 18*(2), 4-8.

Zanis, M. J., Soltis, P. S. Qiu, Y. L., Zimmer, E. & Soltis, D. E. (2003). Phylogenetic analyses and perianth evolution in basal angiosperms. *Pl. Syst. & Evol., 263*, 117-134.

Zavada, M. S. (2007). The identification of fossil angiosperm pollen and its bearing on the time and place of the origin of angiosperms. *Pl. Syst. & Evol., 263*, 117-134.

Zavada, M. S. & Mentis, M. T. (1992). Plant-animal interaction: the effect of Permian megaherbivores on the Glossopterid flora. *Amer. Midland Nat., 127*(1), 1-12.

Zeevart, J. A. D. (1976). Physiology of Flower Formation. *Annual Rev. Pl. Physiol. 27:* 321-348.

Zhang, H. D. (1995). *Florology* (in Chinese). Guangzhou: Zhongshan University Press.

Zhang, Z. Y. (1988). Longfengshania Du emed. An earliest record of bryophyte-like fossils. *Acta Paleontologica Sinica*, *27*, 416-426.

Zherikhin, V. V. (2002). Ecological history of terrestrial insects. In A. P. Rasnitsyn & D. I. J Quicke (Eds.). *History of Insects*. London: Kluwer Academic Publisher.

Zhou, S. L., Zou, X. H., Zhou, Z. Q., Liu, J., Xu, C., Yu, J., Wang, Q., Zhang, D. M., Wang, X. Q., Ge, S., Sang, T., Pan, K. Y. & Hong, D. Y. (2014). Multiple species of wild tree peonies gave rise to the King of Flowers, *Paeonia suffruticosa* Andrews. *Proc. Roy. Soc. Bot.*, *281*(1797), 20141687.

Zimmerman, W. (1930). *Phylogenie der Pflanzen* [Phylogeny of plants]. Jena: Gustav Fischer Verlag.

Zimmerman, W. (1938). Die Telomtheorie. [The Telom theory.] *Biologe, 7*, 385-391.

Zimmerman, W. (1959). *Die Phylogenie der Pflanzen [Phylogeny of plants]*. Stuttgart: Gustav Fischer Verlag.

ABOUT THE AUTHORS

Subir Ranjan Kundu, PhD

Independent Researcher/Freelance Writer

Mr. Subir Ranjan Kundu- is a freelance non-fiction writer, independent researcher and content creator in the interdisciplinary fields of evolutionary biology, phylogeography, cytogenetics, systematic botany and conservation biology.

He has a PhD degree from the University of Calcutta in the field of Phytogeography, Floristics, and Conservation Biology (Botany). He was deputed at Indian Botanic Garden, Howrah and Industrial section of Indian Museum, Calcutta; during his research tenure under National Fellowship Program, funded by the Ministry of Environment & Forests, Government of India.

His research interests are phyto-endemism, conservation and sustainable utilization of endemic plant resources. He is the author of more than 50 articles presently published in international journals and two books; he is also a life member of Indian Science Congress Association, Kolkata.

Pritha Biswas

Senior Content Writer

R & D Division, ARC Document Solution Inc.

Kolkata, India

Ms. Pritha Biswas has a bachelor's degree in Biological Science (major in Zoology) and has been working as a content developer and editor for various web-based media. She is a regular blogger and content writer on the internet and is presently associated with ARC Document Solutions, as the senior content writer.

She also engaged in writing features on science in the popular English magazine- "Gyan Vitaranam" in India for young readers. Currently, she is pursuing her executive MBA from Institute of Chartered Financial Analysts of India, Kolkata, India.

INDEX

#

A

B

C

D

E

F

G

H

I

J

K

L

M

N

O

P

Q

R

S

T

U

V

W

X

Y

Z